AF779046

Reviews and Advances in Materials Processing

Reviews and Advances in Materials Processing

Editor

Manoj Gupta

MDPI • Basel • Beijing • Wuhan • Barcelona • Belgrade • Manchester • Tokyo • Cluj • Tianjin

Editor
Manoj Gupta
Materials Group, Department
of Mechanical Engineering
National University of
Singapore
Singapore
Singapore

Editorial Office
MDPI
St. Alban-Anlage 66
4052 Basel, Switzerland

This is a reprint of articles from the Special Issue published online in the open access journal *Technologies* (ISSN 2227-7080) (available at: www.mdpi.com/journal/technologies/special_issues/rev_adv).

For citation purposes, cite each article independently as indicated on the article page online and as indicated below:

LastName, A.A.; LastName, B.B.; LastName, C.C. Article Title. *Journal Name* **Year**, *Volume Number*, Page Range.

ISBN 978-3-0365-5694-9 (Hbk)
ISBN 978-3-0365-5693-2 (PDF)

© 2022 by the authors. Articles in this book are Open Access and distributed under the Creative Commons Attribution (CC BY) license, which allows users to download, copy and build upon published articles, as long as the author and publisher are properly credited, which ensures maximum dissemination and a wider impact of our publications.
The book as a whole is distributed by MDPI under the terms and conditions of the Creative Commons license CC BY-NC-ND.

Contents

About the Editor

Manoj Gupta

Prof. Dr. Manoj Gupta received his Ph.D. degree from the University of California, Irvine, USA, in 1992, and is currently an Associate Professor at the Department of Mechanical Engineering at the National University of Singapore. In August 2017, he was highlighted among the Top 1% Scientist of the World Position by The Universal Scientific Education and Research Network, as well as among the top 2.5% by ResearchGate and the top 1% in Stanford's list of researchers. To his credit are (i) the disintegrated melt deposition technique and (ii) hybrid microwave sintering technique, an energy efficient solid-state processing method used to synthesize alloys/microcomposites/nanocomposites. He has published over 600 peer-reviewed journal papers, co-authored 8 books, and owns 2 US patents.

technologies

MDPI

Editorial

Editorial for the Special Issue "Reviews and Advances in Materials Processing"

Manoj Gupta

Department of Mechanical Engineering, National University of Singapore, 9 Engineering Drive 1, Singapore 117576, Singapore; mpegm@nus.edu.sg

In the area of Materials Science and Engineering, the tetrahedron comprising of processing, microstructure, properties and performance as four vertex corners is always key to develop new materials and to convert them to a useful shape for end application with the best properties possible. As can be envisaged, processing is an integral part of this tetrahedron and the proper development and usage of processing for a given end goal is a must to take the product efficiently to the consumer. The processing of the materials comprises of primary processing techniques and secondary processing techniques. Primary processing of materials can be broadly but not limited to classified into four sub-categories irrespective of type of materials. These include: (a) liquid-phase processing, (b) solid-phase processing, (c) two phase processing, and (d) vapor phase processing. Each of these subcategories have different types depending on a number of governing factors. Similarly, the secondary processing of materials is also very important for giving a desired shape to the materials and these include, for example, (a) forging, (b) rolling, (c) extrusion, (d) drawing, (e) joining and (f) machining. With the advancement in computing tools and the availability of advanced software, modelling and simulation has also almost become an integral part of the processing process so as to realize the best output from them through optimizing the processing/operating parameters.

In view of the intrinsic importance of 'Materials Processing', especially for the researchers in the area of materials science and manufacturing, this issue was conceived with a specific aim to bring to the notice of readers the latest advances in various processing types currently used for a wide spectrum of materials. A total of 14 papers were accepted and published in this issue. Among them, 3 are review papers and 11 are research papers targeting different research goals.

Among the three review papers, two of the articles address electro-discharge machining of non-conductive ceramics and a combination of materials written by Volosova et al. [1,2]. The critical issues and the way forward in each case is intelligently described by authors. In the third review paper written by TC Yap [3], the effect of cryogenic cooling on the turning of superalloys, ferrous materials and viscoelastic polymers is described along with the challenges faced by the industry in using cryogenic machining.

Eleven research papers written under this thematic issue address a spectrum of topics related to machining, additive manufacturing, oil processing, nanomaterials processing, metallic processing and polymers processing. The article written by Grigoriev et al. [4] described tool electrode behaviour and wear under discharge pulses during electrical discharge machining. Important insight is provided on electrical erosion wear fundamentals. Another paper written by Melnik et al. [5] presents the research on the parameters of vibroacoustic emission for the development of the monitoring and adaptive control system for electrical discharge machining.

Two of the research papers focussed on the secondary processing (joining) of metal-based materials. Tamadon et al. [6] investigated the influence of WC-based pin tool profile and processing parameters during friction stir welding on the microstructure and properties of AA1100 weld. In the second paper, Tamadon et al. [7] used Bobbin Friction Stir Welding

Citation: Gupta, M. Editorial for the Special Issue "Reviews and Advances in Materials Processing". *Technologies* **2022**, *10*, 77. https://doi.org/10.3390/technologies10040077

Received: 14 June 2022
Accepted: 23 June 2022
Published: 24 June 2022

Publisher's Note: MDPI stays neutral with regard to jurisdictional claims in published maps and institutional affiliations.

Copyright: © 2022 by the author. Licensee MDPI, Basel, Switzerland. This article is an open access article distributed under the terms and conditions of the Creative Commons Attribution (CC BY) license (https://creativecommons.org/licenses/by/4.0/).

and investigated the micro-flow patterns within aluminium weld structure. They used Atomic Force Microscopy as the main tool for characterization.

In the area of processing of non-metallic materials and liquid to be specific, Mazumder et al. [8] presented desulfurization technologies in the context of fuel and fuel cells. They focussed on the efficacy of different types of hydrotalcite absorbents.

Two papers in this issue address the advances in polymer processing. T. Aizawa [9] highlighted the CO_2 assisted polymer compression method to prepare porous polymer materials. These authors provided interesting insight in process development. The paper written by Terekhina et al. [10] presented their work on polymer processing using fused filament fabrication method. Different processing variations were attempted and strength was characterized to understand the effects of these variations. Hartmann et al. [11] highlighted the importance of the design of the additive manufacturing process in the context of industrial adaptability. In their article, they proposed a data-driven geometrical compensation approach.

The paper written by Vasiliev et al. [12] reported a set of computational algorithms to optimize the design and the optical or magneto-optical spectral analysis of 1D magnetic photonic crystals together with a windows software implementation. They also reported the methods and algorithms to obtain absorption coefficient spectral dispersion datasets for new materials.

Related to nanomaterials, two articles are included in the present issue. The article by Liem et al. [13] highlighted the use of eco-friendly microwave synthesis of silver nanoparticles using Mulberry leaves extract and silver nitrate solution. In the second article, Chen et al. [14] presented their work on facile fabrication of macroscopic self-standing Ni or Co-doped MnO_2 architectures targeted for enhancing catalytic activities for propane oxidation.

In a nutshell, the papers included in this issue address various processing challenges faced by researchers and industry for liquid, metallic, polymers, ceramics and nanomaterials processing (both primary and secondary). The methods described and the solution proposed are of current relevance and of significant importance to move forward for better efficiency and productivity. It is certain that readers will gain much useful knowledge from these articles.

To conclude, I would like to thank all the contributors to this issue for publishing their immensely useful work with '*Technologies*'. I also thank reviewers for critically going through these papers and making useful comments. Their selfless contribution is much appreciated. I would also like to thank the administrative staff for processing these articles at their earliest and providing much needed support to authors, reviewers and editors.

Funding: This research received no external funding.

Institutional Review Board Statement: Not applicable.

Informed Consent Statement: Not applicable.

Data Availability Statement: Not applicable.

Conflicts of Interest: The author declares no conflict of interest.

References

1. Volosova, M.A.; Okunkova, A.A.; Fedorov, S.V.; Hamdy, K.; Mikhailova, M.A. Electrical Discharge Machining Non-Conductive Ceramics: Combination of Materials. *Technologies* **2020**, *8*, 32. [CrossRef]
2. Volosova, M.; Okunkova, A.; Peretyagin, P.; Melnik, Y.A.; Kapustina, N. On Electrical Discharge Machining of Non-Conductive Ceramics: A Review. *Technologies* **2019**, *7*, 55. [CrossRef]
3. Yap, T.C. Roles of Cryogenic Cooling in Turning of Superalloys, Ferrous Metals, and Viscoelastic Polymers. *Technologies* **2019**, *7*, 63. [CrossRef]
4. Grigoriev, S.N.; Volosova, M.A.; Okunkova, A.A.; Fedorov, S.V.; Hamdy, K.; Podrabinnik, P.A.; Pivkin, P.M.; Kozochkin, M.P.; Porvatov, A.N. Wire Tool Electrode Behavior and Wear under Discharge Pulses. *Technologies* **2020**, *8*, 49. [CrossRef]
5. Melnik, Y.A.; Kozochkin, M.P.; Porvatov, A.N.; Okunkova, A.A. On Adaptive Control for Electrical Discharge Machining Using Vibroacoustic Emission. *Technologies* **2018**, *6*, 96. [CrossRef]

6. Tamadon, A.; Baghestani, A.; Bajgholi, M.E. Influence of WC-Based Pin Tool Profile on Microstructure and Mechanical Properties of AA1100 FSW Welds. *Technologies* **2020**, *8*, 34. [CrossRef]
7. Tamadon, A.; Pons, D.J.; Clucas, D. AFM Characterization of Stir-Induced Micro-Flow Features within the AA6082-T6 BFSW Welds. *Technologies* **2019**, *7*, 80. [CrossRef]
8. Mazumder, M.; Das, R.; Sajib, M.S.J.; Gomes, A.J.; Islam, M.; Selvaratnam, T.; Rahman, A. Comparison of Different Hydrotalcite Solid Adsorbents on Adsorptive Desulfurization of Liquid Fuel Oil. *Technologies* **2020**, *8*, 22. [CrossRef]
9. Aizawa, T. Process Development of CO_2-Assisted Polymer Compression for High Productivity: Improving Equipment and the Challenge of Numbering-Up. *Technologies* **2019**, *7*, 39. [CrossRef]
10. Terekhina, S.; Skornyakov, I.; Tarasova, T.; Egorov, S. Effects of the Infill Density on the Mechanical Properties of Nylon Specimens Made by Filament Fused Fabrication. *Technologies* **2019**, *7*, 57. [CrossRef]
11. Hartmann, C.; Lechner, P.; Himmel, B.; Krieger, Y.; Lueth, T.C.; Volk, W. Compensation for Geometrical Deviations in Additive Manufacturing. *Technologies* **2019**, *7*, 83. [CrossRef]
12. Vasiliev, M.; Alameh, K.; Nur-E-Alam, M. Analysis, Optimization, and Characterization of Magnetic Photonic Crystal Structures and Thin-Film Material Layers. *Technologies* **2019**, *7*, 49. [CrossRef]
13. Liem, L.N.; The, N.P.; Nguyen, D. Microwave Assisted Green Synthesis of Silver Nanoparticles Using Mulberry Leaves Extract and Silver Nitrate Solution. *Technologies* **2019**, *7*, 7. [CrossRef]
14. Chen, L.; Song, X. Facile Fabrication of Macroscopic Self-Standing Ni or Co-doped MnO_2 Architectures as Catalysts for Propane Oxidation. *Technologies* **2019**, *7*, 81. [CrossRef]

technologies

MDPI

Review

Electrical Discharge Machining Non-Conductive Ceramics: Combination of Materials

Marina A. Volosova [1], Anna A. Okunkova [1,*], Sergey V. Fedorov [1], Khaled Hamdy [1,2] and Mariya A. Mikhailova [1]

1 Department of High-Efficiency Processing Technologies, Moscow State University of Technology STANKIN, Vadkovskiy per. 3A, 127055 Moscow, Russia; m.volosova@stankin.ru (M.A.V.); sv.fedorov@icloud.com (S.V.F.); eng_khaled2222@mu.edu.eg (K.H.); m.mikhailova@stankin.ru (M.A.M.)

2 Production Engineering and Mechanical Design Department, Faculty of Engineering, Minia University, Minia 61519, Egypt

* Correspondence: a.okunkova@stankin.ru; Tel.: +7-909-913-1207

Received: 24 April 2020; Accepted: 26 May 2020; Published: 28 May 2020

Abstract: One of the promising processing methods for non-conductive structural and functional ceramics based on ZrO_2, Al_2O_3, and Si_3N_4 systems is electrical discharge machining with the assistance of an auxiliary electrode that can be presented in the form of conductive films with a thickness up to 4–10 μm or nanoparticles - granules, tubes, platelets, multidimensional particles added in the working zone as a free poured powder the proper concentration of which can be provided by ultrasound emission or by dielectric flows or as conductive additives in the structure of nanocomposites. However, the described experimental approaches did not reach the production market and industry. It is related mostly to the chaotic development of the knowledge and non-systematized data in the field when researchers often cannot ground their choice of the material for auxiliary electrodes, assisting powders, or nano additives or they cannot explain the nature of processes that were observed in the working tank during experiments when their results are not correlated to the measured specific electrical conductivity of the electrodes, particles, ceramic workpieces or nanocomposites but depends on something else. The proposed review includes data on the main electrophysical and chemical properties of the components in the presence of heat when the temperature in the interelectrode gap reaches 10,000 °C, and the systematization of data on ceramic pressing methods, including spark plasma sintering, the chemical reactions that occur in the interelectrode gap during sublimation of primary (brass and copper) and auxiliary electrodes made of transition metals Ti, Cr, Co, and carbon, auxiliary electrodes made of metals with low melting point Zn, Ag, Au, Al, assisting powder of oxide ceramics TiO_2, CeO_2, SnO_2, ITO, conductive additives Cu, W, TiC, WC, and components of Al_2O_3 and Zr_2O workpieces in interaction with the dielectric fluid - water and oil/kerosene medium.

Keywords: structural ceramic; oxide ceramic; EDM; ZrO_2; Al_2O_3; electrode; thin films; electrical conductivity; white layer; electro physics; chemical reactions; sublimation

1. Introduction

One of the widespread machining methods is electrical discharge machining (EDM) that allows producing parts regardless of physical and mechanical properties. The only usability condition for EDM is the electrical conductivity of the workpiece material: the higher specific electrical conductivity provides the higher material removal rate [1–3].

In addition, it is one of most precise machining methods as the diameter of the electrode-tool on EDM equipment can vary from 0.6 up to 0.05 mm. The accuracy of electrode positioning on the industrial equipment under certain conditions can reach 1–2 μm (80–100 nm in some cases) when the achievable roughness *Ra* is about 0.4 μm after roughing. It also allows removing significant volumes of material by a single pass of a wire tool. That can be a considerable advantage in the conditions of the tool industry in comparison with other machining methods based on the mechanical nature of removal.

The electrical discharge machining allows for obtaining complex spatial geometry with a slope of the generatrix up to 30° relative to the Z-axis and making holes with an arcuate or spiral-shaped generatrix on die-sink equipment. The most modern machines are equipped with computer numerical control with additional rotary mechanisms for the workpiece fixture that allows moving both rotationally and indexally according to the developed program in full coordination with movements of other axes, expanding the available geometry to six axes (Figure 1).

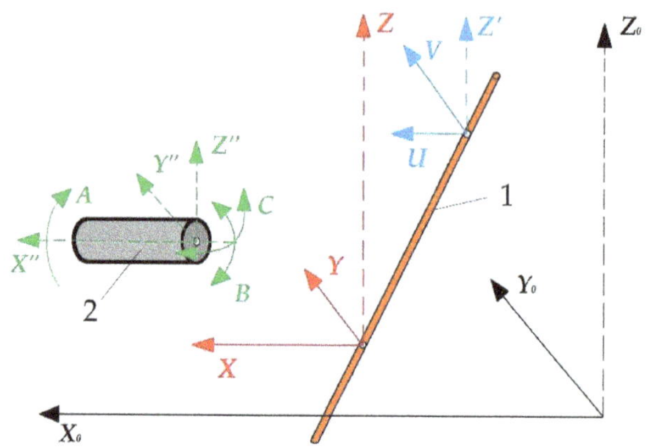

Figure 1. Additional rotary axes of wire electrical discharge machining: (**1**) is a tool, (**2**) is a part, $X_0Y_0Z_0$ is an absolute coordinate system of the machine, *XYZ* is a relative coordinate system of the lower wire attachment point, *UVZ'* is an auxiliary coordinate system of the upper wire attachment point, *X″Y″Z″* is a relative coordinate system of the part, *ABC* is an additional rotary axes.

The process of electrical erosion of the material consists of following stages:

- the initiation of an electric pulse in the interelectrode gap,
- the breakdown of the dielectric medium by a series of the discharge of pulses,
- the formation of a discharge channel with a temperature of more than 10,000 °C of a cloud of low-temperature plasma,
- the sublimation of material from the surface of the electrodes,
- the interruption of the pulses and washing off the cooled-down erosion products out of the interelectrode gap by dielectric medium flows,
- the restoration of dielectric tightness of the medium.

Due to the high temperatures that are fulminantly reached at the time of the formation of the discharge channel, the electrode material bypasses melting to the liquid phase. It sublimates directly from the solid phase to the condition of low-temperature gas plasma [4–7]. The ions of electrodes and fluid components interact with each other during subsequent pulses interruption and form sediment in the form of gases and droplets or dust that are washed out from the interelectrode gap by dielectric medium, which is, in most cases, water, oil, kerosene. The material of the electrodes and medium in the form of solid substances are partially adsorbed by the eroded surface of the electrodes that usually have a structure of nano-frames of the more resistant to high temperatures components.

For example, during processing steels in water, a subsurface layer of electrodes that often saturated with carbon and oxides is formed [8–10]. This layer is known as a recast or white layer. During machining stainless steel 12Kh18N10T (AISI 321), the nickel of the workpiece reacts with zinc of the brass electrode with the formation of the intermetallide accompanies processing by sparks and a black cloud of erosion dust [11,12]. At the same time, a natural oxide film is formed during processing fusible aluminum, copper, chrome, titanium, and their alloys [13–18]. When attempts to machine aluminum contained materials in oil or kerosene, the aluminum interacts with carbon from the organic medium in the presence of high temperatures when the oxide film is sublimated, with the formation of extremely dangerous aluminum carbide Al_4C_3 that may damage the filtration system of the equipment [19–22]. The experiments and chemical analyses of the samples showed that the components of the wire tool penetrate to a depth of 4–10 μm (up to 35 μm). A thermal influence zone is observed at the same depth and characterized by the formation of an austenitic layer in steels [4,10,11,18]. The erosion products depose as well on the walls of the working tank and in filters. After electrical erosion, a crater-like surface topology on the surface of the workpiece is observed [23–26].

The dielectric properties of ceramics and nanoceramics, in combination with excellent physical and mechanical properties, limit their machinability - material removal volumes per tool pass, the complexity of the resulting geometry, and, accordingly, the scope of their potential application [27–30]. It restrains the growth of the economy and prevents the transition to the sixth technological paradigm associated with the concept of "nano", the development of innovative materials and technologies, and its introduction into the industry [31–36].

Today, there are three fundamental approaches to solving this problem (Figure 2). One of them is related to the introduction of a conducting secondary phase to the ceramic matrix of a composite or nanocomposite. It involves the use of powder metallurgy methods, including innovative hybrid sintering using assisting currents in graphite dies. The powder metallurgy methods of production conductive ceramics are known as industrial methods for creating composites from the mid-80s and continuously improves [37]. The idea of creating an electrically conductive ceramic nanocomposite has been developing since the late 80s [38], and many scientists have achieved remarkable results [21,22]. Still, more research is needed before the industrial introduction of this approach, since ceramics, when modified by conducting and non-conducting nanoparticles and platelets, do not always retain their original physical and mechanical properties, whiteness or even transparency, and after modification, even with confirmed improved properties, the material cannot always be electroerosively processed, even with proven conductivity much higher than the percolation threshold.

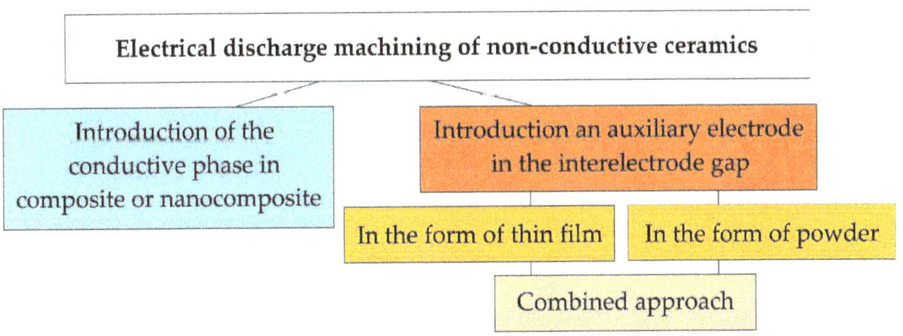

Figure 2. Electroerosive methods for processing non-conductive ceramics.

The second method for solving the problem of electrical discharge machining ceramics with high values of dielectric impermeability is the approach associated with the introduction of an auxiliary electrode in the interelectrode gap. The auxiliary electrode can be introduced in the form of a thin film produced of a conductive material (aluminum, copper, silver, etc.) on the surface of the dielectric workpiece, to which electric current discharges will be initiated. It is assumed that the erosion of the surface of the ceramic workpiece will be carried out simultaneously with the electrically conductive film. For the first time, the idea of introducing the auxiliary electrode was presented in the schematic diagram of electrical discharge machining dielectrics by soviet scientists [39]. Later similar experiments were carried out by many other scientists, and serious results were obtained - the ceramics were processed electroerosively to a depth of 1.5–2.0 mm [40]. In the described case, after the first operational iterations, when the film-electrode sublimates entirely, the question of reinitiation of current pulses in the interelectrode gap remains open.

The third method can be called an approach in which current discharges are directed to electrically conductive particles uniformly weighted in the volume of a liquid dielectric (water, oil, kerosene) in the interelectrode gap. In this case, it is necessary to monitor the constant concentration of conductive particles in the discharge gap. The first experiments to improve the dielectric constant of the working medium have been carried out since the creation of the method of electrical discharge machining materials by Professors Boris and Natalya Lazarenko [41–45]. The most outstanding works on changing the components of the working fluid during conductive materials processing appeared from the beginning of the 2000s [46]. It was related to the possibility of obtaining a finely dispersed substance and the ability to control the morphology of the used powders. The most relevant studies were obtained using carbon nanotubes and other forms of graphene in the discharge gap [4,47].

The direction of research related to the development of innovative methods of processing super-hard, heat- and wear-resistant materials without losing their physical and mechanical properties, with the possibility of obtaining complex spatial geometry and with minimal losses on material, on operating and auxiliary time, with minimum possible tool consumption, while maintaining high volumes of material removal by a pass of tool, is an extremely relevant task of modern science, as evidenced by multiple publications.

The novelty of the work is in the systematization of ceramic pressing methods, electrophysical and chemical properties of electrodes' and dielectrics' components in the presence of heat, data on possible interactions between components, suitable combinations of the materials for processing structural oxide ceramics as alumina and zirconia that can be applied for functional ceramics and nanocomposites.

This analytical study is designed to answer the question of why some materials should not be combined, why certain phenomena are observed during processing, why the workability of the workpiece is often not dependent on the electrical conductivity and temperature stability of the materials of the primary and auxiliary electrodes, assisting powder and conductive additives.

The tasks of the study include the analytical research on the conductivity of materials in the presence of heat and chemical composition of the subsurface layer of some materials, the investigation of physical properties of structural ceramics in the presence of heat and pressing methods, the sintering cycles under pressure and features of spark plasma sintering, main chemical interaction of alumina and zirconia components with materials of electrodes, assisting powder, additives, and working medium.

The analytical research was conducted, taking into account the basic principles of electrophysics and physical chemistry, the laws of the structures of substances, their structure and transformations, and available theoretical and practical information.

2. Conductivity of Materials and Saturation of Components in Recast Layer

All conductive materials are subject to electrical erosion to one degree or another. The group of materials to be electroerosively machined include aluminum, chromium, cobalt, copper, beryllium, molybdenum, nickel, niobium, titanium, tungsten, and their alloys including bronze and brass, high-temperature and magnetic cast alloys, cast iron and steels, graphite, hard alloys (carbide ceramics), titanium nitride and boride, conductive ceramic composites and nanocomposites, and some other materials (Figure 3). However, not only the specific electrical conductivity determines the possibility of using electrical discharge machining methods to process all practically conductive materials. The resistance to electrical erosion of metals is determined as well by a combination of thermophysical properties of materials such as a melting or sublimation point, heat capacity, density, structural and microstructural homogeneity, and thermal conductivity [34,48].

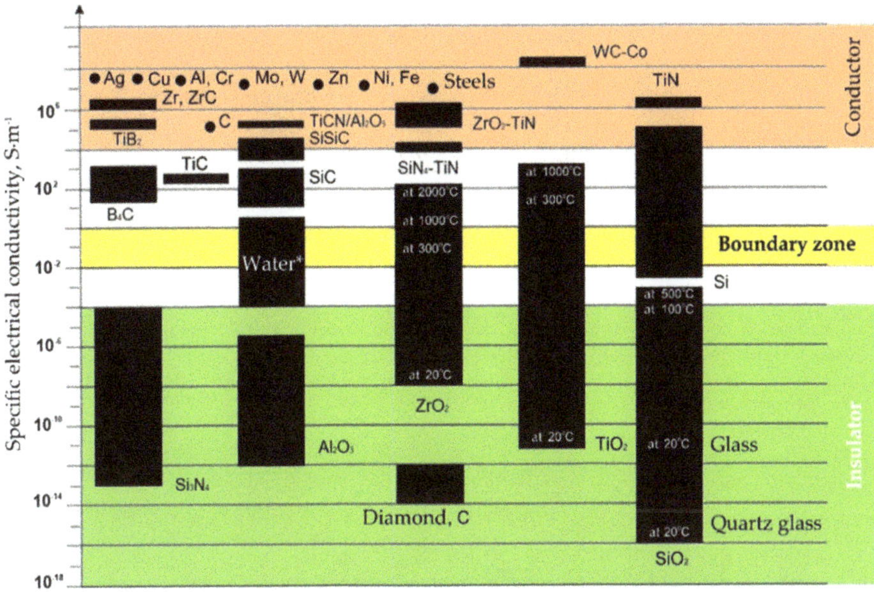

Figure 3. Specific electrical conductivity (ς) of some materials. (* The electrical conductivity depends on impurities and suspensions, for distilled water is about 10^{-5} S·m^{-1}).

Some dielectric materials change their electrical conductivity when heated and become conductors or semiconductors when metals have another tendency to the opposite phenomenon. It can have a positive effect on the electrical discharge machining a part whose chemical composition contains non-conductive at room temperature components. The conductivity is also affected by the presence of impurities in the material composition. In this case, conductive components increase the specific conductivity of the material, and non-conductive ones decrease. As can be seen, for materials such as carbon and silicon dioxide, the electrical conductivity often depends as well on the structure of the crystal lattice. Carbon the form of graphite calls a conductor when the carbon in the form of a diamond is an insulator, despite its anisotropic conductivity associated as well with the structure of the crystal lattice. Another example, the silicon dioxide in the crystalline form (silica glass) is a better dielectric than in the amorphous form of domestic glass. The same trend is observed for titanium dioxide.

One of the promising areas of research is the introduction into the interelectrode gap of a powder or nanosized powder. It can be some conductive metal or titanium dioxide, cerium dioxide, other ceramics in the powder phase [49,50]. In this case, the ceramics in the presence of heat would acquire conductive properties and provoke more dense discharges of pulses in processing main conductive material—usually metals and metal alloys such as extremely popular nickel alloys, construction and tool steels. The experiments showed that this approach gave a higher performance and improved the quality of machined surfaces related to the even topology.

Typically, the electroerosively machined surfaces have a matte appearance, similar to the surface fired by multiple shots. It contains disordered distributed tiny craters as a result of the material removal by separate discharges. The surface structure with the presence of craters and tubercles can be a favorable indicator for the retention of lubricants in the mechanisms and the formation of working cavities of injection molds. This type of surface is called "shagreen leather" [46]. It is easily distinguishable on any molded plastic products, mainly on computer and television case products, but also in the products for the transportation and sale of foods and drinks, toys for outdoor games.

There remains a layer of adsorbed substance, consisting mainly of components of the electrode-part and sometimes of components sublimated from the electrode-tool and decay products of the dielectric medium on the electroerosively machined surfaces. Mostly carbides are formed during processing in the organic dielectric as oil or kerosene and oxides during processing in deionized water [13,14,19,26]. The carbides are formed as well from the composition of steel during processing steels, as carbon is one of the most refractory materials. It counts as well for other alloys and complex materials when refractory components form the recast or white layer [11,14,46,49–58]. In the case of processing ceramic composites and nanocomposites, the surface layer is saturated with low-melting components in a nano-frame of refractory components [4,29,52]. It was observed that this recast or white layer is more hard and brittle than the main material of the workpiece. The subsurface layer can be easily distinguishable on cross-sections of steels, nanocomposites based on oxide ceramics, other materials and alloys using scanning electron microscopy (Figure 4). In the case of machining steels, carbon saturated layer has a thickness of 2.5–50 μm. It does not exceed 4–10 μm for pure 99% chromium. The recast layer is considered favorable for shaping the cavities of injection molds. In cases where high fatigue strength of the product is required, the white layer can be removed mechanically or electrochemically [51].

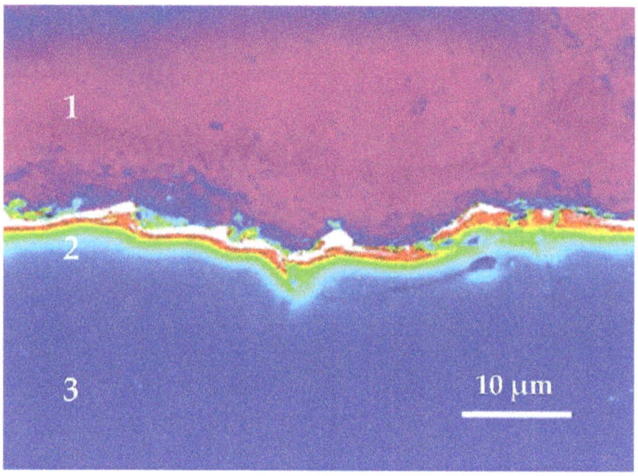

Figure 4. White layer obtained by electric discharge machining 99.9% pure chromium by a brass-wire electrode in a water medium: (**1**) is an epoxy resin that contains a sample for SEM analysis, (**2**) is a white layer of thermal exposure, a saturated component wire electrode—copper and zinc, (**3**) is a cut of the machined part made of chromium.

Obviously, the white layer and the erosion products are formed from the components of ceramics, dielectric medium, and auxiliary electrode during processing ceramics, the components of the primary phase and conductive additives of ceramic composites and nanocomposites also interact with other components in processing.

In some cases, the formed erosion products may have properties incompatible with the concept of safety-toxic or explosive gas or reactive sediment. Therefore, it is necessary to do carefully analyze the composition of all used main and auxiliary materials in the interelectrode gap, especially before designing the processing conditions for new materials.

3. Properties of Structural Ceramics in the Presence of Heat and Pressing Methods

The ceramics are non-metallic and non-organic materials based on crystal compounds of non-metals and metals synthesized and consolidated by various methods to impart specific properties and geometry. At the same time, ceramics is a group of phase-sensitive materials—the properties of ceramics with a similar chemical compound may significantly differ depending on precursors, methods and technology parameters, structure, and phase composition of sintered materials, intermediate processing methods, surface quality, properties estimation methods, exploitation conditions.

The hard-to-melt compounds, which form the basis of structural ceramics, are characterized by low operational ability; in other words, they are difficult in forming blanks, sintering to high-density values, and processing to the required dimensions.

All used ceramics can be divided into structural and functional types. In this context, the structural ceramics provide the integrity and load-bearing capacity of the designs of various products, when physical and mechanical properties are of primary importance, when functional ceramics are aimed at specialized products and devices.

Although WC-Co system cermets still dominate the market of technical ceramics, some oxide ceramics can become a proper alternative in many production areas in the coming years [59–61]. Ceramic cutting tools based on oxide ceramics have gained popularity due to their ability to work in extreme conditions, when high wear resistance, hardness, and low chemical activity are required under operating conditions [62,63]. However, the scope of oxide ceramics as a structural material is still limited due to the low fracture toughness [64–66]. Cracks easily spread and often lead to uncontrolled fracture of specimens [67,68].

There are various methods for improving the mechanical properties of oxide ceramic parts. For example, as was mentioned, the ability to reinforce Al_2O_3, ZrO_2 by adding secondary phases made of carbide or nitride ceramics, metals, intermetallic compounds, or other materials to the ceramic matrix of composites or nanocomposites. The secondary phase in the form of powders and tubes, nano or micro platelets and fibers is helpful to increase strength and fracture toughness of ceramic matrix composites [69,70]. These methods have proven to be effective as experimental and innovative for increasing the properties of oxide ceramics as reinforced zirconia with Ti[C,N] or CeO_2 particles or reinforced alumina with multidimensional ceramic particles of SiC, WC, TiC, TiN [27,28,59,60,71–73].

Silicon nitride is also one of the most promising materials for critical applications because of its high strength, thermal conductivity, low coefficient of thermal expansion, high resistance to thermal shock, which are superior to oxide ceramics. Silicon nitride and Silicon alumina nitride (SiAlON) can be called the most suitable ceramics for machining superalloys. At the same time, silicon nitride Si_3N_4 has low hardness and wear resistance [74,75].

There is a few ceramics compaction technique, sometimes combined with sintering that influences the physical and mechanical properties of blanks [76,77]. Cold static pressing is a unique and straightforward technique for pressing the products of different shapes from any compounds that allow realizing it in the conditions of mass production [78]. However, it does not allow distributing the properties of the product in the volume evenly.

The hot-pressing method solves the problems of the cold pressing method partly since it provides better strength characteristics of the products, the minimum tolerances on the size of the workpieces, reduces of sintering time due to the combination of the pressing and sintering processes. The disadvantage of this method is the rapid wear of the molds and the low productivity of the process [79,80]. Applying a hot-pressing method improves the strength and macrostructural characteristics of the workpieces. However, ferroelectric and ferromagnetic properties deteriorate. That can be a disadvantage for electroceramics but an advantage for structural ceramics. Significant grain growth in zirconia ceramics made by hot-pressing can be observed. In addition, ceramic ferroelectrics obtained by hot-pressing, in comparison with conventional annealing, deteriorate performance due to the presence of residual stresses and violation of stoichiometry of the molar masses of the components.

The hot-pressing method allows obtaining products with a more uniform distribution of density throughout the volume and effective in the production of some types of nanoceramics with superplasticity at elevated temperatures, which can significantly reduce the cost of products due to the elimination of machining. However, the use of the hot-pressing method is limited by high demands on the material of the mold, which must be inert concerning the pressed powders at elevated temperatures, heat-resistant, not having the property of superplasticity, economical.

Hot isostatic pressing is one of the varieties of the method that implemented using various techniques based on the use of elastic media—liquids, gases, polyurethane, etc. by a disposable mold, which avoids the friction forces of the powder granules on the stationary parts of the mold, which is indispensable for the conditions of mass production [81,82].

Other pressing methods are usually aimed at eliminating the drawbacks of previous methods, for example, dynamic pressing methods as isothermal stamping, magnetic pulse (with amplitude up to 5 GPa, when the pressure of conventional pressing methods does not exceed 1 GPa), explosive, hydrodynamic, impact or shock (up to 35 GPa), shear compaction, electrical consolidation, various ultrasound vibration moldings (in the range of 16–22 kHz), pressureless packing [83–89]. They are aimed at improving the density of the workpieces and more uniform density distribution, relieving stresses, preventing grain growth during subsequent sintering. Their use is always limited by the maximum achievable dimensions of the workpiece, the need to use methods of preliminary destruction of agglomerates, and the removal of sorbed gases.

In addition, when choosing ceramics, designers are often guided not only by their physical and mechanical properties but also by their appearance, which is extremely important for some applications and depends not only on the composition of ceramics but also on the methods of their pressing and sintering [77]. For example, in the production of samples by hot-pressing, preforms are predominantly black-gray, while white is predominant in ceramics obtained using cold pressing methods. That correlates to the fundamentals of physical chemistry when other chemical transformations occur in the presence of heat [4,14,29].

The relationship of the basic physical and mechanical properties of some materials to develop requirements for the material is shown in Figure 5. It should be noted that in the presented figure, the range of parameter values for Al_2O_3 and ZrO_2 refers to samples obtained by isostatic pressing from powder, and Elbor refers to any trademarks of cubic boron nitride (c-BN).

It should be noted that oxide ceramics as alumina Al_2O_3 and zirconia ZrO_2 among the mentioned above properties including low electrical conductivity at room temperatures show other highly relevant for some application properties as chemical inertness even at elevated temperatures or in the presence of an oxidizing agent, hot hardness (or redness, high hardness, and wear resistance to temperatures of red heat), low coefficient of thermal conductivity, and low coefficient of thermal expansion in comparison with more widespread construction materials (Table 1).

As can be seen, despite the quite high thermal conductivity and hardness and low thermal expansion coefficient, alumina Al_2O_3 has lower flexural strength and crack growth resistance in comparison with zirconia ZrO_2.

It should be noted that electrical conductivity of ceramics that contain vitreous (glassy) phase (amorphous) increases significantly with an increase in temperatures since the concentration and mobility of charge carriers are exponentially dependent on temperature (t):

$$\varsigma = \varsigma_0 e^{\beta t}, \tag{1}$$

$$\chi = \chi_0 e^{-\beta t}, \tag{2}$$

where ς_0 and χ_0 are values of specific electrical conductivity and volume resistivity at 0 °C, β is the temperature coefficient.

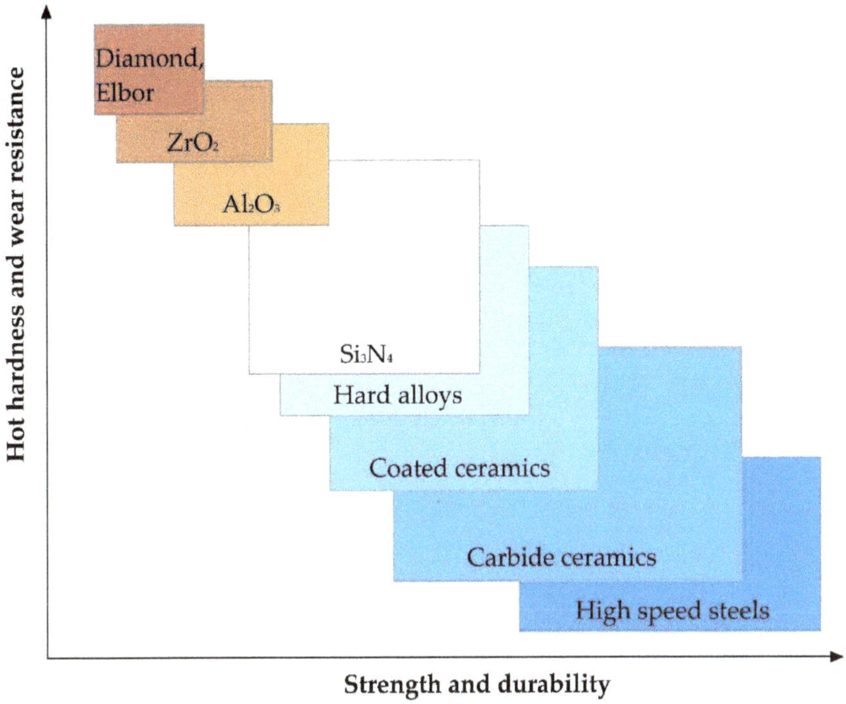

Figure 5. The relationship of the basic physical and mechanical parameters for some materials.

Table 1. The main properties of some materials [90–94].

Physical, Mechanical and Electro Properties	Al_2O_3	ZrO_2	Si_3N_4	Stainless Steel *
Density (ρ), g/cm^3	3.8–4.0	6.0–6.05	2.37–3.25	7.6–7.95
Melting point, °C	2044	2715	1900	1420
Flexural strength (σ), MPa	300–350	750–1050	650–800	110–550
Vickers hardness (HV), GPa	19–21	12–13	16–19.6	0.129
Fracture strength (k_{1c}), Pa·m$^{\frac{1}{2}}$	3.0–3.5	8.0–10.0	6.5–7.2	-
Thermal conductivity (k), W·m^{-1}·K^{-1}	25–30	2–3	10–43	30–45
Thermal expansion coefficient (α), 10^{-6}·K^{-1}	8.0–9.0	10.0–11.0	1.4–3.7	9.9–17.3

Table 1. *Cont.*

Physical, Mechanical and Electro Properties	Al_2O_3	ZrO_2	Si_3N_4	Stainless Steel *
Dielectric permittivity (ε), F/m	9.5–10	~25	~7–7.5	-
Specific electrical conductivity ($\varsigma_{20°}$) at room temperatures (20 °C), $S \cdot m^{-1}$	1.0×10^{-10}–1.0×10^{-12}	1.0×10^{-6}–1.0×10^{-8}	1.0×10^{-4}–1.0×10^{-13}, minimal volume electrical resistivity (χ, Ω·m) 19 at 20 °C	0.5–0.8×10^{7}
Specific electrical conductivity ($\varsigma_{1000°}$) in the presence of high temperatures (1000 °C), $S \cdot m^{-1}$	No more than 1.0×10^{-6}, electrical conductivity is higher in powder materials due to impurities, at low pressures - n-type semiconductor	1.0–10.0	Workability up to 1350 °C, min volume electrical resistivity (χ, Ω·m) 15 at 200 °C, 12 at 600 °C; that corresponds to 6.7×10^{-2} and 8.3×10^{-2} [95–97]	0.81–0.86×10^{-10}

* Given for reference [98,99].

The electrical conductivity of crystalline ceramics (for titanium(IV) dioxide TiO_2, it matches the form of anatase-metastable mineral), changes slower and retain quite low values at very high temperatures.

It should be noted that the porous ceramics significantly increases specific electrical conductivity in the presence of moisture, even in the smallest quantities [95–97].

The study devoted to the evaluation of ceramics workability by electrical discharge machining with an auxiliary electrode [100] describes the properties of non-conductive ceramics as ZrO_2, Si_3N_4, and SiC subjected to electrical discharge machining. The Vickers hardness (*HV*0.3), surface roughness (*Sq*), and flexural strength of samples were compared before and after machining since the use of the methods based on electrical erosion of the materials traditionally causes a decrease in these parameters due to the nature of the material destruction and the zone of thermal influence (or heat-affected zone, HAZ). It was found that the effect shows changes three times higher than the parameters before processing for ZrO_2 samples, and twice for in Si_3N_4. It can correlate with specific electrical conductivity in the presence of high temperatures ($\varsigma^{1000°}$), thermal expansion coefficient (α), and chemical transactions of components that occurred in the presence of heat.

4. Features of Structural Ceramics Sintering

The ceramics in the form of billets obtained by powder pressing and sintering are usually processed by grinding using abrasive materials or solid tools made of technical diamond or c-BN to give them a more functional shape depending on their applications [6,101–105]. Processing ceramics of complex shapes using traditional processing methods is a complicated and time-consuming process due to their exceptional physical and mechanical properties.

The technological process of powder metallurgy for ceramic sintering consists of four main stages: production of powders, their mixing and homogenizing, pressing or molding, and subsequent sintering [106–112]. In all similar conditions, the method of powder compaction of billets often determines the properties of the final product and plays a decisive role in the formation of the internal structure as uniformity, absence of pores, and cracks [113–115]. Forming billets is often carried out by cold pressing under high working pressure up to 1000 MPa in metal dies of a press that is oriented vertically. A mixture of powders is free-poured into the cavity of the dies between the upper and lower punches or several punches in the case of production with several transitions [116]. The stroke of the lower punch controls the dosage volume. The formed workpiece is pushed out of the cavity of the lower punch. For molding, specialized press equipment with mechanical, hydraulic, or pneumatic drives is used. The resulting billet has the dimensions and shape of the final product and the strength is sufficient for unloading and transporting the pre-form into the furnace for subsequent sintering. The workpiece is sintered to increase its strength and ductility. For the production of high-precision parts, sintering is carried out in furnaces with a reducing (hydrogen, dissociated ammonia) or protective (nitrogen, argon, endothermic gas) atmosphere. Forming is also often carried out using the molding method in combination with sintering by electric current pulses [117].

Traditional sintering methods based on an isostatic process form a ceramic preform in the presence of high temperatures (up to 1100 °C). The sintering process can take 1–5 h to obtain a denser structure of the workpiece. At the same time, it is believed that high temperature and prolonged sintering time lead to undesirable grain growth, which may adversely affect the operational properties of the material [38,118,119]. Some chemical transactions can occur with impurities in the ceramic powder during sintering. It leads to the formation of a precipitate that has properties exceeded main material properties and gas evolution. Thus, excessive porosity can be obtained at high temperatures of sintering, which obviously also reduces the physical and mechanical properties of the ceramics.

Spark plasma sintering (SPS) is often used as an alternative method of consolidating powder billets to reduce sintering temperatures. The use of this technology in combination with the use of additional currents facilitates, the rapid consolidation of the powder, makes it possible to produce sufficiently dense ceramic composites and nanocomposites, and prevents grain growth [120–122]. The technology also offers significant advantages, such as a faster and shorter sintering time compared to conventional methods of forming powder billets, which is primarily associated with the complex effect of mechanical pressure on the billet and electrical impulses during processing [123–125].

Spark plasma sintering is an advanced technology with high potential for material processing. Its principles are similar to hot-pressing and differ in the source of heat. SPS uses pulsed direct current (DC), which is passed through the dies and acts in such a way as a heat source that corresponds to the Joule effect. It is used to minimize grain growth and obtain a dense nanostructured material.

A machine for spark plasma sintering consists of a press sintering machine with a vertical uniaxial pressurization mechanism, specially designed electrodes with adjustable water pumping, a cooler, a water-cooled vacuum chamber, a vacuum/air control mechanism, an argon-gas atmosphere control mechanism, a DC pulse generator, a water cooling control unit, a control unit for the position of the upper punch along the Z-axis, a temperature measuring unit (Figure 6).

During sintering, the compacted powder is compressed between the die and punches to which the electrodes are connected. After, the spark discharge is supplied, with the help of which high temperatures are obtained. It leads to thermal and electrolytic diffusion between the powder particles that depend on their turn on the electrical properties of the powder material in the presence of heat. Sintering occurs in the process at temperatures between 1000 and 2500 °C. The heat distribution and the temperature difference between the core and the edges of the sintered billet are determined by the thermal conductivity of the billet material, which, for example, is ten times lower for zirconia than for alumina (Table 1). Cycle time and sintering power consumption depend on the thermal conductivity of the material [126]. The sintering cycle is selected in such a way as to ensure minimal grain growth of the sintered billets, which should provide higher physical and mechanical properties.

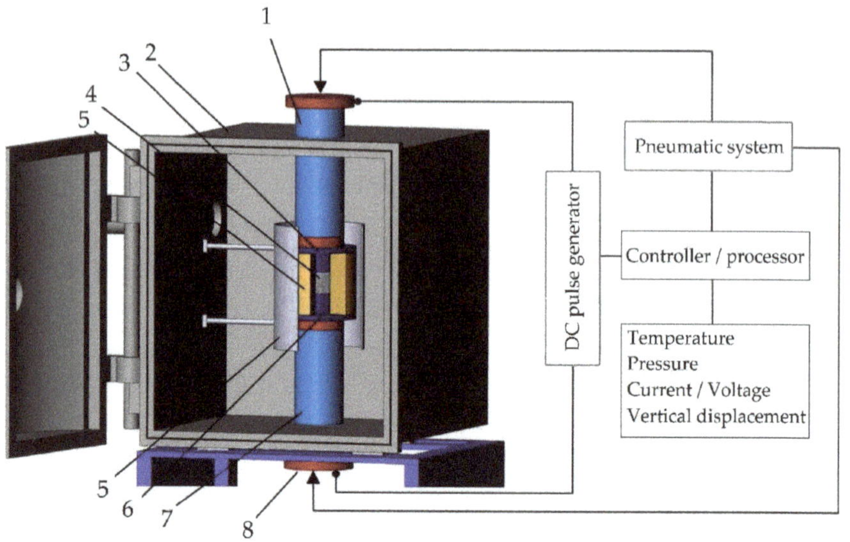

Figure 6. The configuration of the spark plasma sintering system developed in MSTU STANKIN: (**1**) is an upper electrode, (**2**) is a vacuum chamber, (**3**) is an upper punch, (**4**) is pressed powder, (**5**) is a die block, (**6**) is a heater, (**7**) is a lower punch, (**8**) is a lower electrode.

Conventional electric hot press processes using direct current or power are controlled by the basic process parameters that provide Joule sintering. These parameters or factors are related to the characteristics of the power supply, or high-frequency induction of the elements in combination with the smooth loading compacted powder materials through hydraulic or mechanical pressure.

The phenomena generated by spark plasma and pressure exclude the influence of gas adsorption, the formation of oxide films, and interactions with impurities and suspended particles that remain on the surface of the powder when a high-temperature field occurs. The action of the electromagnetic field is enhanced, and a high diffusion rate is provided by the rapid migration of ions. The local high-temperature state caused by impulse voltage is accompanied by vapors, solidification, and recrystallization.

SPS at high pressures is another version of the well-established method of spark sintering for shaping certain materials, such as superhard polycrystalline diamonds, cubic boron nitride, ceramic composites, nanocomposites, including those having refractory properties. Ceramics can also be synthesized into metastable phases or intermetallic alloys using the entire variety of SPS methods [127,128].

Since the sintering process is accompanied by uniaxial compression, the applied force is always limited by the high-temperature tensile strength of a graphite press tool, which is around 100–150 MPa. High pressure prevents the growth of grains in billets. It should be noted that there is a dependence between the physical properties of the sintered workpiece and the particle size of the ceramic powder, which is also determined by the different dispersion of the composition of the powder [129].

SPS consists of the following stages (Figure 7). In essence, vacuum conditions are created inside the working chamber, the workpiece is warmed up and is subjected to pressure, while at the point of contact of the powder particles there arises a spark discharge with the formed zones of local overheating, with continued heating, the surface of the powder grains reaches melting points, isthmuses form between particles, the mold is cooled down together with the workpiece.

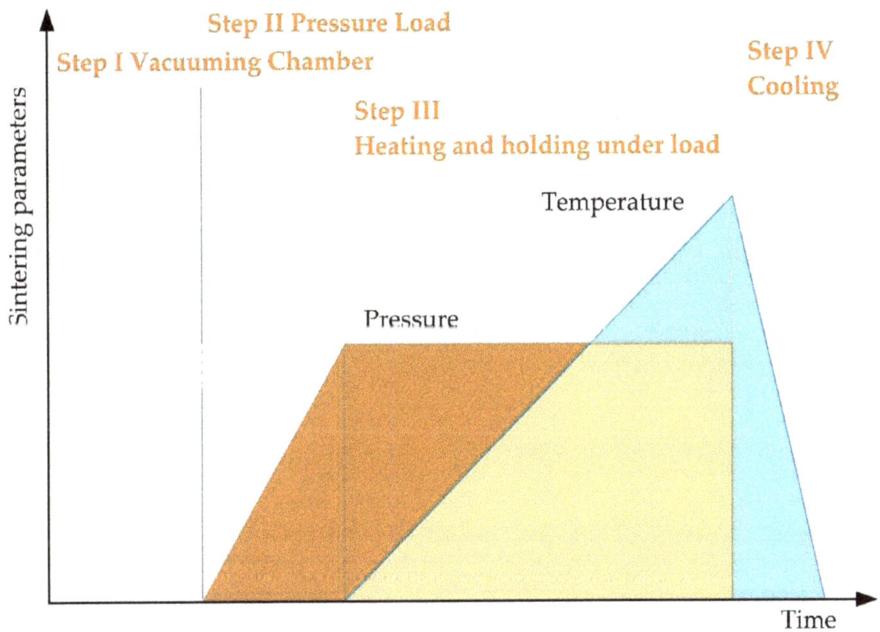

Figure 7. The main stages of spark plasma sintering.

The mold is heated using a sufficiently low voltage of ~10 V and a current of up to 10 kA. The maximum achievable temperature in graphite molds is 2400 °C. The cooling rate reaches 150 °C/min but can be increased to 400 °C/min using additional gas cooling.

The sintering temperature is usually controlled either inside the wall of the graphite dies using a thermocouple, or superficially using a pyrometer during sintering. It should be noted that there are always differences in temperatures between the measured value and the actual sintering temperature in the mold. It has been experimentally established that the measured temperature during sintering metals and ceramic composites can vary between 50–250 °C in comparison with the actually achievable temperature in the center of the mold. The temperature measured on the surface is always lower than the actual value in the center of the preform.

The heterogeneity of heating can lead to increased porosity of the samples, which reduces not only physical and mechanical properties but also thermal and electrical conductivity of the final product [4,5,38,126]. The nature of electrical phenomena during sintering depends on the electrical properties of the raw material.

5. Chemical Interaction of Ceramics Components with Electrodes, Powders, and Medium

Many materials change their electrical conductivity when heated and become semiconductors and conductors from insulators. A reverse transition is also possible for metals. Such properties during the design of the technology can have a positive effect on the electroerosive machinability of a part whose structure contains dielectric components that changes its conductivity in the presence of heat when the material surface forming the interelectrode gap reaches sublimation temperatures in the conditions of low-temperature plasma when forming the discharge channel to the auxiliary electrode.

The introduction of a powder or nanosized powder of titanium dioxide TiO_2, cerium dioxide CeO_2, and other ceramic components into the interelectrode gap is one of the most spread approaches in electrical discharge machining of ceramics [4,24,25]. When the discharge channel temperatures reach a certain level to provoke sublimation of the surface to be machined, some of the ceramics acquire conductive properties as it was mentioned above. That causes denser discharges of pulses in the interelectrode gap during processing the main conductive material, leading to higher performance and improved quality of the processed surfaces

Thus, the erosion products and the surface layer of the processed surfaces are formed directly from the components of ceramics, dielectric medium, auxiliary electrode, assisting powder, secondary phase of materials in the case of nanocomposites in the presence of heat, some of the possible chemical interactions for alumina Al_2O_3 and zirconia ZrO_2 are provided in Table 2. It is necessary to carefully analyze the composition of all used primary and auxiliary materials before designing the processing conditions [130–137].

Table 2. The analytical chemical composition of erosion products and the subsurface layer of the processing surfaces during electrical erosion of some ceramics using a brass or copper electrode tool.

<table>
<tr><th>Auxiliary Electrode</th><th>Dielectric Medium</th><th>Chemical Properties in the Presence of Heat</th></tr>
<tr><td></td><td></td><td>Al_2O_3 ceramics [19–22,138–143]</td></tr>
<tr><td rowspan="2">Any</td><td>Water</td><td>It is stable up to 2044 °C, chemically stable, insoluble in water, at 1000 °C it can interact with alkali metals to form aluminates as $NaAlO_2$, which react with water:
$NaAlO_2 + 2H_2O \rightarrow NaOH + Al(OH)_3$
During fusion, it can form anhydrous aluminates, for example, $Ca(AlO_2)_2$, which with water can form calcium hydroaluminates, $CaO{\cdot}Al_2O_3{\cdot}H_2O$ therefore it can only be processed in distilled and deionized water</td></tr>
<tr><td>Oil/Kerosene</td><td>Aluminum carbide Al_4C_3 is obtained by direct reaction:
$4Al + 3C \rightarrow Al_4C_3$
Or in the reaction of alumina Al_2O_3 with carbon C in the presence of heating to 1800 °C:
$2Al_2O_3 + 9C \rightarrow Al_4C_3 + 6CO\uparrow$
It is steady up to 1400 °C, reacts with water:
$Al_4C_3 + 12H_2O \rightarrow 4Al(OH)_3\downarrow + 3CH_4\uparrow$
With hydrogen at 2200 °C:
$Al_4C_3 + 6H_2\uparrow \rightarrow 4Al + 3CH_4\uparrow$
With oxygen at 650–700 °C:
$Al_4C_3 + 6O_2\uparrow \rightarrow 2\ Al_2O_3 + 3CO_2\uparrow$
It is used in pyrotechnics</td></tr>
</table>

Table 2. *Cont.*

Auxiliary Electrode	Dielectric Medium	Chemical Properties in the Presence of Heat
Ti, Cr, Co, C	Water	Titanium(IV) dioxide TiO_2 becomes a semiconductor at 1000 °C with specific electrical conductivity up to 10^3 S·m^{-1}
		Chromium(II) oxide CrO decomposes into metallic chromium and chromium(III) oxide at 697 °C: $3CrO \rightarrow Cr_2O_3 + Cr$ Hydrogen is reduced to metallic chromium at 1000 °C: $CrO + H_2\uparrow \rightarrow Cr + H_2O$ Chromium(II) oxide Cr_2O_3 is resistant up to 2440 °C, toxic and causes dermatitis on contact with skin, is used as abrasive and catalyst in a number of organic reactions; in a strongly acidic environment, a reaction may occur: $Cr_2O_3 + 6H^+ + 9H_2O \rightarrow 2[Cr(H_2O)_6]^{3+}$ Strong reducing agents reduce it: $Cr_2O_3 + 2Al \rightarrow Al_2O_3 + 2Cr$ Chromium(IV) oxide CrO_2 is obtained by heating hydrated chromium(III) oxide Cr_2O_3 in oxygen at 350–400 °C: $Cr_2O_3{\cdot}nH_2O + O_2\uparrow \rightarrow 4CrO_2 + 2nH_2O$ It decomposes on heating at 420–510 °C: $4CrO_2 \rightarrow 2Cr_2O_3 + O_2\uparrow$ It reacts with water at 100 °C: $3CrO_2 + 2H_2O \rightarrow 2CrO(OH) + H_2CrO_4$ $3CrO_2 \rightarrow Cr_2O_3 + CrO_3$ Chromium(VI) oxide CrO_3 has the melting point at 196 °C, when it is dissolved in water, chromic acid is formed with a lack of CrO_3: $CrO_3 + H_2O \rightarrow H_2CrO_4$ It forms dichromic acid in water with an excess of CrO_3: $2CrO_3 + H_2O \rightarrow H_2Cr_2O_7$ It decomposes with the formation of chromium(III) oxide and oxygen at 250 °C: $4CrO_3 \rightarrow 2Cr_2O_3 + 3O_2\uparrow$
		Cobalt oxide Co_3O_4 is a complex oxide having a spinel structure and stable at room temperature; it decomposes with the formation of cobalt monoxide CoO at temperatures above 900 °C The α-form or β-form of cobalt monoxide CoO can be obtained at high temperatures All cobalt oxide Co_3O_4 is reduced with hydrogen: $Co_3O_4 + 4H_2\uparrow \rightarrow 3Co + 4H_2O$
		Aluminum carbide Al_4C_3 can be formed
Zn, Ag, Au, Al	Water	Zinc oxide ZnO sublimates at a temperature of 1800 °C, when fused with silicon dioxide forms a refractory glassy silicate $ZnSiO_3$
		Silver(I) oxide Ag_2O decomposes when heated above 280 °C Silver(I,III) oxide Ag_2O_2 decomposes when heated above 100 °C
		Gold(I) oxide Au_2O is unstable under normal conditions, decomposes in Au and Au_2O_3 at 225 °C Gold(II) oxide Au_2O_3 is stable, has excellent conductive properties, is used in microelectronics
		Alumina Al_2O_3 is insoluble in water, is an n-type dielectric or semiconductor at nominal pressure

Table 2. *Cont.*

Auxiliary Electrode	Dielectric Medium	Chemical Properties in the Presence of Heat
		ZrO_2 ceramics [144–150]
Any	Any	Becomes a semiconductor in the presence of temperatures
		The released zirconium has the property to burn in oxygen at a self-ignition temperature of 250 °C with high speed and smokeless, which is why it has become widespread in pyrotechnics
		The released zirconium actively absorbs hydrogen forming hydrides ZrH_x at 250–300 °C [151] that can develop super-conducting properties under pressure of 150 GPa at 10.6 K [152]; the hydrides dissociate at 1200–1300 °C
	Water	Does not interact with water
Ti, Cr, Co, C		Most of the possible components' interactions are similar to the mentioned above
		The sublimation point of carbon C is 3642 °C; it does not exist in liquid form at normal pressures
		Zirconium forms zirconium carbide ZrC with carbon C, which is an excellent conductor, at temperatures above 900 °C; its electrical properties are similar to those of pure zirconium
Zn, Ag, Au, Al		Most of the components' interactions are similar to the mentioned above
Any	Oil/Kerosene	It forms zirconium carbide ZrC at temperatures above 900 °C
Ti, Cr, Co, C		The formed titanium carbide TiC has a slightly higher electrical resistance than titanium nitride, a semiconductor, becomes a conductor with increasing temperature
		Chromium carbide Cr_3C_2 has the specific electrical conductivity * of 1.3×10^6 $S{\cdot}m^{-1}$; chromium carbides have melting point in the range of 1520–1890 °C; chromium carbide Cr_7C_3 turns into $Cr_{23}C_6$ after prolonged heating at 730–870 °C; Cr_3C_2 carbide insoluble in water but it can interact with zinc Zn at a temperature of 940 °C; the oxidation of chromium carbide begins at temperatures of 700–1100 °C
		Tricobalt carbide Co_3C in the presence of 3Co + C at a temperature of 1800 °C decomposes upon crystallization. At temperatures above 230 °C, tricobalt carbide reacts with hydrogen to form methane at 250 °C, while methane and metallic cobalt are released
		Copper(I) acetylenide Cu_2C_2 in dry form is explosive, detonates when heated or impact, forms hydrates in the presence of water $Cu_2C_2{\cdot}H_2O$ Copper(II) acetylenide CuC_2 forms hydrates in the presence of water $CuC_2{\cdot}\frac{1}{2}H_2O$, insoluble in water, but decomposes with explosion upon drying and loss of water
Zn, Ag, Au, Al		Zinc carbide ZnC_2 reacts with water to form acetylene (colorless flammable gas): $ZnC_2 + 2H_2O \rightarrow Zn(OH)_2 + C_2H_2\uparrow$
		Silver acetylide Ag_2C_2 (CAg≡CAg) is a very unstable and explosive compound, explodes when heated and mechanically exposed
		Gold carbide Au_2C_2 is insoluble in water and explosive, with ammonia forms an explosive adduct $Au_2C_2{\cdot}NH_3$

* At 20 °C.

As can be seen in Table 2, titanium Ti and chromium Cr are most suited for electrical discharge machining of zirconia ZrO_2 in oil or kerosene from all considered transition metals as they provoke the formation of conductive particles in the interelectrode gap.

EDM of zirconia ZrO_2 in oil or kerosene and with assisting carbon particles or nanotubes can form conductive zirconium carbide ZrC by interacting with oil or kerosene decomposition components and elements of the auxiliary electrode or assisting powder. Then the erosion products can contribute to a denser erosion in the interelectrode gap.

Zirconia ZrO_2 tends to change its electrical properties in the presence of heat up to 2000 °C (Figure 3), but it is inert to the water medium. The zirconium starts actively to absorb hydrogen H_2 at temperatures around 250–300 °C forming solid solution and hydrides ZrH_x (x = 0.05–2) representing fragile sulfur black powder. The hydride powder is stable at room temperature but ignites at 430 °C on the air when the hydride crystals are stable up to 600–750 °C and then they decompose with the release of hydrogen and dissociate to the pure metal at 1200–1300 °C in a vacuum [153,154]. Thus, the released hydrides may not cause any difficulties during electrical discharge machining.

Processing alumina Al_2O_3 in water should not provoke the appearance of new chemically active substances in the interelectrode gap, while processing in oil or kerosene can cause of damage of the filtration system of the equipment due to the formation of explosive and chemically active aluminum carbide Al_4C_3.

It seems that the use of copper Cu, silver Ag, gold Au, or aluminum Al in the form of the film is the most suitable for EDM of alumina Al_2O_3 in water. In normal conditions, alumina of the workpiece, an oxide film of the auxiliary electrode made of aluminum, and formed alumina particles in the process of electrical erosion are inert to water. Ag and Au films and their oxide films do not form compounds that are resistant to high temperatures: the formed gold(III) oxide Au_2O_3 is a thermally unstable conductor.

It should be noted that, usually, EDM in oil or kerosene provides more uniform morphology than in water due to more even heat removal from the treatment zone and the damping effect of a more viscous medium to compensate for forced oscillations of the wire electrode during processing [4,11,26,49].

The use of cerium(IV) dioxide CeO_2 (Table 3) as an assisting powder with brass wire or profiled electrode is not highly recommended. It has a similar but brighter nature as the uses of the brass tool during EDM of nickel-containing steel or Ni-coated workpieces. The cerium Ce and nickel Ni react with zinc Zn of brass very actively with the formation of intermetallides. The reaction of nickel Ni and zinc Zn at temperatures above 1000 °C has an explosive character that can be observed as a series of sparks in the interelectrode gap during processing [26,44,53,155,156]. The reaction of cerium Ce and zinc Zn has a more violent character that is typical for cerium. The reaction at high temperatures during local heating and consequent fusion of two powders provide a bright flash and a powerful explosion [157,158].

One of the useful properties of titanium carbide TiC assisting powder can be its interaction with nitrogen N at a temperature above 2500 °C. It forms conductive titanium nitride TiN that can be an advantage in EDM of non-conductive silicon nitride Si_3N_4 [27,28].

Table 3. The analytical chemical composition of erosion products and the subsurface layer of the processing surface during electrical erosion of some ceramics using a brass or copper electrode tool [129,159–163].

Assisting Powder	Dielectric Medium	Chemical Properties in the Presence of Heat
		Al_2O_3 ceramics
		+ conductive additives
W	Water	Tungsten W is heat- and chemically resistant under standard conditions; the sublimation point is 3422 °C; it exists in liquid form only at high pressures; it is slowly oxidized to tungsten(VI) trioxide WO_3 at a temperature of red heat; oxidation of tungsten W in an atmosphere of oxygen occurs at temperatures above 500 °C: $2W + 3O_2\uparrow \rightarrow 2WO_3$ It is reduced to metallic tungsten W by hydrogen at a temperature of 700–900 °C
Cu		Melting point is 1083.4 °C, it is oxidized to copper(I) oxide Cu_2O, which does not react with water and dissociates to a small degree, with oxygen deficiency at a temperature of 200 °C: $4Cu + O_2\uparrow \rightarrow 2Cu_2O$ and to copper(II) oxide CuO with oxygen excess at temperatures of about 400–500 °C or by heating metallic copper in the air at temperatures below 1100 °C: $2Cu + O_2\uparrow \rightarrow 2CuO$ Copper(II) oxide CuO is reduced to metallic copper by hydrogen (exothermic reaction): $CuO + H_2\uparrow \rightarrow Cu + H_2O + Q\uparrow$ It also decomposes into copper Cu and oxygen O, when heated to 1100 °C
		+ oxide ceramics
TiO_2	Water	As was mentioned above, titanium(IV) dioxide TiO_2 becomes a semiconductor at 1000°C with electrical conductivity up to 10^3 $S{\cdot}m^{-1}$
CeO_2		Cerium(IV) dioxide CeO_2 has a relatively high ionic conductivity of oxygen at 500–800°C; it also exhibits high electronic conductivity at low oxygen partial pressures; melting point is 2400 °C; refectory material up to 2300 °C in an oxidizing or inert atmosphere; cerium(IV) oxide reduces to cerium(III) oxide Ce_2O_3 with hydrogen gas at about 1400 °C: $2CeO_2 + H_2\uparrow \rightarrow Ce_2O_3 + H_2O$ And, on the contrary, cerium(III) oxide is oxidized to cerium(IV) oxide with an excess of oxygen: $2Ce_2O_3 + O_2\uparrow \rightarrow 4CeO_2$ Cerium(III) oxide Ce_2O_3 is stable on air, but cerium Ce in powder is pyrophoric and unstable at room temperature; presence cerium as an alloying element for aluminum alloy reduces its conductivity; the melting point is 2177 °C Cerium Ce reacts fulminant with some metals as zinc Zn at higher temperatures and forms intermetallides with heat and light emission; thus using a brass tool is not recommended for EDM in the presence of cerium Ce
SnO_2		Tin(IV) oxide is a wide-gap n-type semiconductor, electrical resistivity 3.4 × 10^3 Ω·cm (electrical conductivity ~2.94 × 10^2 $S{\cdot}m^{-1}$); doping with elements of group V (for example, antimony Sb) increases electrical conductivity by three-five orders of magnitude; the melting point is 1630 °C; it evaporates with the decomposition of tin monoxide SnO (and its di-, tri- and tetramers) and oxygen at high temperatures; it is restored by hydrogen to metal tin at 500–600 °C: $SnO_2 + 2H_2\uparrow \rightarrow Sn + 2H_2O$
ITO		Indium tin oxide (ITO) is semiconductor material with n-type conductivity comparable to metallic, where tin ions serve as electron donors; it is a solid solution of 90% indium(III) oxides and 10% tin(IV) $(In_2O_3)_{0.9}-(SnO_2)_{0.1}$; insoluble in water and extremely expensive in applications due to its transparency

Table 3. *Cont.*

Assisting Powder	Dielectric Medium	Chemical Properties in the Presence of Heat
		ZrO_2 ceramics
		+ conductive additives
WC	Any	The specific electrical conductivity * is ~5.22×10^2 S·m^{-1}; significant and rapid oxidation of tungsten carbide WC in air begins at 500–700 °C, and it is completely oxidized due to the high volatility of tungsten oxide at temperatures above 800 °C: $WC + 2O_2 \rightarrow WO_3 + CO\uparrow$ Tungsten(VI) trioxide WO_3 is reduced as well to metallic tungsten W by hydrogen H as it was mentioned above The temperature stability range for W_2C is up to 2750 °C; it is up to 2600 °C for WC; WC weakly interacts with a zinc melt at a temperature of 940 °C
	Oil/Kerosene	Tungsten(VI) trioxide WO_3 is reduced as well to metallic tungsten W by carbon C at a temperature of 1000 °C
TiC	Any	Melting point is ~3260 °C; it begins to react with nitrogen N at high temperatures above 2500 °C; it is decarburized during interacting with hydrogen H; the temperature of the active oxidation of titanium carbide is 1100–1200 °C; the temperature stability range of titanium carbide TiC is below 3140 °C; it is highly resistant to molten low-melting metals and metals such as copper, aluminum, brass, cast irons and steels
	Oil/Kerosene	It is oxidized by carbon dioxide CO_2 at temperatures above 1200 °C
Cu	Oil/Kerosene	It should be noted to the mentioned above that copper(II) oxide CuO is reduced as well to copper Cu by carbon(II) monoxide CO and carbon C: $2CuO + C \rightarrow 2Cu + CO_2\uparrow$
		+ oxide ceramics
TiO_2	Oil/Kerosene	As it was mentioned above, the formed titanium carbide TiC has a slightly higher electrical resistance than titanium nitride TiN (less electrically conductive); it is a semiconductor; it becomes a conductor with increasing temperature
CeO_2	Oil/Kerosene	Cerium(IV) oxide is reduced by carbon(II) monoxide CO to cerium(III) oxide, when there is not enough oxygen: $4CeO_2 + 2CO \rightarrow 2Ce_2O_3 + 2CO_2\uparrow$ The contrary reaction as presented in (27)
SnO_2	Oil/Kerosene	It should be added to the mentioned above, it is restored as well by carbon to metal tin at 800–900 °C: $SnO_2 + 2C \rightarrow Sn + 2CO\uparrow$

* At 20 °C.

6. Conclusions

It was shown that the erosion products and subsurface layer of the machined surfaces are formed directly from the components of ceramics, dielectric medium, auxiliary electrode, assisting powder during electrodes sublimation in the presence of heat. The chemical composition of the subsurface layer of the machined surfaces and the machinability of new materials not always depends on the electrophysical properties of the material but also on the combination of the materials of the primary and auxiliary electrode, conductive additives, assisting powder and workpiece.

All possible component interactions should be taken into account when developing the technology for electrical discharge machining non-conductive ceramics since the formation of certain chemicals in the form of insoluble or pyrotechnically dangerous sludge and gas can have dramatic consequences for the quality of the machined surfaces, the service life of the equipment and its units, and even for personnel.

The ceramic pressing methods, electrophysical and chemical properties of components in the presence of heat, data on possible interactions, and suitable combinations of the materials for processing structural oxide ceramics were presented for the most spread aluminum oxide and zirconium dioxide but can be applied for functional ceramics and nanocomposites.

According to the conducted analytical study, electrical discharge machining in oil or kerosene medium gives better qualities of the machined surface. A uniform surface morphology is formed due to more uniform heat removal and the damping effect of a more viscous medium to compensate for forced oscillations of the wire tool.

Titanium and chromium are most suited from the considered transition metals, taking into account the basic principles of physical chemistry, the laws of the structures of substances, their structure and transformations, available theoretical and practical data for electric discharge machining in oil or kerosene. They allow obtaining electrically conductive substances as erosion products, which can contribute to more dense erosion in the interelectrode gap.

During processing in a water medium, the use of silver, gold, aluminum as an auxiliary electrode are more suitable for copper or brass electrode, since aluminum is inert to water, silver and gold do not form compounds that are resistant at high temperatures. Moreover, the formed Au_2O_3 is the conductor but not heat-resistant.

Machining alumina in water should not provoke the appearance of chemically active substances in the interelectrode gap, while processing in oil or kerosene can have a negative influence on the filtration system of the equipment due to the formation of explosive and chemically active Al_4C_3 in the treatment zone. It is also evident that sintering alumina cannot be implemented in carbon dies, as the consequences of this combination can be non-electrical nature of the workpiece destruction and rapid wear of die paddings.

Probably, applying the brass tool during machining nickel alloys should be revised to the direction of using more chemically neutral to nickel materials.

Cerium dioxide cannot be used as assisting powder in combination with a brass tool as well.

During the development of the technology for ceramic machining, the preferred materials for the auxiliary electrodes, assisting powder and conductive particles should be the materials that provide conductive erosion dust in the interelectrode gap.

The developed approaches in electrical discharge machining structural ceramics can have an impact on the industry if the obtained data will be taken into account.

Author Contributions: Conceptualization, M.A.V.; methodology, A.A.O.; software, M.A.M.; validation, K.H.; formal analysis, K.H.; investigation, S.V.F.; resources, S.V.F.; data curation, M.A.M.; writing—original draft preparation, K.H., M.A.M.; writing—review & editing, A.A.O.; visualization, A.A.O.; supervision, S.V.F.; project administration, M.A.V.; funding acquisition, M.A.V. All authors have read and agreed to the published version of the manuscript.

Funding: This research was financially supported by the Russian Foundation for Basic Research, grant number 19-08-00899.

Acknowledgments: The research was done at the Department of High-Efficiency Processing Technologies of MSTU Stankin.

Conflicts of Interest: The authors declare no conflict of interest.

References

1. Grigoriev, S.N.; Gurin, V.D.; Volosova, M.A.; Cherkasova, N.Y. Development of residual cutting tool life prediction algorithm by processing on CNC machine tool. *Mater. Werkst.* **2013**, *44*, 790–796. [CrossRef]
2. Qudeiri, J.E.A.; Zaiout, A.; Mourad, A.-H.I.; Abidi, M.H.; Elkaseer, A. Principles and characteristics of different EDM processes in machining tool and die steels. *Appl. Sci.* **2020**, *10*, 2082. [CrossRef]
3. Mouangue Nanimina, A.; Abdul-Rani, A.M.; Ahmad, F.; Zainuddin, A.; Jason Lo, S.H. Effects of Electro-discharge Machining on Aluminium Metal Matrix Composite. *J. Appl. Sci.* **2011**, *11*, 1668–1672. [CrossRef]
4. Grigoriev, S.N.; Kozochkin, M.P.; Porvatov, A.N.; Volosova, M.A.; Okunkova, A.A. Electrical discharge machining of ceramic nanocomposites: Sublimation phenomena and adaptive control. *Heliyon* **2019**, *5*, e02629. [CrossRef] [PubMed]

5. Grigoriev, S.N.; Melnik, Y.A.; Metel, A.S.; Panin, V.V. Broad beam source of fast atoms produced as a result of charge exchange collisions of ions accelerated between two plasmas. *Instrum. Exp. Tech.* **2009**, *52*, 602–608. [CrossRef]
6. Grigoriev, S.; Metel, A. Plasma and beam-assisted deposition methods. In *Nanostructured Thin Films and Nanodispersion Strengthened Coatings*; Voevodin, A.A., Shtansky, D.V., Levashov, E.A., Moore, J.J., Eds.; Kluwer Academic Publishers: New York, NY, USA; Boston, MA, USA; Dordrecht, The Netherlands; London, UK; Moscow, Russia, 2004; pp. 147–154.
7. Grigoriev, S.; Melnik, Y.; Metel, A. Broad fast neutral molecule beam sources for industrial-scale beam-assisted deposition. *Surf. Coat. Tech.* **2002**, *156*, 44–49. [CrossRef]
8. Wijaya, H.; Wahyudi, S.; Soenoko, R.; Setyarini, P.H.; Yasid, S.; Gapsari, F. The effect of power supply current on recast layer in S45C steel using wire EDM. *IOP Conf. Ser. Mater. Sci. Eng.* **2019**, *494*, 012102. [CrossRef]
9. Muthuramalingam, T. Measuring the influence of discharge energy on white layer thickness in electrical discharge machining process. *Measurement* **2019**, *131*, 694–700. [CrossRef]
10. Tsai, D.C.; Lui, T.S.; Chen, L.H. Effect of graphite nodules on the EDM machinability of ferritic SG cast irons. *Mater. Trans. JIM* **2000**, *41*, 293–299. [CrossRef]
11. Grigoriev, S.N.; Kozochkin, M.P.; Kropotkina, E.Y. Okunkova, A.A. Study of wire tool-electrode behavior during electrical discharge machining by vibroacoustic monitoring. *Mech. Ind.* **2016**, *17*, 717. [CrossRef]
12. Betsofen, S. X-ray diffraction methods for the evaluation of residual stresses in the surface layers with gradient structure. *Mater. Sci.* **2006**, *42*, 367–375. [CrossRef]
13. Markopoulos, A.P.; Papazoglou, E.-L.; Karmiris-Obratański, P. Experimental study on the influence of machining conditions on the quality of electrical discharge machined surfaces of aluminum alloy Al5052. *Machines* **2020**, *8*, 12. [CrossRef]
14. Rahman, S.S.; Ashraf, M.Z.I.; Bashar, M.S.; Kamruzzaman, M.; Nurul Amin, A.K.M.; Hossain, M.M. Crystallinity, surface morphology, and chemical composition of the recast layer and rutile-TiO2 formation on Ti-6Al-4V ELI by wire-EDM to enhance biocompatibility. *Int. J. Adv. Manuf. Technol.* **2017**, *93*, 3285–3296.
15. Umar Farooq, M.; Pervez Mughal, M.; Ahmed, N.; Ahmad Mufti, N.; Al-Ahmari, A.M.; He, Y. On the investigation of surface integrity of Ti6Al4V ELI using Si-mixed electric discharge machining. *Materials* **2020**, *13*, 1549. [CrossRef] [PubMed]
16. Thangaraj, M.; Annamalai, R.; Moiduddin, K.; Alkindi, M.; Ramalingam, S.; Alghamdi, O. Enhancing the surface quality of micro titanium alloy specimen in WEDM process by adopting TGRA-based optimization. *Materials* **2020**, *13*, 1440. [CrossRef]
17. Świercz, R.; Oniszczuk-Świercz, D. Investigation of the influence of reduced graphene oxide flakes in the dielectric on surface characteristics and material removal rate in EDM. *Materials* **2019**, *12*, 943. [CrossRef]
18. Volosova, M.A.; Okunkova, A.A.; Povolotskiy, D.E.; Podrabinnik, P.A. Study of electrical discharge machining for the parts of nuclear industry usage. *Mech. Ind.* **2015**, *16*, 706. [CrossRef]
19. Niamat, M.; Sarfraz, S.; Shehab, E.; Ismail, S.O.; Khalid, Q.S. Experimental characterization of electrical discharge machining of aluminum 6061 T6 alloy using different dielectrics. *Arab. J. Sci. Eng.* **2019**, *44*, 8043–8052. [CrossRef]
20. Obrosov, A.; Gulyaev, R.; Zak, A.; Ratzke, M.; Naveed, M.; Dudzinski, W.; Weiß, S. Chemical and morphological characterization of magnetron sputtered at different bias voltages Cr-Al-C coatings. *Materials* **2017**, *10*, 156. [CrossRef]
21. Bains, P.S.; Singh, S.; Sidhu, S.S.; Kaur, S.; Ablyaz, T.R. Investigation of surface properties of Al–SiC composites in hybrid electrical discharge machining. In *Futuristic Composites*, 1st ed.; Sidhu, S., Bains, P., Zitoune, R., Yazdani, M., Eds.; Springer: Singapore, 2018; pp. 181–196.
22. Vozniakovskii, A.A.; Kidalov, S.V.; Kol'tsova, T.S. Development of composite material aluminum-carbon nanotubes with high hardness and controlled thermal conductivity. *J. Compos. Mater.* **2019**, *53*, 2959–2965. [CrossRef]
23. Razeghiyadaki, A.; Molardi, C.; Talamona, D.; Perveen, A. Modeling of material removal rate and surface roughness generated during electro-discharge machining. *Machines* **2019**, *7*, 47. [CrossRef]
24. Schubert, A.; Zeidler, H.; Hackert-Oschaetzchen, M.; Schneider, J.; Hahm, M. Enhancing micro-EDM using ultrasonic vibration and approaches for machining of nonconducting ceramics. *Stroj. Vestn. J. Mech. Eng.* **2013**, *59*, 156–164.

25. Grigoriev, S.N.; Masterenko, D.A.; Teleshevskii, V.I.; Emelyanov, P.N. Contemporary state and outlook for development of metrological assurance in the machine-building industry. *Meas. Tech.* **2013**, *55*, 1311.
26. Grigor'ev, S.N.; Kozochkin, M.P.; Fedorov, S.V.; Porvatov, A.N.; Okun'kova, A.A.; Kozochkin, M. Study of electroerosion processing by vibroacoustic diagnostic methods. *Meas. Tech.* **2015**, *58*, 878–884.
27. Volosova, M.A.; Grigor'ev, S.N.; Kuzin, V.V. Effect of titanium nitride coating on stress structural inhomogeneity in oxide-carbide ceramic. Part 4. Action of heat flow. *Refract. Ind. Ceram.* **2015**, *56*, 91–96.
28. Volosova, M.A.; Gurin, V.D. Influence of vacuum-plasma nitride coatings on contact processes and a mechanism of wear of working surfaces of high-speed steel cutting tool at interrupted cutting. *J. Frict. Wear* **2013**, *34*, 183–189.
29. Fominski, V.Y.; Grigoriev, S.N.; Romanov, R.I.; Volosova, M.A.; Grunin, A.I.; Teterina, G.D. The formation of a hybrid structure from tungsten selenide and oxide plates for a hydrogen-evolution electrocatalyst. *Tech. Phys. Lett.* **2016**, *42*, 555–558.
30. Volosova, M.A.; Grigoriev, S.N.; Ostrikov, E.A. Use of laser ablation for formation of discontinuous (discrete) wear-resistant coatings formed on solid carbide cutting tool by electron beam alloying and vacuum-arc deposition. *Mech. Ind.* **2016**, *17*, 720.
31. Glaziev, S.Y. The discovery of regularities of change of technological orders in the central economics and mathematics institute of the soviet academy of sciences. *Econ. Math. Methods* **2018**, *54*, 17–30.
32. Korotayev, A.V.; Tsirel, S.V. A spectral analysis of world GDP dynamics: Kondratiev waves, Kuznets swings, Juglar and Kitchin cycles in global economic development, and the 2008–2009 economic crisis. *Struct. Dyn.* **2010**, *4*, 3–57.
33. Schumpeter, J.; Becker, M.C.; Knudsen, T. The fundamental phenomenon of economic development. *Am. J. Econ. Sociol.* **2002**, *61*, 405–437.
34. Perez, C. Technological revolutions and techno-economic paradigms. *Camb. J. Econ.* **2010**, *34*, 185–202.
35. Bobovich, B.B. Glass-fiber reinforced plastics—construction materials of the sixth technological paradigm? *Glas. Ceram.* **2019**, *76*, 38–41.
36. Wonglimpiyarat, J. Towards the sixth Kondratieff cycle of nano revolution. *Int. J. Nanotechnol. Mol. Comput.* **2011**, *3*, 87–100.
37. Lee, M.; Borom, M.P. Rapid rate sintering of Al2O3-TiC composites for cutting-tool applications. *Adv. Ceram. Mater.* **1988**, *3*, 38–44.
38. Gao, L.; Hong, J.S.; Miyamoto, H.; Torre, S.D.D.L. Bending strength and microstructure of Al2O3 ceramics densified by spark plasma sintering. *J. Eur. Ceram. Soc.* **2000**, *20*, 2149–2152.
39. Lukashenko, S.V.; Kovtun, A.V.; Dashuk, P.N.; Sokolov, B.N. The Method of Electrical Discharge Machining of Dielectrics. Patent 1,542,715, 10 December 1986.
40. Mohri, N. EDM of advanced ceramics—From finish machining to machining insulating ceramics. *CIRP Ann. Manuf. Technol.* **1996**, *45*, 289–296.
41. Lazarenko, B.R.; Lazarenko, N.I. A Method of Working Metals and Other Electro-Conductive Materials and Means for Applying Same. United Kingdom Patent GB637793A, 3 April 1943.
42. Lazarenko, B.R.; Gitlevich, A.E.; Tkachenko, V.N. Certain characteristics of the discharge and plasma flow during the deposition of powdered coatings (Nekotorye Kharakteristiki Razryada I Plazmennogo Potoka Pri Nanesenii Pokrytii Iz Poroshkovykh Materialov). *Surf. Eng. Appl. Electrochem. (Elektronnaya Obrabotka Materialov)* **1974**, *60*, 29–31.
43. Lazarenko, B.R.; Mikhailov, V.V.; Gitlevich, A.E.; Verkhoturov, A.D.; Anfimov, I.S. Distribution of elements in surface layers during electric spark alloying [Raspredelenie Elementov V Poverkhnostnykh Sloyakh Pri Elektroiskrovom Legirovanii]. *Surf. Eng. Appl. Electrochem. (Elektronnaya Obrabotka Materialov)* **1977**, *3*, 28–33.
44. Lazarenko, B.R.; Duradzhi, V.N.; Bryantsev, I.V. Effect of Incorporating an additional inductance on the characteristics of anode and cathode processes (O Vliyanii Vklyucheniya Dopolnitel'noi Induktivnosti Na Kharakteristiki Anodnogo I Katodnogo Protsessov). *Surf. Eng. Appl. Electrochem. (Elektronnaya Obrabotka Materialov)* **1979**, *5*, 8–13.
45. Lazarenko, B.R.; Lazarenko, N.I. Electric spark machining of metals in water and electrolytes. (Elektroiskrovaya Obrabotka Metallov V Vode I Elektrolitakh). *Surf. Eng. Appl. Electrochem. (Elektronnaya Obrabotka Materialov)* **1980**, *1*, 5–8.

46. Tzeng, Y.F.; Lee, C.Y. Effects of powder characteristics on electrodischarge machining efficiency. *Int. J. Adv. Manuf. Technol.* **2001**, *17*, 586–592.
47. Ivanov, R.; Hussainova, I.; Aghayan, M.; Drozdova, M.; Perez-Coll, D.; Carvajal Rodriguez, M.A.; Rubio-Marcos, F. Graphene-encapsulated aluminium oxide nanofibers as a novel type of nanofillers for electroconductive ceramics. *J. Eur. Ceram. Soc.* **2015**, *35*, 4017. [CrossRef]
48. Liu, J.; Li, J.; Xu, C. Interaction of the cutting tools and the ceramic-reinforced metal matrix composites during micro-machining: A review. *CIRP J. Manuf. Sci. Technol.* **2004**, *7*, 55–70.
49. Melnik, Y.A.; Kozochkin, M.P.; Porvatov, A.N.; Okunkova, A.A. On adaptive control for electrical discharge machining using vibroacoustic emission. *Technologies* **2018**, *6*, 96. [CrossRef]
50. Baseri, H.; Sadeghian, S. Effects of nanopowder TiO2-mixed dielectric and rotary tool on EDM. *Int. J. Adv. Manuf. Technol.* **2016**, *83*, 519–528.
51. Zhang, Y.; Wang, C.; Wang, Y.; Ni, Q.; Ji, L. Geometric Accuracy Improvement by Using Electrochemical Reaming with a Helical Tube Electrode as Post-Processing for EDM. *Materials* **2019**, *12*, 3564.
52. Kozochkin, M.P.; Grigor'ev, S.N.; Okun'kova, A.A.; Porvatov, A.N. Monitoring of electric discharge machining by means of acoustic emission. *Russ. Eng. Res.* **2016**, *36*, 244–248.
53. Yue, X.; Yang, X.; Tian, J.; He, Z.; Fan, Y. Thermal, mechanical and chemical material removal mechanism of carbon fiber reinforced polymers in electrical discharge machining. *Int. J. Mach. Tools Manuf.* **2018**, *133*, 4–17.
54. Zasimchuk, E.; Baskova, O.; Gatsenko, O.; Turchak, T. Universal mechanism of viscoplastic deformation of metallic materials far from thermodynamics equilibrium. *J. Mater. Eng. Perform.* **2018**, *27*, 4183–4196.
55. Ehle, L.C.; Schneider, S.; Schwedt, A.; Richter, S.; Klink, A.; Mayer, J. Electron microscopic characterization of surface zones thermo-chemically modified by electrical discharge machining. *J. Mater. Process. Tech.* **2020**, *280*, 116596.
56. Faisal, N.; Kumar, K. Optimization of machine process parameters in EDM for EN 31 using evolutionary optimization techniques. *Technologies* **2018**, *6*, 54. [CrossRef]
57. Gavrin, V.N.; Kozlova, Y.P.; Veretenkin, E.P.; Logachev, A.V.; Logacheva, A.I.; Lednev, I.S.; Okunkova, A.A. Reactor target from metal chromium for "pure" high-intensive artificial neutrino source. *Phys. Part. Nucl. Lett.* **2016**, *13*, 267–273.
58. Porvatov, A.N.; Kozochkin, M.P.; Fedorov, S.V.; Okunkova, A.A. About possibility of vibroacoustic diagnostics of electrical discharge machining and characterization of defects. *Mech. Ind.* **2015**, *16*, 707. [CrossRef]
59. Kumar, A.S.; Durai, A.R.; Sornakumar, T. Machinability of hardened steel using alumina-based ceramic cutting tools. *Int. J. Refract. Met. Hard Mater.* **2003**, *21*, 109–117.
60. Gevorkyan, E.; Lavrynenko, S.; Rucki, M.; Siemiatkowski, Z.; Kislitsa, M. Ceramic cutting tools out of nanostructured refractory compounds. *Int. J. Refract. Met. Hard Mater.* **2017**, *68*, 142–144.
61. Wang, D.; Xue, C.; Cao, Y.; Zhao, J. Fabrication and cutting performance of an Al2O3/TiC/TiN ceramic cutting tool in turning of an ultra-high-strength steel. *Int. J. Adv. Manuf. Technol.* **2017**, *91*, 1967–1976.
62. Laine, R.M.; Marchal, J.C.; Sun, H.P.; Pan, X.Q. Nano-α-Al2O3 by liquid-feed flame spray pyrolysis. *Nat. Mater.* **2006**, *5*, 710–712.
63. Kolodiazhnyi, T.; Annino, G.; Spreitzer, M.; Taniguchi, T.; Freer, R.; Azough, F.; Panariello, A.; Fitzpatrick, W. Development of Al2O3–TiO2 composite ceramics for high-power millimeter-wave applications. *Acta Mater.* **2009**, *57*, 3402–3409.
64. Evans, A.G. Perspective on the development of high-toughness ceramics. *J. Am. Ceram. Soc.* **1990**, *73*, 187–206.
65. Boatemaa, L.; Bosch, M.; Farle, A.-S.; Bei, G.-P.; Van der Zwaag, S.; Sloof, W.G. Autonomous high-temperature healing of surface cracks in Al2O3 containing Ti2AlC particles. *J. Am. Ceram. Soc.* **2018**, *101*, 5684–5693.
66. Ercenk, E.; Guven, B.; Yilmaz, S. Crystallization kinetics of machinable glass ceramics produced from volcanic basalt rock. *J. Non-Cryst. Solids* **2018**, *498*, 262–271.
67. Song, Q.; Zhang, Z.-H.; Hu, Z.-Y.; Yin, S.-P.; Wang, H.; Wang, H.; Cheng, X.-W. Fully dense B4C ceramics fabricated by spark plasma sintering at relatively low temperature. *Mater. Res. Express* **2018**, *5*, 105201. [CrossRef]
68. Li, L.; Pu, S.; Liu, Y.; Zhao, L.; Ma, J.; Li, J. High-purity disperse alpha-Al2O3 nanoparticles synthesized by high-energy ball milling. *Adv. Powder Technol.* **2018**, *29*, 2194–2203.
69. Li, L. Effect of cyclic fatigue loading on matrix multiple fracture of fiber-reinforced ceramic-matrix composites. *Ceramics* **2019**, *2*, 327–346.

70. Dassios, K.G.; Matikas, T.E. Assessment of fatigue damage and crack propagation in ceramic matrix composites by infrared thermography. *Ceramics* **2019**, *2*, 393–406.
71. Ghouli, S.; Ayatollahi, M.R.; Bushroa, A.R. Fracture characterization of ceria partially stabilized zirconia using the GMTSN criterion. *Eng. Fract. Mech.* **2018**, *199*, 647–657.
72. Ighodaro, O.L.; Okoli, O.I. Fracture toughness enhancement for alumina systems: A Review. *Int. J. Appl. Ceram. Technol.* **2008**, *5*, 313–323.
73. Szutkowska, M.; Cygan, S.; Podsiadlo, M.; Laszkiewicz-Lukasik, J.; Cyboron, J.; Kalinka, A. Properties of TiC and TiN Reinforced Alumina-Zirconia Composites Sintered with Spark Plasma Technique. *Metals* **2019**, *9*, 1220.
74. Zheng, G.; Zhao, J.; Gao, Z.; Cao, Q. Cutting performance and wear mechanisms of Sialon-Si3N4graded nano-composite ceramic cutting tools. *Int. J. Adv. Manuf. Technol.* **2012**, *58*, 19–28.
75. Grigoriev, S.N.; Metel, A.S.; Melnik, Y.A.; Volosova, M.A. Equipment and Technology for Combined Ion–Plasma Strengthening of Cutting Tools. *Machines* **2018**, *6*, 58. [CrossRef]
76. Orlova, A.I.; Ojovan, M.I. Ceramic mineral waste-forms for nuclear waste immobilization. *Materials* **2019**, *12*, 2638.
77. Kachaev, A.A.; Grashchenkov, D.V.; Lebedeva, Y.E.; Solntsev, S.S.; Khasanov, O.L. Optically transparent ceramic (review). *Glass. Ceram.* **2016**, *73*, 117–123.
78. Lyamina, G.V.; Ilela, A.E.; Dvilis, E.S.; Petyukevich, M.A.; Tolkachev, O.S. Medical ceramics from powders of the system Al2O3–ZrO2–Y2O3 obtained on an installation of nanospray drying. *Nanotechnol. Russ.* **2018**, *13*, 337–343.
79. Gommeringer, A.; Kern, F. Mechanical properties and electrical discharge machinability of alumina-10 vol% zirconia-28 vol% titanium nitride composites. *Ceramics* **2020**, *3*, 199–209.
80. Boulesteix, R.; Chevarin, C.; Belon, R.; Maître, A.; Cochain, L.; Sallé, C. Manufacturing of large size and highly transparent Nd: YAG ceramics by pressure slip-casting and post-sintering by HIP: An experimental and simulation study. *Materials* **2020**, *13*, 2199.
81. Maksimov, R.N.; Khrustov, V.R.; Shitov, V.A.; Yurovskikh, A.S. Effect of the thermal shrinkage behavior of Yb:Lu2O3 nanopowder compacts on the structural and optical characteristics of ceramics. *Inorg. Mater.* **2019**, *55*, 634–639.
82. Grishaev, V.V.; Tostanovskii, V.I. Properties of oxide structural ceramics made from mechanically milled powders. *Refract. Ind. Ceram.* **1993**, *34*, 98–102.
83. Annenkov, Y.M.; Ivanov, V.V.; Ivashutenko, A.S.I.; Kondratyuk, A.A.; Sivkov, A.A. Efficiency of different compaction methods for corundum-zirconia powders. *Refract. Ind. Ceram.* **2008**, *49*, 461–465.
84. Gavrilkin, S.M.; Kopaneva, L.I.; Batsanov, S.S. Anomalous phase transition in shock-compressed PbO. *Combust. Explos. Shock Waves* **2003**, *39*, 723–726.
85. Grigoriev, S.N.; Dmitriev, A.M.; Korobova, N.V.; Fedorov, S.V. A cold-pressing method combining axial and shear flow of powder compaction to produce high-density iron parts. *Technologies* **2019**, *7*, 70. [CrossRef]
86. Lazar, A.; Kosmač, T.; Zavašnik, J.; Abram, A.; Kocjan, A. TiN-nanoparticulate-reinforced ZrO2 for electrical discharge machining. *Materials* **2019**, *12*, 2789. [CrossRef]
87. Morozova, L.V.; Kalinina, M.V.; Drozdova, I.A.; Shilova, O.A. Preparation and characterization of nanoceramics for solid oxide fuel cells. *Inorg. Mater.* **2018**, *54*, 79–86.
88. Khasanov, O.L.; Pokholkov, Y.P.; Sokolov, V.M.; Dvilis, É.S.; Slosman, A.I.; Matrenin, S.V. Ultrasonic compacting of zirconium ceramics from ultradispersed powders. *Glass. Ceram.* **1995**, *52*, 177–180.
89. Gadow, R.; Kern, F. Pressureless sintering of injection molded zirconia toughened alumina nanocomposites. *J. Ceram. Soc. Jpn.* **2006**, *114*, 958–962.
90. Morozova, L.V.; Kovalko, N.Y.; Kalinina, M.V.; Shilova, O.A. Production of chemically pure zirconia-based nanoceramics in the ZrO2(Y2O3)-Al2O3 system for restorative dentistry. *Theor. Found. Chem. Eng.* **2019**, *53*, 848–854.
91. Smirnov, V.V.; Smirnov, S.V.; Obolkina, T.O.; Antonova, O.S.; Goldberg, M.A.; Barinov, S.M. The influence of manganese oxide on the sintering and properties of the eutectic ceramics of the ZrO2-Al2O3-SiO2 system. *Dokl. Chem.* **2019**, *486*, 160–163.
92. Morozova, L.V. Mechanochemical Activation of precursor powders for the preparation of dense Al2O3-ZrO2 <Y2O3> nanoceramics. *Inorg. Mater.* **2019**, *55*, 295–301.

93. Novikov, Y.N. *The FLASH Memory Based on Silicon Nitride (SONOS)*, 1st ed.; Nova Publisher: New York, NY, USA, 2011; pp. 35–66.
94. Roman, O.V.; Kovalevskaya, A.V.; Fomikhina, I.V.; Grigoriev, S.V. Development of carbide and nitride ceramics of increased strength (Razrabotka karbidnoy i nitridnoy keramiki povyshennoy prochnosti). *Lit'e Metall.* **2005**, *4*, 157–162.
95. Podbolotov, K.B.; Volochko, A.T.; Khort, N.A.; Gusarov, S.V. Refractory materials based on secondary resources and phosphate compounds. *Refract. Ind. Ceram.* **2019**, *59*, 579–582.
96. Podbolotov, K.B.; Dyatlova, E.M.; Volochko, A.T. Synthesis and reinforcement of heat-resistant cordierite-mullite ceramic structure with introduction of a fiber filler. *Refract. Ind. Ceram.* **2016**, *57*, 151–154.
97. Volochko, A.T.; Podbolotov, K.B.; Dyatlova, E.M. *Refractory and Infusible Ceramics (Ogneupornye I Tugoplavkie Keramicheskie Materialy)*, 1st ed.; Belorus. Nauka: Minsk, Belarus, 2013; pp. 79–82.
98. Makarov, A.V.; Skorynina, P.A.; Volkova, E.G.; Osintseva, A.L. Effect of heating on the structure, phase composition, and micromechanical properties of the metastable austenitic steel strengthened by nanostructuring frictional treatment. *Phys. Met. Met.* **2018**, *119*, 1196–1203.
99. Borisov, M.; Lobanov, D.; Yanyushkin, A.; Skeeba, V. Investigation of the process of automatic control of current polarity reversal in the conditions of hybrid technology of electrochemical processing of corrosion-resistant steels. *Obrab. Met.-Met. Work. Mater. Sci.* **2020**, *22*, 6–15.
100. Ojha, N.; Zeller, F.; Mueller, C.; Reinecke, H. The effect of the electrical discharge machining process on the material properties of nonconductive ceramics. *J. Micro Nano-Manuf.* **2016**, *4*, 1–5.
101. Kumar, R.; Chaubey, A.K.; Maity, T.; Prashanth, K.G. Mechanical and tribological properties of Al2O3-TiC composite fabricated by spark plasma sintering process with metallic (Ni, Nb) binders. *Metals* **2018**, *8*, 50. [CrossRef]
102. Godin, N.; Reynaud, P.; R'Mili, M.; Fantozzi, G. Identification of a critical time with acoustic emission monitoring during static fatigue tests on ceramic matrix composites: Towards lifetime prediction. *Appl. Sci.* **2016**, *6*, 43. [CrossRef]
103. Grigoriev, S.N.; Krapostin, A.A. Multilayer composite nanoscale coatings as a method to increase reliability and tool life of cutting tools made of mixed ceramic Al2O3-TiC. *Mech. Ind.* **2016**, *17*, 704.
104. Volosova, M.; Grigoriev, S.; Metel, A.; Shein, A. The Role of Thin-Film Vacuum-Plasma Coatings and Their Influence on the Efficiency of Ceramic Cutting Inserts. *Coatings* **2018**, *8*, 287. [CrossRef]
105. Khodaee, A.; Melander, A. Numerical and experimental analysis of the gear size influence on density variations and distortions during the manufacturing of PM gears with an innovative powder processing route incorporating HIP. *J. Manuf. Mater. Process.* **2018**, *2*, 49. [CrossRef]
106. Jayalakshmi, S.; Singh, R.A.; Gupta, M. Metallic glasses as potential reinforcements in Al and Mg matrices: A review. *Technologies* **2018**, *6*, 40.
107. Balog, M.; Ibrahim, A.M.H.; Krizik, P.; Bajana, O.; Klimova, A.; Catic, A.; Schauperl, Z. Bioactive Ti plus Mg composites fabricated by powder metallurgy: The relation between the microstructure and mechanical properties. *J. Mech. Behav. Biomed.* **2019**, *90*, 45–53.
108. Nicoara, M.; Buzdugan, D.; Locovei, C.; Bena, T.; Stoica, M. About thermostability of biocompatible Ti-Zr-Ag-Pd-Sn amorphous alloys. *J. Therm. Anal. Calorim.* **2018**, *133*, 189–197.
109. Veronesi, P.; Rosa, R.; Colombini, E.; Leonelli, C. Microwave-Assisted preparation of high entropy alloys. *Technologies* **2015**, *3*, 182–197.
110. Sezer, N.; Evis, Z.; Kayhan, S.M.; Tahmasebifar, A.; Koc, M. Review of magnesium-based biomaterials and their applications. *J. Magnes. Alloy.* **2018**, *6*, 23–43.
111. Xanthopoulou, G.; Thoda, O.; Boukos, N.; Krishnamurthy, S.; Dey, A.; Roslyakov, S.; Vekinis, G.; Chroneos, A.; Levashov, E. Effects of precursor concentration in solvent and nanomaterials room temperature aging on the growth morphology and surface characteristics of ni–nio nanocatalysts produced by dendrites combustion during SCS. *Appl. Sci.* **2019**, *9*, 4925. [CrossRef]
112. Volykhov, A.A.; Sanchez-Barriga, J.; Sirotina, A.P.; Neudachina, V.S.; Frolov, A.S.; Gerber, E.A.; Kataev, E.Y.; Senkovsky, B.; Khmelevsky, N.O.; Aksenenko, A.Y.; et al. Rapid surface oxidation of Sb2Te3 as indication for a universal trend in the chemical reactivity of tetradymite topological insulators. *Chem. Mater.* **2016**, *28*, 8916–8923.

113. Grigoriev, S.N.; Kozochkin, M.P.; Sabirov, F.S.; Kutin, A.A. Diagnostic systems as basis for technological improvement. *Proc. CIRP* **2012**, *1*, 599–604.
114. Ferreira, S.C.; Conde, A.; Arenas, M.A.; Rocha, L.A.; Velhinho, A. Anodization mechanism on sic nanoparticle reinforced al matrix composites produced by power metallurgy. *Materials* **2014**, *7*, 8151–8167.
115. Xu, C.; Liu, F.; Huang, L.; Jiang, L. Dependence of creep performance and microstructure evolution on solution cooling rate in a polycrystalline superalloy. *Metals* **2018**, *8*, 4. [CrossRef]
116. Iankov, R. Finite element simulation of powder metal compaction processes in container with several punches. *Nato Sci. Ser.* **2001**, *176*, 169–180.
117. Sobol, O.V.; Andreev, A.A.; Grigoriev, S.N.; Gorban', V.F.; Volosova, M.A.; Aleshin, S.V.; Stolbovoy, V.A. Physical characteristics, structure and stress state of vacuum-arc tin coating, deposition on the substrate when applying high-voltage pulse during the deposition. *Probl. Atom. Sci. Tech.* **2011**, *4*, 174–177.
118. Loh, N.J.; Simao, L.; Jiusti, J.; De Noni, A., Jr.; Montedo, O.R.K. Effect of temperature and holding time on the densification of alumina obtained by two-step sintering. *Ceram. Int.* **2017**, *43*, 8269–8275.
119. Tamura, Y.; Moshtaghioun, B.M.; Gomez-Garcia, D.; Rodríguez, A.D. Spark plasma sintering of fine-grained alumina ceramics reinforced with alumina whiskers. *Ceram. Int.* **2017**, *43*, 658–663.
120. Jin, Y.; Han, M.-K.; Kim, S.-J. Na-Doping effects on thermoelectric properties of Cu2−xSe nanoplates. *Appl. Sci.* **2018**, *8*, 12. [CrossRef]
121. Ning, J.; Wu, D.; Zhao, D. Synthesis and thermoelectric properties of TiO2/Cu2SnSe3 composites. *Appl. Sci.* **2017**, *7*, 1043.
122. Alleno, E. Review of the thermoelectric properties in nanostructured Fe2Val. *Metals* **2018**, *8*, 864. [CrossRef]
123. Zhang, H.; Zhang, L.; Liu, X.; Chen, Q.; Xu, Y. Effect of Zr addition on the microstructure and mechanical properties of CoCrFeNiMn high-entropy alloy synthesized by spark plasma sintering. *Entropy* **2018**, *20*, 810. [CrossRef]
124. Viswanathan, V.; Laha, T.; Balani, K.; Agarwal, A.; Seal, S. Challenges and advances in nanocomposite processing techniques. *Mater. Sci. Eng. R* **2006**, *54*, 121–285.
125. Wang, L.; Zhang, J.; Jiang, W. Recent development in reactive synthesis of nanostructured bulk materials by spark plasma sintering. *Int. J. Refract. Met. Hard Mater.* **2013**, *39*, 103–112.
126. Zhang, Y.F.; Wang, L.J.; Jiang, W.; Chen, L.-D. Microstructure and properties of Al2O3-TiC composites fabricated by combination of high-energy ball milling and spark plasma sintering (SPS). *J. Inorg. Mater.* **2005**, *20*, 1445.
127. Oraei, M.; Mostaan, H.; Rafiei, M.; Abbasian, A.R.; Zarezadeh, M. Investigation into microstructural evolutions, mechanical properties and thermal analysis of Al (Zn)/Al2O3 nano-composite fabricated by mechanical milling and SPS method. *Mater. Res. Express* **2019**, *6*, 0865g9.
128. Moriceau, J.; Houizot, P.; Pasturel, M.; Guizouarn, T.; Rouxel, T. A magnetic glass matrix (ZnO-BaO-B2O3) particulate (Fe3O4) nanocomposite obtained by SPS. *J. Non Cryst. Solids* **2019**, *514*, 116–121.
129. Fedorov, S.V.; Pavlov, M.D.; Okunkova, A.A. Effect of structural and phase transformations in alloyed subsurface layer of hard-alloy tools on their wear resistance during cutting of high-temperature alloys. *J. Frict. Wear* **2013**, *34*, 190–198.
130. Chaudhary, T.; Siddiquee, A.N.; Chanda, A.K. Effect of wire tension on different output responses during wire electric discharge machining on AISI 304 stainless steel. *Def. Technol.* **2019**, *15*, 541. [CrossRef]
131. Maradia, U.; Filisetti, E.; Boccadoro, M.; Roten, M.; Dutoit, J.-M.; Hengsberger, S. Increasing the injection moulding productivity through EDM surface modulation. *Proc. CIRP* **2018**, *68*, 58. [CrossRef]
132. Uhlmann, E.; Oberschmidt, D.; Bolz, R. Application of micro structured, boron doped CVD-diamond as mu EDM tool electrodes. *Proc. CIRP* **2018**, *68*, 649.
133. Sahu, A.K.; Chatterjee, S.; Nayak, P.K.; Mahapatra, S.S. Study on effect of tool electrodes on surface finish during electrical discharge machining of Nitinol. *IOP Conf. Ser. Mater. Sci. Eng.* **2018**, *338*, 012033.
134. Prathipati, R.P.; Devuri, V.; Cheepu, M.; Gudimetla, K.; Kiran, R.U. Machining of AISI D2 tool steel with multiple hole electrodes by EDM process. *IOP Conf. Ser. Mater. Sci. Eng.* **2018**, *330*, 012067.
135. Trung, K.H. Development of a WEDM system with high machining efficiency. *Int. Conf. Syst. Sci. Eng.* **2017**, 397–399. [CrossRef]
136. Wu, H.; Wang, T.; Wang, J. Research on discharge state detection of finishing in high-speed wire electrical discharge machine. *Int. J. Adv. Manuf. Technol.* **2019**, *103*, 2301.

137. Takale, A.; Chougule, N. Optimization of process parameters of wire electro discharge machining for Ti49.4Ni50.6 shape memory alloys using the Taguchi technique. *Int. J. Struct. Integr.* **2019**, *10*, 548. [CrossRef]
138. Reshetnikov, S.; Kurzina, I.; Livanova, A.; Meshcheryakov, E.; Isupova, L. Effect of Li, Na and K modification of alumina on its physical and chemical properties and water adsorption ability. *Materials* **2019**, *12*, 4212. [CrossRef]
139. Konishcheva, M.V.; Svintsitskiy, D.A.; Potemkin, D.I.; Rogozhnikov, V.N.; Sobyanin, V.A.; Snytnikov, P.V. Catalytic performance and characterization of highly efficient composite Ni(Cl-x)/CeO2/eta-Al2O3/FeCrAl wire mesh catalysts for preferential CO methanation. *Chemistryselect* **2020**, *5*, 1228–1234.
140. Kholodnaya, G.; Sazonov, R.; Ponomarev, D.; Zhirkov, I. Pulsed plasma chemical synthesis of carbon-containing titanium and silicon oxide based nanocomposite. *Radiat. Phys. Chem.* **2018**, *144*, 132–137.
141. Matveev, V.A.; Kalinnikov, V.T.; Zakharov, V.I.; Maiorov, D.V. Investigation of an effect exerted by methods of production of chromium(III) oxides-hydroxides on their physicochemical properties. *Russ. J. Appl. Chem.* **2011**, *84*, 1524–1528.
142. Yang, J.; Liu, Y.; Deng, J.; Zhao, X.T.; Zhang, K.F.; Han, Z.; Dai, H.X. AgAuPd/meso-Co3O4: High-performance catalysts for methanol oxidation. *Chin. J. Catal.* **2019**, *40*, 837–848.
143. Shiryaev, M.A.; Jin, Y.J.; Bong, H.C.; Baranov, A. ZnO nanoparticle modification by polyethylenimine for biomolecule conjugation. *Nanotechnol. Russ.* **2017**, *12*, 613–619.
144. Gavelova, P.; Halodova, P.; Namburi, H.K.; Prokupkova, I.A.; Miklos, M.; Krejci, J. From micro to nano: Material characterization methods for testing of nuclear core and structural materials. *J. Nucl. Eng. Radiat. Sci.* **2019**, *5*, 030917. [CrossRef]
145. Olenin, A.Y.; Lisichkin, G.V. Surface-modified oxide nanoparticles: Synthesis and application. *Russ. J. Gen. Chem.* **2019**, *89*, 1451–1476.
146. Voroshilov, Y.V.; Gorshkova, L.V.; Popova, A.M.; Fedorov, T.F. Ternary systems Ti-Zr-C and Ti-Hf-C. *Sov. Powder Met.* **1967**, *6*, 403–405.
147. Bulienkov, N.A.; Zheligovskaya, E.A.; Chernogorova, O.P.; Drozdova, E.I.; Ushakova, I.N.; Ekimov, E.A. Nonequilibrium diamond growth during the high-temperature high-pressure synthesis of a composite material made of a mixture of cobalt and fullerene powders. *Russ. Met.* **2018**, *1*, 35–41.
148. Titkov, Y.; Berdnikova, O.; Tyurin, Y.; Kolisnichenko, O.; Polovetskiy, Y.; Kushnaryova, O. Effect of structure on the properties of composite Cr3C2 + NiCr coatings. *Springer Proc. Phys.* **2020**, *240*, 151–159.
149. Zotti, G.; Cattarin, S.; Mengoli, G.; Fleischmann, M.; Peter, L.M. Photoelectrochemistry of copper(I) acetylide films electrodeposited onto copper electrodes. *J. Electroanal. Chem.* **1986**, *200*, 341–351.
150. Denisov, E.T.; Sarkisov, O.M.; Likhtenshtein, G.I. *Chemical Kinetics: Fundamentals and Recent Developments*, 1st ed.; Elsevier Science: Amsterdam, The Netherlands, 2008; pp. 472–486.
151. Zirconium (Zirconiy). *Chemical Encyclopedia (Khimicheskaya Enciklopediya)*, 1st ed.; Zefirov, N.S., Kulov, N.N., Eds.; Great Russian Encyclopedia (Bolshaya Rossiyskaya Enciklopediya): Moscow, Russia, 1998; Volume 5.
152. Li, X.F.; Hu, Z.Y.; Huang, B. Phase diagram and superconductivity of compressed zirconium hydrides. *Phys. Chem. Chem. Phys.* **2017**, *19*, 3538–3543. [PubMed]
153. Sukhushina, I.S. Phase transitions of the ZrNiH x hydride. *Russ. J. Phys. Chem. A* **2007**, *81*, 1595–1600.
154. Besedin, D.V.; Ustynyuk, L.Y.; Ustynyuk, Y.A.; Lunin, V.V. Hydrogenolysis and hydroisomerization of neopentane on titanium and zirconium hydrides stabilized on the surface of SiO2: A theoretical study by density functional theory. *Russ. J. Phys. Chem. A* **2008**, *82*, 193–200.
155. Grigoriev, S.; Kozochkin, M.; Porvatov, A.; HtuM, T.; Zhavoronsky, P.; Jiang, X.; Pivkin, P. Dynamic model of electrical discharge machining and algorithm of extreme control through acoustic signal. *EPJ Web. Conf.* **2019**, *224*, 05002. [CrossRef]
156. Grigoriev, S.N.; Fedorov, S.V.; Hamdy, K. Materials, properties, manufacturing methods and cutting performance of innovative ceramic cutting tools—A review. *Manuf. Rev.* **2019**, *6*, 19.
157. Fedotov, P.S.; Ermolin, M.S.; Karandashev, V.K.; Ladonin, D.V. Characterization of size, morphology and elemental composition of nano-, submicron, and micron particles of street dust separated using field-flow fractionation in a rotating coiled column. *Talanta* **2014**, *130*, 1–7.
158. Makarova, O.V.; Palatnikov, M.N.; Biryukova, I.V.; Sidorov, N.V. Impact of a dopant impurity electronic structure on physical properties, defect structure, and features of lithium niobate doping technology. *Tech. Phys.* **2019**, *64*, 1872–1878.

159. Ikim, M.I.; Spiridonova, E.Y.; Belysheva, T.V.; Gromov, V.F.; Gerasimov, G.N.; Trakhtenberg, L.I. Structural properties of metal oxide nanocomposites: Effect of preparation method. *Russ. J. Phys. Chem. B* **2016**, *10*, 543–546.
160. Brauer, G.; Baudler, M.; Fehér, F. *Handbuch der Präparativen Anorganischen Chemie*, 3rd ed.; Enke: Stuttgart, Germany, 1978.
161. Kovtunenko, P.V.; Nesterova, I.L. Tin oxides. In *Chemical Encyclopedia (Khimicheskaya Enciklopediya)*, 1st ed.; Knunyants, I.L., Ed.; Great Russian Encyclopedia (Bolshaya Rossiyskaya Enciklopediya): Moscow, Russia, 1992; Volume 3, pp. 515–548.
162. Gommeringer, A.; Kern, F.; Gadow, R. Enhanced mechanical properties in ED-Machinable Zirconia-Tungsten carbide composites with Yttria-Neodymia Co-Stabilized zirconia matrix. *Ceramics* **2018**, *1*, 26–37.
163. Samsonov, G.V.; Vinitskiy, I.M. *Refractory Compounds (Handbook) (Tugoplavkie Soedineniya (Spravochnik))*, 1st ed.; Mettalurgiya: Moscow, Russia, 1976; p. 560.

© 2020 by the authors. Licensee MDPI, Basel, Switzerland. This article is an open access article distributed under the terms and conditions of the Creative Commons Attribution (CC BY) license (http://creativecommons.org/licenses/by/4.0/).

technologies

Review

On Electrical Discharge Machining of Non-Conductive Ceramics: A Review

Marina Volosova, Anna Okunkova *, Pavel Peretyagin, Yury A. Melnik and Natalya Kapustina

Department of High-Efficiency Machining Technologies, Moscow State University of Technology STANKIN, Vadkovskiy per. 3A, 127055 Moscow, Russia

* Correspondence: a.okunkova@stankin.ru; Tel.: +7-909-913-12-07

Received: 4 July 2019; Accepted: 5 August 2019; Published: 8 August 2019

Abstract: The inability of ceramic and nanoceramic processing without expensive diamond tools and with a high-material-removal rate hampers the scope of its potential applications and does not allow humanity to make a full shift to the sixth technological paradigm associated with Kuhn scientific revolutions and Kondratieff's waves and restrains the growth of the economy. The authors completed a review on the research state of ceramic and nanoceramic processing by electrical discharge machining, which is possibly solved by two principal approaches associated with the usage of standard commercially available machine tools. The first approach is related to the introduction of expensive secondary phase; the second approach proposes initiate processing by adding auxiliary electrodes in the form of coating, suspension, aerosol, or 3D-printed layer based on the components of silver, copper, or graphite in combination with an improved dielectric oil environment by introducing graphite or carbon nanoparticles, which is hugely relevant today.

Keywords: electrical discharge machining; nanoceramics; coatings; auxiliary electrode; electrical conductivity; oxides; nitrides; carbon particles; oil medium

1. Introduction

By the beginning of the 21st century, ceramics and hard alloys were entirely developed and widely spread in the field of engineering [1–3]. It is difficult to imagine modern production in almost all machinery industries without cutting tools or other products made of ceramics and hard alloys. The scope of their applications is extensive but has one distinction—products of ceramic and hard alloys are used in extreme operating conditions under heat and ultimate loads. In this condition, no other material can provide the necessary performance, heat and wear resistance [4,5].

Due to the excellent exploitation properties of ceramics [6,7] and according to the summary analysis of experts, the volume of the world ceramic market grows at an average annual rate of 9.2%, and by 2020, the market volume will reach $103 billion [8–11]. The trends for nanoceramics can be even more significant in further perspective. According to the results of recent studies [12,13], the samples of nanoceramics obtained by improved hybrid spark plasma sintering can exceed in main operational parameters (such as hardness, crack and wear resistance) of technical ceramics sintered by traditional isostatic pressing with subsequent vacuum sintering by 30%. The specialists of the subject area associate this phenomenon with the Hall–Petch relation, which describes the growth of the yield strength of a polycrystalline material with a decrease in the grain size [14,15].

However, the introduction of nanoceramics into real life is limited because of their high costs and low workability. The manufacture of each product requires high costs and is very labor-intensive. It requires fewer allowances in processing and the purchase of precision cutting tools based on artificial diamond, as cutting tools made of traditional alloys are worn extremely fast. In its turn, the geometry of cutting tools limits the resulting geometry of the product. Other machining methods such as vibroacoustic machining, laser and electron beam scribing, selective laser sintering with a binder, find their specific place in the processing of ceramics. However, they do not allow the receiving of all geometrically possible shapes of the product without losses on the operational parameters of ceramics [16–18].

One of the most popular methods for the production of parts, regardless of their hardness, is electrical discharge machining (EDM). However, it requires the material to be conductive [19–21]. The process of electrical erosion of the material consists of initiating electric pulses in the interelectrode gap, in the breakdown of the dielectric medium in the gap, in establishing a discharge channel, where the temperature reaches more than 10,000 °C (low-temperature plasma). Due to such temperatures, the material is ablated, the drops of cooled material as eroded debris are washed away by the flow of the dielectric medium, and the electrodes again approach to resume the cycle. Thus, a unique crater-like topology on the surface of the workpiece is created [22–25].

The inability of ceramics and nanoceramics to conduct current hampers the scope of its potential applications and restrains the growth of the progress and the switch to the next technological paradigm that most of the scientists associate with the term "nano" [26–32].

Thus, it can be concluded that the study and development of a method for electrical discharge machining of non-conductive ceramics and nanoceramics are incredibly relevant today.

The authors propose to apply the dialectical approach of cognition as a scientific approach to solve the formulated scientific problem. The dialectical approach includes the systematization and updating of existing knowledge and new data about the object of research.

2. Scientific Problem

The scientific problem of the physical impossibility of superhard ceramic materials machining (as the exception is the diamond grinding method) is considered because of their nearly zero conductivity since the electrical conductivity of ceramics, as for all solids, depends on the concentration of charge carriers, their magnitude, and mobility. The material conductivity is composed of each its constituent phase electrical conductivity: The measurement of the electrical conductivity of especially pure oxide ceramics becomes a complex scientific and technical problem. In many cases, the problem is the impossibility of obtaining the necessary complex spatial geometric shape of surfaces of functional products from high-strength ceramics by traditional methods of processing. There is a strong need of the development of a new inexpensive way of ceramic nanocomposites machining.

There are two main approaches to solving this problem to achieve this goal:

- The creation of a new class of nanocomposite ceramics with the inclusion of the conductive phase allowing the machining of blanks in order to obtain the finished product with the use of minimal technological effort and the absence of direct contact between the material and tool, i.e., using the electric discharging methods of machining [33–39];
- The modification of the surface layer of the nonconducting workpiece, the geometric shape of which is already close to the geometry of the final product, by applying a conductive coating (auxiliary electrode) with a thickness of ~20 μm, which could become an intensifier of electrical discharge machining of non-conductive ceramics [40–44].

The second approach is the less expensive and may have a high level of technological adaptation to the conditions of real machine-building production, and is capable of producing positive results on a large scale. First of all, these conclusions are related to favorable technological conditions for the sintering of single-phase nanoceramics from the relatively low cost of ZrO_2 and Al_2O_3 nanopowders in comparison with the cost of the nanoscale modified electrically conductive phases (for example, graphene, graphene oxide). The other reason is the relatively low cost of materials for the production of a profile tool for EDM.

The design of constructive and technological features of the product allows the withdrawal of the application of superhard materials such as nanoceramics based on ZrO_2 and Al_2O_3 on a new production level [45–48]. It contributes to their spread in the industry for creating a new class of wear-resistant parts, which are capable of operating under extreme operating loads without losing their performance. That is extremely important in solving tasks of the transition to advanced intellectual production, new materials, and methods.

3. Current State of Research

3.1. The Main Scientific Competitors

The research subject is connected to the research of the method of applying EDM to the processing of nanoceramics made by advanced sintering technologies. The analysis of the leading world scientific competitors in this domain shows that the world community is separated as it was mentioned above. One group of scientists is focused on the modification of nanoceramics with conductive inclusions to create systems such as ZrO_2-Ta, ZrO_2-TiC, ZrO_2-TiCN, Si_3N_4-TiC, Si_3N_4-TiCN, Al_2O_3-TiC, Al_2O_3-TiCN [49–59], as well as inclusions based on graphene and oxide graphene and graphene nanotubes [23,60–64]. Another group of scientists has focused their research on the processing capabilities of existed ceramics and non-conductive ceramic composites. The processing ability (ability to EDM) of such ceramics can be modified by introducing an auxiliary electrode based on copper, brass, steel, carbon, etc. in the form of particles, nanoparticles or thin-film in the process of electrical discharge machining. In this case, an auxiliary electrode will play the role of an initiator and intensifier of technology.

One of the most successful examples of such work is the work of the scientists at the Universities of Chemnitz and Freiburg and their colleagues at the Toyota Concern [65–68]. The research team at Chemnitz Technical University (Germany) under the guidance of Prof. Andreas Schubert have worked on the processing of insulating materials [23,69]. In particular, they propose replacing expensive diamond cutting with micro electrical discharge machining (μEDM) for specific technical applications. The researchers obtained a result on the μEDM of technical oxide and nitride ceramics (ZrO_2, Si_3N_4-TiN, and others). In this case, the copper electrode moved progressively along the surface, following the developed control program of CNC-machine. The electrode tool motion looks like kinematic motion of milling cutter during 3-axis milling of grooves and pockets.

The search for alternative methods of processing non-conductive materials was also conducted at the University of Freiburg (Germany) [70]. The authors described two technologies for forming an auxiliary electrode in the form of a coating: Applying a suspension of gas soot and polymethyl methacrylate (PMMA, organic glass or acrylic) by hand and applying the suspension by direct printing carbon conductive ink (C-W) on a commercially available Ekra M2 printer. The coating was carried out on cylindrical samples of non-conductive technical ceramics ZrO_2. As a result, the fundamental possibility of initiating EDM was proved; eroded debris were involved in processing. However, the authors proposed to use both methods exclusively for super-precision low-cost final finishing since the specific material removal rate could be characterized as modest in comparison with traditional EDM of metals. In their work, the authors criticized the use of plasma-vacuum coating as an initiator of EDM because of its high cost and inaccessibility. The authors of the article, in turn, referred to the successful experience of their Japanese colleagues, in particular, the results of Prof. Mohri.

In Japan, with the cooperation of such scientific schools as the Technological Institute of Toyota, The Technological University in Nagaoka and the Technological College in Tsukuba, in the 1990s, a team of scientists was formed under the guidance of the eminent Prof. Naotake Mohri, with the participation of Prof. Katsushi Furutani. In 1996, this group [71] proposed a new method of EDM of dielectric ceramics, by applying an auxiliary electrode in the form of a metal grid on the surface being machined. The EDM was carried out with a brass or copper electrode using kerosene as a working fluid. Prof. Mohri continues to publish the results of research as part of the same research team [72], in which work is underway to develop a method for applying a conductive coating for ceramics based on crystallized carbon or carbides and working oil, and explores the phenomenon and mechanisms of initiation and acceleration of electrical erosion of a non-conductive substance.

Their colleagues from the Council for Scientific and Technical Research of Turkey, under the guidance of Prof. Can Çoğun, obtained other impressive results: The process of coarse electrical discharge and electrochemical machining of non-conductive materials KTU-1, TiO_2, glass, ZrO_2 (with the addition of Y_2O_3), and Al_2O_3 by adding carbon particles no larger than 30 microns in a dielectric liquid (kerosene) [73–75]. The treatment was carried out using an electrode in the form of a hollow copper tube (a diameter of 3.5 mm). Thus, it was possible to drill the rough holes in workpieces of various shapes, their geometry was obviously faulted, perhaps due to the lack of reliable fixation of spherical blanks on the machine's worktable; the experiment with Al_2O_3 turned out to be especially unsuccessful.

The research on the processing of non-conductive materials was also conducted at several research centers in Malaysia. The work published by the research team under the supervision of Dr. Muhammed Abdul Maleque at the International Islamic University of Malaysia (Kuala Lumpur) [76] demonstrated the results of the processing of technical ZrO_2, wholly covered with a casing of platinum copper. The obtained result proved the ability and effectiveness of EDM processing in kerosene; copper residues from the surface of the ceramic workpiece were successfully removed. Another Malaysian group of scientists under the supervision of Dr. Mohd Amri Lajis at the Technical University of Malaysia (Malacca) in cooperation with the University of Malaysia described the result [77,78] where an acrylic blank was machined by placing an electrode inside a steel sleeve of 15 mm in diameter. Thus, the authors obtained a conical hole in the non-conducting material following the geometry of the electrode.

The research in the field of EDM processing of dielectric materials is carried out at the Perm National Research Polytechnic University under the supervision of Dr. Timur Rizovich Ablyaz. The authors theoretically and practically investigated the principal ability of EDM processing of non-conductive materials as Al_2O_3+ZrO_{15} ceramics by initiating a discharge channel by placing the erosion products in the interelectrode gap [79–81]. The authors provided an analysis of the EDM methods for dielectric materials and considered modern approaches to optimize technologies proposed by previously mentioned colleagues. Their ideas deserve attention and detailed experimental approbation.

3.2. Production of Nanoceramics by Advanced SPS

Today the most sought-after researchers are focused on creating new nanomaterials by advanced spark plasma sintering (SPS) [37,39,82–84]. The development and creation of new progressive nanocomposite materials based on ceramic-ceramic and metal-ceramic compositions strengthened with metal nanoparticles and carbides, carbon nanotubes and nanofibers, graphene and graphene oxide are popular due to improved physical and mechanical properties and chemical inertness. The mathematical models developed for SPS assist in the controlling of properties of new materials [85–87].

A technical concept for creating a new class of nanocomposites by spark plasma sintering [88–92] includes the principles of interfacing the components of an innovative spark plasma sintering system, components of a high-current source node equipped with a pressure and temperature controller, and calibration.

Structural nanoceramics can have gradient properties according to particular customer requirements. The developed concept of creating nanocomposite gradient materials consists of creating a new functional mold and using low sintering temperature to minimize residual stresses [93]. The nanostructured powders for this kind of material are prepared according to the known scheme by colloidal processing [37,39].

As is known, one of the critical parameters for sintering materials is the distribution of temperature fields over the entire volume of the sintered product [94,95]. For this reason, the achievement of a uniform distribution of temperature fields in volume is a critical point in the design of tooling and die. Thus, a new model of semi-molds for sintering was designed [37]. The device contains a die of heat-resistant conductive material, insulating a sleeve made of heat-resistant heat-conducting and non-conductive material, a lower punch made of heat-resistant conductive material, an upper punch consisting of two heat-resistant conductive concentric parts, separated by an insulating material. At the same time, punches are processed, and the composition of powder materials is separated through gaskets. This new design allows changing the flow of electrons flowing through the die and, thereby, ensuring its heating. When the percolation of the sintered material is reached, the electrons begin to flow through both the die and the powder material to be processed. Thus, uniform heating of the product throughout its volume is achieved.

Simulated physical processes in order to optimize the technology can help to study the process of sintering and develop solutions for the problems related to the creation of new materials. As part of the study of the problems of technological development for the synthesis of new nanostructured materials by the method of spark plasma sintering, a theoretical description of the sintering process using methods of mathematical and theoretical physics, based on the theory of density functions and nucleation theory, the Keldysh method for electron transfer in a medium consisting of a conglomerate of nanocatrices (tungsten, aluminum and titanium alloys), with additions in the form of crystalline and amorphous mesostructures, causes genuine interest. This method is currently developed in detailed mathematical models of heat and mass transfer in areas of axisymmetric complex shapes of rotation bodies (in connection with the problems of modeling nozzles and nozzles of rockets, aircraft engines). However, with the parameters of the spark plasma sintering process—pulse duration in the range from 1 to 250 ms, average sintering pulse power up to 150 kW, and heating of the mold with powder material up to 2000 °C—new approaches are needed to solve the mass transfer equations, vibrational equations, and temperature distributions, with singular parameters of the substance, thermal conductivity and heat capacity, and time modulated by short pulses of sintering current of extra-large power. In this regard, the problem can be solved by simulating a substance in a cylindrical sintering area by directly pulsing the current through a mold with powder material and additional induction heating and obtaining the microphysical parameters of powder materials using the method of asymptotic approximation and singular generalized functions.

Since in the process of pulsed sintering in the mold, nanoparticles of metals (W, Ti, Al) and crystalline nanostructures (nanotubes, graphene, graphene oxides, nano-diamonds) with a particle size of 50 to 500 nm pass through a characteristic time of several tens of phase transformations, with diffusion and threshold processes for the transfer of molecules and electrons, the result of sintering and the physical properties of the substance obtained are not predictable without precisely setting and modeling the initial data and the dynamic behavior of the physicochemical properties of particles during the sintering process.

Preliminary experimental models were developed that use a Langmuir probe, an optical probe–waveguide to remove the spectral characteristics of the plasma during sintering, for measuring the phases of the voltage and current of the sintering pulse to study these properties. The main theory explaining the physicochemical properties of the material obtained after the action of a series of pulses is the theory of the growth of germ clusters and nucleation (Frenkel–Landau).

In the course of the work for the investigation of the process of pulsed sintering, new effects that were not previously considered in international literature were found: Accelerated nucleation, electronic nanojet, nanofilament, and photon nanojet in the field of thermal photons. At present, an important physicochemical process of reducing the threshold power of electronic breakdown for sintering conglomerates of dielectric nanoparticles by introducing additional impurities from metallic conductive nanoparticles was found and studied. It should be noted that for these processes there are important optical analogs. In the visible and infrared range, there is a new effect, not previously considered as an application to the process of a pulse, as well as a laser sintering effect—the photon nanojet (photonic nanojet). This effect is used as an analog of the electronic jet, the spatial effect of high electron density on the surface of nanoparticles, due to the excitation of the Frohlich modes to simulate the sintering processes. In addition to the theoretical study of nonlinear effects, in the process of pulsed sintering, there is the possibility of the practical application of a photon jet for a wide range of technical applications as a working medium for microlasers, microlenses, radiation detectors, and metamaterials. Interest in the considered processes takes place in the microwave field, in the calculation and design of metamaterials for the microwave coatings of airplanes and rockets, as well as in the calculations of thermal protective coatings.

The properties of sintered Al_2O_3-TiC-Ni nanocomposite in comparison with the properties of a sample of aluminum oxide (Al_2O_3) obtained by hot pressing—the traditional sintering method exceed them due to smaller alumina grains [39]. The average particle size of the aluminum oxide powder in the nanocomposite is approximately 0.3 ± 0.1 μm, while in a monolithic material this size is 3 ± 1 μm. The structure of the Al_2O_3-TiC-Ni nanocomposite contains Ni and TiC nanoparticles. Although the process of sintering occurs at a temperature of 1375 °C, the nanoparticles do not significantly increase in size. The hardening of the nanocomposite was expected due to the presence of Ni nanoparticles. The measured relative density of the samples was >98%. The difference in wear resistance between Al_2O_3 and Al_2O_3-TiC-Ni nanocomposite under dry slip conditions depends on the hardness of the material, as well as the size of the aluminum oxide grains in the nanocomposite. Both parameters are most important for improving the tribological properties in working conditions.

3.3. Classification of Technical Ceramics and Its Workability by EDM

Technical ceramics are classified based on functionality such as biocompatibility, electrical properties, optical properties, mechanical strength, and resistance to high temperatures according to the following types: Medical ceramics; electroceramics; optical ceramics; structural ceramics and ceramics for extreme conditions; ceramics for energy transfer, storage and conversion; and other types of ceramics [96,97].

It should be noted that high-tech industries are experiencing an acute shortage in the production of ceramic materials and high-tech products made from them, which have high performance properties.

On the one hand, this is due to the use of outdated technologies in production and the lack of knowledge in obtaining advanced high-strength ceramic nanomaterials and composites based on them, and on the other hand, using outdated technologies for processing ceramics. The most advanced production is directly dependent on suppliers of equipment and limited in the choice of technological processing modes. This undoubtedly endangers the adequate response to modern challenges.

Electrical discharge machining is a highly efficient method of processing electrically conductive materials regardless of their mechanical properties (Figure 1). The schematic presentation of the electrical erosion steps under discharge pulses is shown in Figure 2. The regular discharge gap forms on the surface of the workpiece during the formation of the cavity to be processed: Two electrodes approach up to the moment of first working discharges, which transforms into the discharge channel (Figure 1a). Then the tool electrode advances in the cavity formed by the first discharges (Figure 1b) and creates a regular cavity in the workpiece (Figure 1c). On this stage, the pulses acting between the two electrodes balance each other (Figure 1d), and electrical discharge machining becomes stable.

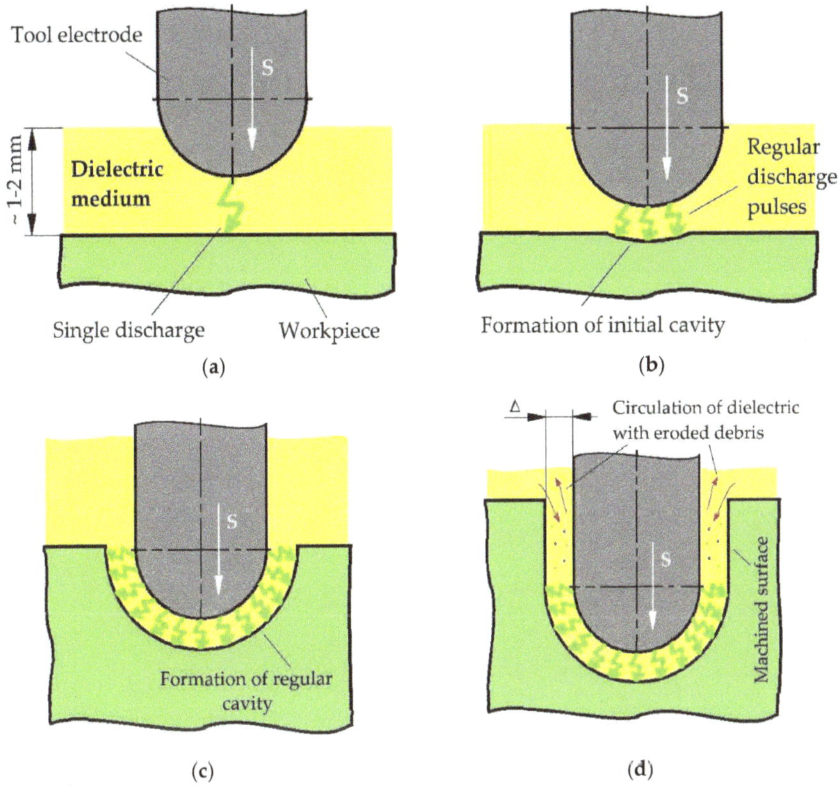

Figure 1. Diagram of the electrical discharge machining (EDM) process by example of cavity formation on the surface of conductive nanoceramics: (**a**) Dielectric breakdown; (**b**) stabilization of discharge pulses; (**c**) formation of regular cavity; (**d**) ongoing processing; S is electrode feed speed; Δ is a discharge gap.

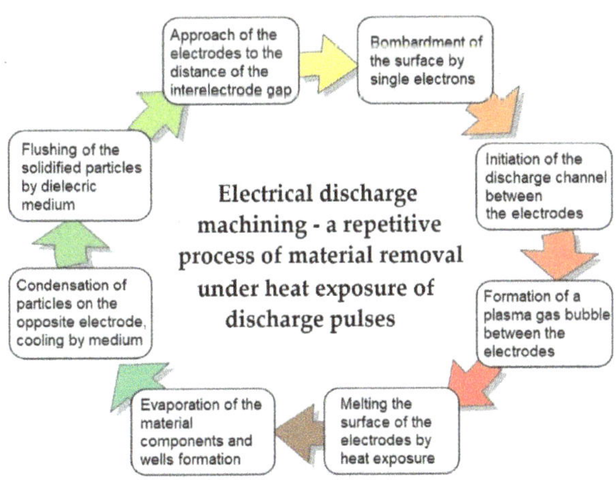

Figure 2. A schematic cycle of the material erosion under discharge pulses.

Modern scientists divide the main processing factors that influence the results of electrical discharge machining into four main groups [24,25,98,99] (Figure 3):

- factors related to the material of electrodes (e.g., homogeneity, conductivity for a tool and a workpiece, heat resistance of the material components);
- factors related to the environment (e.g., temperature, dielectric composition, a grade of workpiece immersion into the fluid);
- factors related to machining process and controlled by CNC-system (e.g., operation voltage, strength of the operational current, auxiliary voltage, strength of the auxiliary current, operational and auxiliary impulses rate, electrode feed, the dielectric pressure in nozzles);
- factors related to technological features of processing (e.g., scheme and system of workpiece fastening, configuration of the workpiece, complexity of surface to be processed).

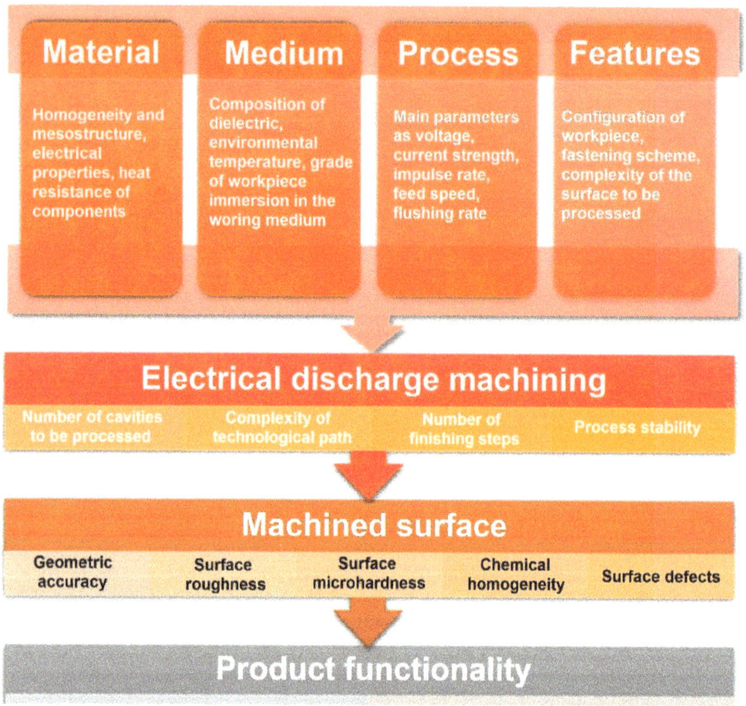

Figure 3. The main factors of EDM affecting the functionality of the final product.

All these factors influence the stability of EDM. The quality of the processed surface can be controlled by measuring final geometrical accuracy, roughness, microhardness, change of chemical content in subsurface layer (up to the depth of ~4–10 μm), homogeneity of the processed surface, and an absence of the visible defects, which can influence the final product appearance and functionality. In this case, the product functionality can be presented as a complex, which consists of its operational ability as part of a more sophisticated unit such as a mechanism or machine and its appearance following customer requirements.

EDM is possible only if the minimum electrical conductivity of materials is about 10^{-2} $\Omega \cdot cm^{-1}$. The electrical conductivity of the material is a critical factor in the success of the process of electrical erosion. All metals fulfill the condition in terms of minimum electrical conductivity, which is not the case with ceramics and its composites. Figure 5 shows the main ceramic materials, depending on their conductivity. Electrically conductive ceramics such as SiC, TiB_2, TiN, and TiC are treated with EDM, while the electrical conductivity of Al_2O_3, ZrO_2, Si_3N_4 is not enough to initiate processing. As it was mentioned before, at present, two different ways of processing non-conductive ceramics by EDM are used:

1) Doping ceramics with an additional electrically conductive phase;
2) Deposition of an electrically conductive coating on the surface of the non-conductive ceramics, initiating the start of interaction and initiation of the first sparks. The authors propose to use the second approach in the research and development of a method for processing non-conductive ceramics and composites as it can help to introduce available up-to-date solutions for the machining of ceramics without the additional costs of a second conductive phase of ceramics, which can reduce the exploitation properties of ceramics (Figure 4).

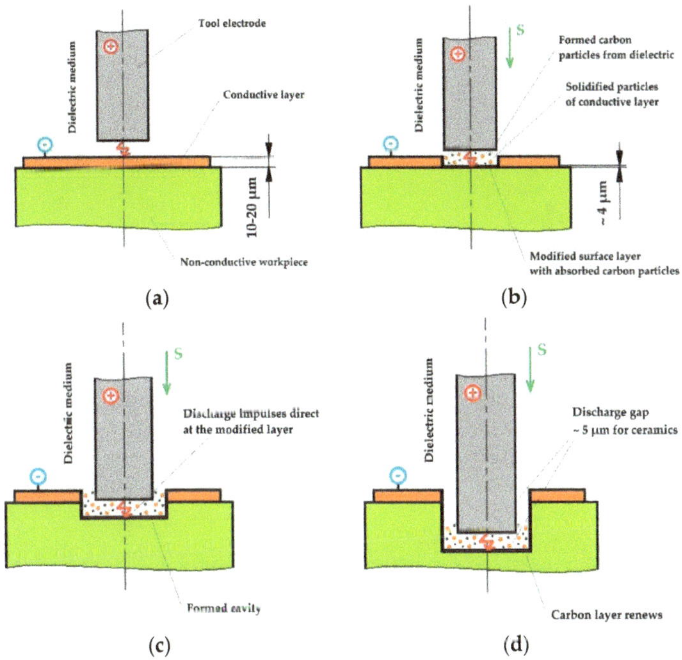

Figure 4. Schematic diagram of the EDM of non-conductive materials using a specially developed electrically conductive coating: (**a**) Initiating discharge impulses between the coating and the tool electrode; (**b**) electrical erosion of the coating, formation of the erosion products between the conductive film and the electrode tool; (**c**) electrical erosion of the non-conductive workpiece with absorbed erosion products by the workpiece; (**d**) reinitiating electrical erosion of the workpiece that keeps constant presences of the erosion product in sub layer of the workpiece.

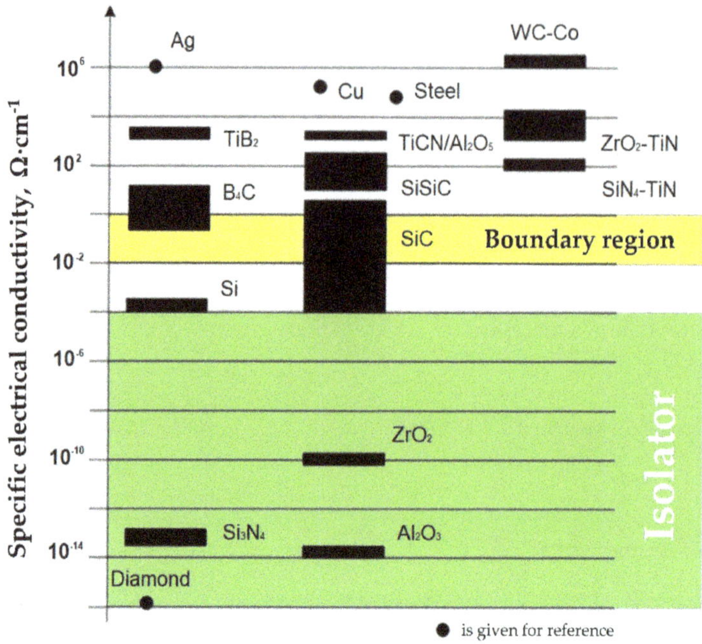

Figure 5. Main ceramic materials in dependence of its electrical conductivity [100,101].

4. Discussion

An analysis of the current state of research on this issue and the main guidelines of research in world science showed that the world community is focused mainly on the modification of nanoceramics with current-conducting inclusions to create systems of the type "oxide or nitride ceramics—conductive additives". As well, the most popular but extremely expensive inclusions today for the production of a material system with advanced exploitation properties are nanotubes of graphene and graphene oxide [28,56].

Despite the efforts of the world scientific community to solve the global problem of processing non-conductive ceramics and nanoceramics due to the inclusion of the secondary phase, the issue remains unresolved [34,37,47,48,50,51,88,102]. The main issue that hampers the advances is related to the material of the secondary phase, which often is not widely available for use in the conditions of real production [33,34,36,37,40,44,89,103]. Besides, the material of the secondary phase can significantly change the appearance of ceramics, such as transparency and color, which directly affects and restricts the areas of its potential application [7,8,39,104].

In parallel with the mentioned approach, the world's leading scientists work on the feasibility of processing technical ceramics by including an initiator and intensifier of the process in the form of an auxiliary electrode based on copper, brass, steel, carbon inclusions, etc. [15,23,35,43]. The additional electrode can be presented in the form of coating deposited by different techniques (including thermal spray and coating technologies or by aerosol, 3D-printing, or drying suspension disposed at a workpiece) or in the form of carbon particles introduced into the processing zone. Suspension can be based on the components of silver, copper, or graphite. A dielectric oil environment can be improved by adding graphite particles and nanoparticles (carbon, graphene nanotubes) in combination with a conductive coating or thin film (up to ~4 μm) or aerosol nanoparticles, which is exceptionally relevant today [15,38,42–44,47].

The developed method of the precision shaping of functional products will subsequently contribute to the introduction and distribution of structural nanoceramics in the machine-building industry. It can significantly vary available shapes and properties of nanoceramics regardless of their conductive ability.

The problem of EDM of ceramics needs a fundamentally new approach to find a way of processing by using an economically available solution for direct individual production. Today, the usage of a specially developed conductive coating or thin film with a thickness of ~10-20 μm based on the suspension can be one of the technically available solutions.

The presence of this coating will initiate discharges at non-conductive material. The temperature in the discharge gap can reach 10,000 °C, which is suitable for processing any matter. The presence of carbon particles in the oil medium and the eroded debris, which contain the carbon formed from the organic medium, will contribute to processing.

For example, in the case of processing naturally widely available oxide ceramics such as ZrO_2 and Al_2O_3, their melting points are of 2715 °C and 2054 °C, respectively [33,49,50,68]. The introduction of an additional process initiator can result in the physical ablation of the ceramics with the formation of the conductive form ZrC and unstable compound Al_4C_3, which can theoretically assist processing.

This method can be called superior to the proposals of its competitors in terms of its use for processing structural nanoceramics. The development and application of the technique can contribute to widespread structural nanoceramics in such industries as aircraft and aerospace, where resistance to the heat and wear is relevant. The processing of nanoceramics with the usage of standard mechanical equipment such as EDM without adding an expensive secondary phase can significantly reduce the cost in manufacturing to surpass any potential manufacture competitors by several times.

5. Conclusions

5.1. Research Work

Based on the results of the research and obtained experimental and analytical results described above, a conductive coating with a thickness of ~10-20 μm for use as an initiator and intensifier of electric erosion corresponds to general world trends. Its application to technical ceramics and nanoceramics in addition to the refusal to modify nanoceramics itself using an expensive conductive secondary phase, which hampers the switch to the next technological order, will allow an advanced and accessible development, in all respects, of a way of processing structural and non-conductive nanomaterials without the loss of their original operational properties such as high resistance to the heat and abrasive wear. In perspective, it can allow the mechanical engineering industry to shift to a new paradigm of production related to the application of nanomaterials, which is unique in its performance properties in terms of its thermal stability and wear-resistance.

5.2. Future Work

Further experiments with the processing of innovative nanoceramics can give a fundamentally new development for the presented idea. It will reveal the principal ability of processing non-conductive oxide nanoceramics and allow the identification of specific conditions and technological modes, and will draw conclusions about the fundamental laws of the main EDM parameters' influence on the workability of nanoceramic workpieces. That is necessary to form a unique geometry of the typical product sample, for which the solution of the global problem of the operational ability for non-conductive materials is critical. It is important to promote the switch to the next technological paradigm with the elaboration of a new accessible method to process innovative ceramics with equipment, which is widely available at modern mechanical production enterprises.

Author Contributions: Conceptualization, M.V.; Methodology, A.O.; Software, N.K.; Validation, P.P.; Formal Analysis, Y.A.M.; Investigation, A.O.; Resources, Y.A.M.; Data Curation, N.K.; Writing-Original Draft Preparation, P.P.; Writing-Review & Editing, A.O.; Visualization, P.P. and N.K.; Supervision, M.V.; Project Administration, M.V.; Funding Acquisition, Y.A.M.

Funding: This research was funded by the Ministry of Education and Science of Russian Federation, grant number No. 9.7453.2017/6.7.

Acknowledgments: The research was done at the Department of High-Efficiency Machining Technologies of MSTU Stankin.

Conflicts of Interest: The authors declare no conflict of interest.

References

1. Grigoriev, S.N.; Kozochkin, M.P.; Sabirov, F.S.; Kutin, A.A. Diagnostic Systems as Basis for Technological Improvement. *Procedia CIRP* **2012**, *1*, 599–604. [CrossRef]
2. Grigoriev, S.; Metel, A. Plasma- and beam-assisted deposition methods. *NATO Sci. Ser.* **2004**, *155*, 147–154.
3. Kozochkin, M.P.; Porvatov, A.N.; Sabirov, F.S.; Kozochkin, M. The fitting of technological equipment with data-measuring systems. *Meas. Tech.* **2012**, *55*, 530–534. [CrossRef]
4. Hameed, S.; Rojas, H.A.G.; Benavides, J.I.P.; Alberro, A.N.; Egea, A.J.S. Influence of the Regime of Electropulsing-Assisted Machining on the Plastic Deformation of the Layer Being Cut. *Materials* **2018**, *11*, 886. [CrossRef]
5. Volosova, M.; Grigoriev, S.; Metel, A.; Shein, A. The Role of Thin-Film Vacuum-Plasma Coatings and Their Influence on the Efficiency of Ceramic Cutting Inserts. *Coatings* **2018**, *8*, 287. [CrossRef]
6. Gallo, L.S.; Boas, M.O.V.; Rodrigues, A.C.; Melo, F.C.; Zanotto, E.D. Transparent glass–ceramics for ballistic protection: materials and challenges. *J. Mater. Res. Technol.* **2019**, *8*, 3357–3372. [CrossRef]
7. Naveau, A.; Rignon-Bret, C.; Wulfman, C. Zirconia abutments in the anterior region: A systematic review of mechanical and esthetic outcomes. *J. Prosthet. Dent.* **2019**, *121*, 775. [CrossRef]
8. Panzani, A. Globalisation Scenarios in the World Ceramic Tile Market Extract of the speach given by Alfonso Panzani at QUALICER *CFI Ceram. Forum Int.* **2014**, *91*, E20.
9. Pascariu, P.; Homocianu, M. ZnO-based ceramic nanofibers: Preparation, properties and applications. *Ceram. Int.* **2019**, *45*, 11158–11173. [CrossRef]
10. Roata, I.C.; Croitoru, C.; Pascu, A.; Stanciu, E.M. Photocatalytic coatings via thermal spraying: A mini-review. *AIMS Mater. Sci.* **2019**, *6*, 335–353. [CrossRef]
11. Xie, S.; Xu, J.; Chen, Y.; Tan, Z.; Nie, R.; Wang, Q.; Zhu, J. Indentation Behavior and Mechanical Properties of Tungsten/Chromium co-Doped Bismuth Titanate Ceramics Sintered at Different Temperatures. *Materials* **2018**, *11*, 503. [CrossRef]
12. Killinger, A.; Gadow, R.; Kern, F. Manufacturing of nanocomposite structural ceramic materials and coatings. *Int. J. Mater. Prod. Technol.* **2009**, *35*, 334–345.
13. Volosova, M.A.; Grigor'ev, S.N.; Kuzin, V.V. Effect of Titanium Nitride Coating on Stress Structural Inhomogeneity in Oxide-Carbide Ceramic. Part Action of Heat Flow. *Refract. Ind. Ceram.* **2015**, *56*, 91–96. [CrossRef]
14. Guo, D.; Song, S.; Luo, R.; Goddard, W.A.; Chen, M.; Reddy, K.M.; An, Q. Grain Boundary Sliding and Amorphization are Responsible for the Reverse Hall-Petch Relation in Superhard Nanocrystalline Boron Carbide. *Phys. Rev. Lett.* **2018**, *121*, 145504. [CrossRef]
15. Balokhonov, R.R.; Zinoviev, A.V.; Romanova, V.A.; Batukhtina, E.E. A Mesomechanical Analysis of the Deformation and Fracture in Polycrystalline Materials with Ceramic Porous Coatings. *AIP Conf. Proc.* **2015**, *1683*, 20020.
16. Murzin, S.P.; Balyakin, V.B. Microstructuring the surface of silicon carbide ceramic by laser action for reducing friction losses in rolling bearings. *Opt. Laser Technol.* **2017**, *88*, 96–98. [CrossRef]
17. Kuzin, V.V.; Seleznev, A.E.; Fedorov, S.Y. Effect of conditions of diamond grinding on tribological behavior of alumina-based ceramics. *J. Frict. Wear* **2016**, *37*, 371–376. [CrossRef]
18. Grigoriev, S.N.; Gurin, V.D.; Volosova, M.A.; Cherkasova, N.Y.; Volosova, M. Development of residual cutting tool life prediction algorithm by processing on CNC machine tool. *Materialwissenschaft und Werkstofftechnik* **2013**, *44*, 790–796. [CrossRef]
19. Boswell, B.; Islam, M.N.; Davies, I.J. A review of micro-mechanical cutting. *Int. J. Adv. Manuf. Technol.* **2018**, *94*, 789–806. [CrossRef]
20. Pratap, A.; Patra, K.; Dyakonov, A.A. On-machine texturing of PCD micro-tools for dry micro-slot grinding of BK7 glass. *Precis. Eng.* **2019**, *55*, 491–502. [CrossRef]

21. Isaev, A.V.; Kozochkin, M.P.; Kozochkin, M. Use of a Measurement Information System to Increase the Precision with which Thin-Walled Parts are Machined on Numerically Controlled Milling Machines. *Meas. Tech.* **2014**, *56*, 1155–1161. [CrossRef]
22. Razeghiyadaki, A.; Molardi, C.; Talamona, D.; Perveen, A. Modeling of Material Removal Rate and Surface Roughness Generated during Electro-Discharge Machining. *Machines* **2019**, *7*, 47. [CrossRef]
23. Schubert, A.; Zeidler, H.; Hackert-Oschaetzchen, M.; Schneider, J.; Hahm, M. Enhancing Micro-EDM using Ultrasonic Vibration and Approaches for Machining of Nonconducting Ceramics. *Strojniski Vestnik J. Mech. Eng.* **2013**, *59*, 156–164. [CrossRef]
24. Okunkova, A.A.; Povolotskiy, D.E.; Podrabinnik, P.A. Study of electrical discharge machining for the parts of nuclear industry usage. *Mech. Ind.* **2015**, *16*, 706.
25. Grigor'ev, S.N.; Kozochkin, M.P.; Fedorov, S.V.; Porvatov, A.N.; Okun'kova, A.A.; Kozochkin, M. Study of Electroerosion Processing by Vibroacoustic Diagnostic Methods. *Meas. Tech.* **2015**, *58*, 878–884. [CrossRef]
26. Glaziev, S.Y. The Discovery of Regularities of Change of Technological Orders in the Central Economics and Mathematics Institute of the Soviet Academy of Sciences. *Ekonomika i Matematiceskie Metody. Econ. Math. Methods* **2018**, *54*, 17–30. [CrossRef]
27. Chen, H. Sustainable Consumption and Production in China. In *Handbook on Sustainability Transition and Sustainable Peace. Hexagon Series on Human and Environmental Security and Peace*; Brauch, H., Oswald Spring, U., Grin, J., Scheffran, J., Eds.; Springer: Cham, Switzerland, 2016; Volume 10, pp. 571–591.
28. Korotayev, A.V.; Tsirel, S.V. A Spectral Analysis of World GDP Dynamics: Kondratiev Waves, Kuznets Swings, Juglar and Kitchin Cycles in Global Economic Development, and the 2008–2009 Economic Crisis. *Struct. Dyn.* **2010**, *4*, 3–57.
29. Schumpeter, J.; Becker, M.C.; Knudsen, T. The fundamental phenomenon of economic development. *Am. J. Econ. Sociol.* **2002**, *61*, 405–437. [CrossRef]
30. Perez, C. Technological revolutions and techno-economic paradigms. *Camb. J. Econ.* **2010**, *34*, 185–202. [CrossRef]
31. Bobovich, B.B. Glass-Fiber Reinforced Plastics—Construction Materials of the Sixth Technological Paradigm? *Glas. Ceram.* **2019**, *76*, 38–41. [CrossRef]
32. Wonglimpiyarat, J. Towards the Sixth Kondratieff Cycle of Nano Revolution. *Nanotechnol. Microelectron.* **2010**, 87–100. [CrossRef]
33. Hou, J.; Feng, Y.; Liao, J.; Ding, W.; Shui, L.; Li, H.; Wang, Y.; Tang, B.; Umar, A.; Zhou, G. Multiscale Interface Effect on Homogeneous Dielectric Structure of ZrO2/Teflon Nanocomposite for Electrowetting Application. *Polymers* **2018**, *10*, 1119. [CrossRef]
34. Sun, S.; Xu, Q. Fabricating a Novel Intragranular Microstructure for Al2O3/GdAlO3 Ceramic Composites. *Materials* **2018**, *11*, 1879. [CrossRef]
35. Hanzel, O.; Singh, M.A.; Marla, D.; Sedlák, R.; Šajgalík, P. Wire electrical discharge machinable SiC with GNPs and GO as the electrically conducting filler. *J. Eur. Ceram. Soc.* **2019**, *39*, 2626–2633. [CrossRef]
36. Gordeev, Y.I.; Abkaryan, A.K.; Surovtsev, A.V.; Lepeshev, A.A. Investigation into the Peculiarities of Structure Formation and Properties of Copper-Based Powder Pseudoalloys Modified by ZnO and TiN Nanoparticle Additives. *Russ. J. Non Ferrous Met.* **2019**, *60*, 68–75. [CrossRef]
37. Grigoriev, S.N.; Volosova, M.A.; Peretyagin, P.Y.; Seleznev, A.E.; Okunkova, A.A.; Smirnov, A. The Effect of TiC Additive on Mechanical and Electrical Properties of Al2O3 Ceramic. *Appl. Sci.* **2018**, *8*, 2385. [CrossRef]
38. Grigoriev, S.N.; Melnik, Y.A.; Metel, A.S.; Panin, V.V.; Prudnikov, V.V. A compact vapor source of conductive target material sputtered by 3-keV ions at 0.05-Pa pressure. *Instruments Exp. Tech.* **2009**, *52*, 731–737. [CrossRef]
39. Díaz, L.A.; Montes-Morán, M.A.; Peretyagin, P.Y.; Vladimirov, Y.G.; Okunkova, A.; Moya, J.S.; Torrecillas, R. Zirconia–alumina–nanodiamond composites with gemological properties. *J. Nanoparticle Res.* **2014**, *16*, 2257. [CrossRef]
40. Lei, J.; Wu, X.; Wang, Z.; Xu, B.; Zhu, L.; Wu, W. Electrical discharge machining of micro grooves using laminated disc electrodes made of Cu and Sn foils. *J. Mater. Process. Technol.* **2019**, *271*, 455–462. [CrossRef]
41. Arab, J.; Mishra, D.K.; Kannojia, H.K.; Adhale, P.; Dixit, P. Fabrication of multiple through-holes in non-conductive materials by Electrochemical Discharge Machining for RF MEMS Packaging. *J. Mater. Process. Technol.* **2019**, *271*, 542–553. [CrossRef]

42. Urazov, M.N.; Efimov, A.A.; Lizunova, A.A.; Ivanov, V.V.; Mylnikov, D.A. Investigation of electric erosion of silicon electrodes in aerosol nanoparticles synthesis. *AIP Conf. Proc.* **2017**, *1858*, 040007.
43. Metel, A.; Bolbukov, V.; Volosova, M.; Grigoriev, S.; Melnik, Y. Equipment for deposition of thin metallic films bombarded by fast argon atoms. *Instrum. Exp. Tech.* **2014**, *57*, 345–351. [CrossRef]
44. Kuzin, V.V.; Grigor'ev, S.N.; Volosova, M.A. Effect of a TiC Coating on the Stress-Strain State of a Plate of a High-Density Nitride Ceramic Under Nonsteady Thermoelastic Conditions. *Refract. Ind. Ceram.* **2014**, *54*, 376–380. [CrossRef]
45. Wysocka, I.; Hupka, J.; Rogala, A. Catalytic Activity of Nickel and Ruthenium–Nickel Catalysts Supported on SiO2, ZrO2, Al2O3, and MgAl2O4 in a Dry Reforming Process. *Catalysts* **2019**, *9*, 540. [CrossRef]
46. Promakhov, V.; Zhukov, A.; Dubkova, Y.; Zhukov, I.; Kovalchuk, S.; Zhukova, T.; Olisov, A.; Klimenko, V.; Savkina, N. Structure and Properties of ZrO2–20%Al2O3 Ceramic Composites Obtained Using Additive Technologies. *Materials* **2018**, *11*, 2361. [CrossRef]
47. Di, J.; Liu, W.; Xue, L.; Zhou, Q.; Yan, Y. A Dense Cr_2O_3/Al_2O_3 Composite Ceramic Coating Prepared by Electrodeposition and Sealing with Al_2O_3. *Coatings* **2019**, *9*, 14. [CrossRef]
48. Koltsov, I.; Smalc-Koziorowska, J.; Prześniak-Welenc, M.; Małysa, M.; Kimmel, G.; McGlynn, J.; Ganin, A.; Stelmakh, S. Mechanism of Reduced Sintering Temperature of Al2O3–ZrO2 Nanocomposites Obtained by Microwave Hydrothermal Synthesis. *Materials* **2018**, *11*, 829. [CrossRef]
49. Gommeringer, A.; Kern, F.; Gadow, R. Enhanced Mechanical Properties in ED-Machinable Zirconia-Tungsten Carbide Composites with Yttria-Neodymia Co-Stabilized Zirconia Matrix. *Ceramics* **2018**, *1*, 26–37. [CrossRef]
50. Nayebzadeh, H.; Saghatoleslami, N.; Tabasizadeh, M. Application of microwave irradiation for fabrication of sulfated ZrO2–Al2O3 nanocomposite via combustion method for esterification reaction: process condition evaluation. *J. Nanostructure Chem.* **2019**, *9*, 141–152. [CrossRef]
51. Podzorova, L.I.; Il'Icheva, A.A.; Pen'Kova, O.I.; Antonova, O.S.; Baikin, A.S.; Konovalov, A.A. Al2O3-Based Ceramic Composites with a High Brittle Fracture Resistance. *Inorg. Mater.* **2019**, *55*, 628–633. [CrossRef]
52. Deirmina, F.; Almangour, B.; Grzesiak, D.; Pellizzari, M. H13–partially stabilized zirconia nanocomposites fabricated by high-energy mechanical milling and selective laser melting. *Mater. Des.* **2018**, *146*, 86–297. [CrossRef]
53. Bin Mansoor, M.; Köble, S.; Wong, T.W.; Woias, P.; Goldschmidtböing, F. Design, Characterization and Sensitivity Analysis of a Piezoelectric Ceramic/Metal Composite Transducer. *Micromachines* **2017**, *8*, 271. [CrossRef] [PubMed]
54. Jurczyk, K.; Adamek, G.; Kubicka, M.M.; Jakubowicz, J.; Jurczyk, M. Nanostructured Titanium-10 wt% 45S5 Bioglass-Ag Composite Foams for Medical Applications. *Materials* **2015**, *8*, 1398–1412. [CrossRef] [PubMed]
55. Todorova, T.; Gaier, M.; Zwanziger, J.; Plucknett, K. Understanding the elastic and thermal response in TiC-based ceramic-metal composite systems: First-principles and mechanical studies. *J. Alloy. Compd.* **2019**, *789*, 712–719. [CrossRef]
56. Beitollahi, H.; Garkani-Nejad, F.; Tajik, S.; Ganjali, M.R. Voltammetric Determination of Acetaminophen and Tryptophan Using a Graphite Screen Printed Electrode Modified with Functionalized Graphene Oxide Nanosheets Within a Fe3O4@SiO2 Nanocomposite. *Iran. J. Pharm. Res.* **2019**, *18*, 80–90. [PubMed]
57. Ming, K.; Su, Q.; Gu, C.; Xie, D.; Wang, Y.; Nastasi, M.; Wang, J. Influence of Metal Additives on Microstructure and Properties of Amorphous Metal–SiOC Composites. *JOM* **2019**, *71*, 2445–2451. [CrossRef]
58. Balokhonov, R.; Romanova, V.; Schmauder, S.; Emelianova, E. A numerical study of plastic strain localization and fracture across multiple spatial scales in materials with metal-matrix composite coatings. *Theor. Appl. Fract. Mech.* **2019**, *101*, 342–355. [CrossRef]
59. Smolin, A.; Shilko, E.; Astafurov, S.; Konovalenko, I.; Buyakova, S.; Psakhie, S.; Smolin, A. Modeling mechanical behaviors of composites with various ratios of matrix–inclusion properties using movable cellular automaton method. *Def. Technol.* **2015**, *11*, 18–34. [CrossRef]
60. Ojha, N.; Hoesel, T.; Mueller, C.; Reinecke, H. Characterization of The Conductive Layer Formed During Mu - Electric Discharge Machining Of Non-Conductive Ceramics. *Ceram. Trans.* **2013**, *240*, 105–110.
61. Gotoh, H.; Tani, T.; Mohri, N. EDM of Insulating Ceramics by Electrical Conductive Surface Layer Control. *Procedia CIRP* **2016**, *42*, 201–205. [CrossRef]
62. Ali, M.Y.; Moudood, M.A.; Maleque, M.A.; Hazza, M.; Adesta, E.Y.T. Electro-discharge machining of alumina: Investigation of material removal rate and surface roughness. *J. Mech. Eng. Sci.* **2017**, *11*, 3015–3026. [CrossRef]

63. Zhang, G.; Guo, Y.; Wang, L. Experimental Study on the Machining of Inclined Holes for Thermal Barrier-Coated Nickel Superalloys by EDM. *J. Mater. Eng. Perform.* **2016**, *25*, 4574–4580. [CrossRef]
64. Bilal, A.; Jahan, M.P.; Talamona, D.; Perveen, A. Electro-Discharge Machining of Ceramics: A Review. *Micromachines* **2019**, *10*, 10. [CrossRef] [PubMed]
65. Hanaoka, D.; Fukuzawa, Y.; Yamashita, K. Research of Large-Area Electrical Discharge Machining for Insulating Si3N4 Ceramics with the Assisting Electrode Method. *Adv. Mater. Res.* **2014**, *939*, 76–83. [CrossRef]
66. Kaneko, K.; Fukuzawa, Y. Characteristics of Micro EDM for Insulating Aluminum Nitride Ceramics. *Adv. Mater. Res.* **2012**, *579*, 86–91. [CrossRef]
67. Furutani, K.; Tomoto, M. Performance of Wire-Sawing of Glass Assisted by Electro-Chemical Discharge. *Key Eng. Mater.* **2012**, *523*, 299–304. [CrossRef]
68. Hanaoka, D.; Ito, R.; Fukuzawa, Y. Electrical Discharge Machined Surface of the Insulating ZrO2 Ceramics. *J. Adv. Mech. Des. Syst. Manuf.* **2011**, *5*, 372–384. [CrossRef]
69. Schubert, A.; Berger, T.; Martin, A.; Hackert-Oschätzchen, M.; Treffkorn, N.; Kuhn, R. Surface structuring of boron doped CVD diamond by micro electrical discharge machining. *AIP Conf. Proc.* **2018**, *1960*, 080003.
70. Zeller, F.; Hoesel, T.; Mueller, C. Microstructuring of non-conductive silicon carbide by electrical discharge machining. *Microsyst. Technol.* **2014**, *20*, 1875–1880. [CrossRef]
71. Mohri, N. EDM of advanced ceramics—From finish machining to machining insulating ceramics. *CIRP Ann. Manuf. Technol.* **1996**, *45*, 289–296.
72. Takayuki, T.; Tsujita, Y.; Gotoh, H.; Okada, M.; Mohri, N. Observation of Material Removal Process by Single Discharge in Air Gap. *Procedia CIRP* **2018**, *68*, 276–279. [CrossRef]
73. Rona, N.; Yenisey, M.; Kucukturk, G.; Gürün, H.; Cogun, C.; Esen, Z. Effect of electrical discharge machining on dental Y-TZP ceramic-resin bonding. *J. Prosthodont. Res.* **2017**, *61*, 158–167. [CrossRef] [PubMed]
74. Esen, Z.; Genc, A.; Akturk, N.; Cogun, C.; Cogun, F. Effect of powder metallurgy Cu-B 4 C electrodes on workpiece surface characteristics and machining performance of electric discharge machining. *Proc. Inst. Mech. Eng. Part B J. Eng. Manuf.* **2016**, *230*, 2190–2203.
75. Erdem, O.; Cogun, C.; Urtekin, L. The effect of powder mixed and heated dielectric on drilling performance of electric discharge machining (EDM). *J. Fac. Eng. Archit. Gazi Univ.* **2007**, *31*, 531–544.
76. Khan, M.A.R.; Rahman, M.M.; Noor, M.; Kadirgama, K.; Maleque, M. Current Research Trends on Dry, Near-Dry and Powder Mixed Electrical Discharge Machining. *Adv. Mater. Res.* **2011**, *264*, 956–961. [CrossRef]
77. Hosni, N.; Lajis, M. Multi-response optimization of the machining characteristics in electrical discharge machining (EDM) using span-20 surfactant and chromium (Cr) powder mixed. *Materialwissenschaft Und Werkstofftechnik* **2019**, *50*, 329–335. [CrossRef]
78. Lajis, M.; Hosni, N. The influences of various mixed dielectric fluids on the performance electrical discharge machining of AISI D2 hardened steel. *Materialwissenschaft Und Werkstofftechnik* **2018**, *49*, 413–419. [CrossRef]
79. Ablyaz, T.R.; Simonov, M.Y.; Shlykov, E.S. Analysis of the Surface of Deposited Copper After Electroerosion Treatment. *Met. Sci. Heat Treat.* **2018**, *59*, 779–785. [CrossRef]
80. Bains, P.S.; Singh, S.; Sidhu, S.S.; Kaur, S.; Ablyaz, T.R. Investigation of Surface Properties of Al–SiC Composites in Hybrid Electrical Discharge Machining. In *Materials Horizons: From Nature to Nanomaterials*; Springer Singapore: Singapore, 2018; pp. 181–196.
81. Ogleznev, N.; Oglezneva, S.; Ablyaz, T. Perspective Composition Materials for Electrode-Tools Production. In *Materials Horizons: From Nature to Nanomaterials*; Springer Singapore: Singapore, 2018; pp. 319–344.
82. Yang, L.; Ditta, A.; Feng, B.; Zhang, Y.; Xie, Z. Study of the Comparative Effect of Sintering Methods and Sintering Additives on the Microstructure and Performance of Si3N4 Ceramic. *Materials* **2019**, *12*, 2142. [CrossRef]
83. Jia, D.; Ma, J.; Gan, X.; Tao, J.; Xie, M.; Yi, J.; Liu, Y. A Comparison Study of Ag Composites Prepared by Spark Plasma Sintering and Hot Pressing with Silver-Coated CNTs as the Reinforcements. *Materials* **2019**, *12*, 1949. [CrossRef]
84. Lun, H.; Zeng, Y.; Xiong, X.; Zhao, L.; Li, D.; Ye, Z.; Qian, T. The Effect of SiC Content on Microstructure and Microwave Heating Rate of h-BN/SiC Ceramics Fabricated by Spark Plasma Sintering. *Materials* **2019**, *12*, 1909. [CrossRef] [PubMed]
85. Hsieh, C.; Tuan, W. Thermal expansion behavior of a model ceramic–metal composite. *Mater. Sci. Eng. A* **2007**, *460*, 453–458. [CrossRef]

86. Hsieh, C.; Tuan, W. Elastic properties of ceramic–metal particulate composites. *Mater. Sci. Eng. A* **2005**, *393*, 133–139. [CrossRef]
87. Pędzich, Z.; Wajler, C. Slow crack propagation in Y-TZP/metal composites. *J. Eur. Ceram. Soc.* **2006**, *26*, 679–682. [CrossRef]
88. Oraei, M.; Mostaan, H.; Rafiei, M.; Abbasian, A.R.; Zarezadeh, M. Investigation into microstructural evolutions, mechanical properties and thermal analysis of Al(Zn)/Al2O3 nano-composite fabricated by mechanical milling and SPS method. *Mater. Res. Express* **2019**, *6*, 0865g9. [CrossRef]
89. Moriceau, J.; Houizot, P.; Pasturel, M.; Guizouarn, T.; Rouxel, T. A magnetic glass matrix (ZnO-BaO-B2O3) particulate (Fe3O4) nanocomposite obtained by SPS. *J. Non Crystalline Solids* **2019**, *514*, 116–121. [CrossRef]
90. Tukmakova, A.; Novotelnova, A.; Samusevich, K.; Usenko, A.; Moskovskikh, D.; Smirnov, A.; Mirofyanchenko, E.; Takagi, T.; Miki, H.; Khovaylo, V. Simulation of Field Assisted Sintering of Silicon Germanium Alloys. *Materials* **2019**, *12*, 570. [CrossRef]
91. Moskovskikh, D.; Song, Y.; Rouvimov, S.; Rogachev, A.; Mukasyan, A. Silicon carbide ceramics: Mechanical activation, combustion and spark plasma sintering. *Ceram. Int.* **2016**, *42*, 12686–12693. [CrossRef]
92. Grigoriev, S.N.; Melnik, Y.A.; Metel, A.S.; Panin, V.V. Broad beam source of fast atoms produced as a result of charge exchange collisions of ions accelerated between two plasmas. *Instruments Exp. Tech.* **2009**, *52*, 602–608. [CrossRef]
93. Galatanu, M.; Enculescu, M.; Galatanu, A. High temperature thermo-physical properties of SPS-ed W–Cu functional gradient materials. *Mater. Res. Express* **2018**, *5*, 026502. [CrossRef]
94. Kotoban, D.; Grigoriev, S.; Okunkova, A.; Sova, A. Influence of a shape of single track on deposition efficiency of 316L stainless steel powder in cold spray. *Surf. Coatings Technol.* **2017**, *309*, 951–958. [CrossRef]
95. Grigoriev, S.; Fominski, V.Y.; Romanov, R.; Gnedovets, A.; Volosova, M.; Gnedovets, A.; Volosova, M. Shadow masked pulsed laser deposition of WSex films: Experiment and modeling. *Appl. Surf. Sci.* **2013**, *282*, 607–614. [CrossRef]
96. Fook, P.; Berger, D.; Riemer, O.; Karpuschewski, B. Structuring of Bioceramics by Micro-Grinding for Dental Implant Applications. *Micromachines* **2019**, *10*, 312. [CrossRef] [PubMed]
97. Garcia-Sanchez, A.M.; Machado-Moreira, B.; Freire, M.; Santos, R.; Monteiro, S.; Dias, D.; Neves, O.; Dionísio, A.; Miller, A.Z. Characterization of Microbial Communities Associated with Ceramic Raw Materials as Potential Contributors for the Improvement of Ceramic Rheological Properties. *Minerals* **2019**, *9*, 316. [CrossRef]
98. Pramanik, A.; Islam, M.N.; Basak, A.K.; Dong, Y.; Littlefair, G.; Prakash, C. Optimizing dimensional accuracy of titanium alloy features produced by wire electrical discharge machining. *Mater. Manuf. Process.* **2019**, *34*, 1083–1090. [CrossRef]
99. Fedorov, S.V.; Pavlov, M.D.; Okunkova, A.A. Effect of structural and phase transformations in alloyed subsurface layer of hard-alloy tools on their wear resistance during cutting of high-temperature alloys. *J. Frict. Wear* **2013**, *34*, 190–198. [CrossRef]
100. Liu, K.; Lauwers, B.; Reynaerts, D. Crossing barriers in structuring ceramics. *Mikroniek* **2010**, *50*, 28–34.
101. Zhang, H.; Jayaseelan, D.; Bogomol, I.; Reece, M.; Hu, C.; Grasso, S.; Lee, W. A novel microstructural design to improve the oxidation resistance of ZrB2-SiC ultra-high temperature ceramics (UHTCs). *J. Alloy. Compd.* **2019**, *785*, 958–964. [CrossRef]
102. Ghelich, R.; Aghdam, R.M.; Jahannama, M.R. Elevated temperature resistance of SiC-carbon/phenolic nanocomposites reinforced with zirconium diboride nanofibers. *J. Compos. Mater.* **2018**, *52*, 1239–1251. [CrossRef]
103. Chiang, K.-T. Modeling and analysis of the effects of machining parameters on the performance characteristics in the EDM process of Al2O3+TiC mixed ceramic. *Int. J. Adv. Manuf. Technol.* **2008**, *3*, 523–533. [CrossRef]
104. Egea, A.S.; Peiró, J.J.; Signorelli, J.W.; Rojas, H.G.; Celentano, D.J. On the microstructure effects when using electropulsing versus furnace treatments while drawing inox 308L. *J. Mater. Res. Technol.* **2019**, *8*, 2269–2279. [CrossRef]

© 2019 by the authors. Licensee MDPI, Basel, Switzerland. This article is an open access article distributed under the terms and conditions of the Creative Commons Attribution (CC BY) license (http://creativecommons.org/licenses/by/4.0/).

technologies

MDPI

Review

Roles of Cryogenic Cooling in Turning of Superalloys, Ferrous Metals, and Viscoelastic Polymers

Tze Chuen Yap

School of Engineering and Physical Sciences, Heriot-Watt University Malaysia, Putrajaya 62200, Malaysia; t.yap@hw.ac.uk; Tel.: +60-3-8894-3780

Received: 24 June 2019; Accepted: 31 August 2019; Published: 5 September 2019

Abstract: Cryogenic machining is a relatively new technique in machining. This concept was applied on various machining processes such as turning, milling, drilling etc. Cryogenic turning technique is generally applied on three major groups of workpiece materials—superalloys, ferrous metals, and viscoelastic polymers/elastomers. The roles of cryogen in machining different materials are unique and are summarised in this review article. Finally, the challenges in using cryogenic machining in industries are also highlighted.

Keywords: cryogenic machining; review; liquid nitrogen; liquid carbon dioxide

1. Introduction

Conventional cutting fluids are used to reduce the temperature of tool and work in machining. However, conventional cutting fluids are unable to work effectively in high-speed machining of some superalloys such as titanium alloys, inconel alloys and tantalum alloys [1]. The other major problems caused by the conventional cutting fluids are health and environmental problems [2–5]. The conventional cutting fluids can be categorised into semi-synthetic, synthetics, straight, and soluble oils [6]. Such additives are added onto cutting fluids to provide lubricating effects. Some of these additives (aromatic amines, chlorinated paraffines, etc.) might cause a carcinogenic potential [3]. Beside this, Suliman et al. [5] identified several species of bacteria and fungi growth found in cutting fluids used in workshops. The contamination of the cutting fluids by these micro-organisms can cause several types of health problems, such as septic infections, primary allergic bronchopulmonary aspergillosis, dermatomycosis, and neonatal meningitis [4,5].

The conventional cutting fluid can be recycled several times until its quality degrades after the useable period. The storage and proper disposal of these cutting fluids require special processes and extra costs [2,7,8]. Moreover, incorrect disposal of these cutting fluids causes environmental pollution. As such, new machining techniques such as green machining, cryogenic machining, and dry machining have been proposed to substitute conventional machining with cutting fluids.

2. Cryogenics Machining

2.1. Cryogenic Machining in General

Cryogenic machining is a relatively new technique in reducing machining temperature by replacing the conventional cutting fluid with extremely cold or sub-zero (below −150 °C) cutting fluid at the machining interface. Nitrogen (N_2), carbon dioxide (CO_2), oxygen (O_2), helium (He), etc., in compressed forms, are the potential liquid cryogens to be used. These gases are exist in the atmosphere and can be converted to a liquid form. These liquid cryogens evaporate in room temperature and convert to gas again [9]. Nevertheless, only liquid carbon dioxide (LCO_2) and liquid nitrogen (LN_2) are commonly used in cryogenic machining studies [10–12]. The boiling points of CO_2 and N_2 are

−78.5 °C and −196 °C. Clearly, the boiling point of nitrogen is lower than carbon dioxide. Besides, carbon dioxide gas is denser than atmospheric air, and it might accumulate at the ground of the plant and can cause breathing problems to machine operators [11]. Thus, liquid carbon dioxide is not recommended to be used as a cutting fluid in machining, and liquid nitrogen is preferred for use as a cryogenic cutting fluid. In 1953, Bartle used liquid carbon dioxide as the coolant in machining [11]. The earliest investigation on cryogenic machining with liquid nitrogen (LN_2) as cryogen was carried out by Uehara and Kumagai [12]. Cryogenic machining with liquid nitrogen LN_2 reduced cutting forces and improved tool life and surface finish of the machined parts. A similar result was reported by Fillippi and Ippolito [10], who conducted cryogenic face milling. However, the usage of cryogenic technology in early days was limited by the high cost. The idea of cryogenic machining was again re-proposed after the 1990's when the cryogenic technology was improved, and higher production rate (high-speed machining) was required [11]. At the same time, the awareness of health and environment issue was also spreading. From 1995–2007, three main groups of researchers have conducted investigations on cryogenic machining. Most of their works focused on cryogenic turning [1,9,11,13–23] and cryogenic grinding [24–27]. The number of reported works in cryogenic milling and cryogenic drilling researches are insignificant, compared to cryogenic turning.

Previous cryogenic turning explorations investigated the tool wear, surface roughness, and cutting forces. Improvements on cutting tool life under the cryogenic conditions have been reported by Hong and co-workers [11,17,18], Wang and co-workers [1,21–23] and Dhar and co-workers [13–16,19,20]. Beside this, better machined surface roughness is obtained through cryogenic machining, as reported by Dhar et al. [14,15] and Wang et al. [1,22,23].

Cryogenic machining technique is commonly applied on superalloys (titanium alloys, inconel alloy, tantalum alloy, etc.), ferrous metals, viscoelastic polymers and elastomers. In general, positive results are reported, regardless of the type of workpiece materials. Nevertheless, roles of cryogen are slightly different in machining different materials.

2.2. *Cryogenic Turning of Superalloys*

A superalloy has a combination of high mechanical strength and surface stability at high temperature [28]. Examples of superalloys are titanium alloy, inconel alloy, and tantalum alloy. Due to their characteristics, superalloys are used in various engineering applications. However, high mechanical strength of superalloys also causes poor machinability. Conventional cutting fluids are not efficient to reduce the machining temperature in machining superalloys, and cryogenic fluids were proposed to replace them.

Most of the work materials studied in previous cryogenic machining studies are titanium alloys. One of the earliest reported cryogenic liquid nitrogen machining studies was conducted by Wang and Rajurkar [1]. They designed a liquid nitrogen cooling system for their experiments. Liquid nitrogen was circulated through a sealed cap which was placed on the top of the insert to reduce the cutting insert's temperature. They measured and compared the cement carbide tool wear in turning Ti-6Al-4V with conventional cutting fluids (Mobilmet Omicron) and liquid nitrogen cooling. Liquid nitrogen outperformed conventional cutting fluids in terms of reduction of both tools' flank wear and surface roughness of the machined surface. However, no significant difference was observed in the cutting forces.

On the other hand, Hong and his co-workers did a series of systematic studies on cryogenic turning of titanium alloys in term of tool wear, surface roughness, cutting forces, friction force, and friction coefficient. In their studies, liquid nitrogen was injected on the interface of tool and workpiece, i.e., the liquid nitrogen was directly sprayed onto the tool and workpiece. In 2001, Hong and Ding [17] studied several cooling approaches in cryogenic machining of Ti-6Al-4V. They reported that the best cooling approach was simultaneous cooling rake and flank, followed by cryogenic cooling at the rake face and cryogenic cooling at the flank face. Beside this, they also concluded that by applying liquid nitrogen, the tool temperature can be reduced and the cutting speed of conventional machining of titanium alloys can be increased (around 60 m/min). In the same year, Hong et al. [18] conducted a similar study but they changed the position of the chip breaker. At the optimum position of the chip breaker and with cryogenic cooling, the tool life was extended by five times compared to conventional machining. Hong et al. [29] furthered the research in measuring the cutting forces. They reported that cryogenic liquid nitrogen hardened the material, and this caused higher cutting force. However, the friction force between the cutting tool and workpiece was reduced due to two main reasons: the material became less sticky at a lower temperature, and liquid nitrogen formed a fluid cushion between the interfaces. As such, they concluded that liquid nitrogen was able to function as a lubricant. Furthermore, Hong [11] also compared the cost of cryogenic machining with conventional machining and showed that cryogenic machining with his patented nozzle was superior in terms of cost, tool life, surface finish, as well as environmental friendliness.

Cryogenic turning of Ti-6Al-4V was also studied by Venugopal et al. [19,20]. They built a special nozzle which was able to inject liquid nitrogen jets on the crater and flank faces of cutting tools. They conducted the cryogenic turning tests in cutting speed above the recommended speed for titanium alloys, (i.e., 60 m/min) under dry, oil, and cryogenic liquid nitrogen environments. From their study, adhesion-dissolution-diffusion wear are the main tool wear mechanisms at the crater in turning Ti-6Al-4V with uncoated carbide; while abrasion and attrition occurred at the flank. Liquid nitrogen cooling was able to control the cutting temperature and then reduced the tool wear (both flank wear and crater wear) at cutting speed of 70 m/min, but this effect decreased at higher cutting speeds (100 and 117 m/min).

Cryogenic turning of Ti-6Al-4V was reported by Dhananchezian and Kumar [30]. They modified the cutting insert used in their experiment by drilling a hole at the rake surface of the insert and applied liquid nitrogen to the hole (Figure 1). The results showed that cryogenic cooling performed better than wet machining in terms of cutting force, surface roughness, and tool wear. Yap et al. [31] studied the influence of low pressure cryogenic liquid nitrogen to the machining forces, friction, tool wear and surface quality under high-speed machining of Ti-5Al-4V-0.6Mo-0.4Fe. The experiments were conducted in dry and cryogenic liquid nitrogen conditions. The 0.034 MPa internal pressure inside a liquid nitrogen Dewar tank caused the liquefied nitrogen to be injected onto the tool–work interface. The experimental results showed that low pressure liquid nitrogen can reduce the friction force and friction coefficient during machining. In additional, surface roughness of the machined titanium alloy was improved and tool wear was reduced. Reduction in machining forces and improvement in surface roughness of machined parts can be attributed to better heat removal by injection of liquid nitrogen. While most of the researchers used only one cryogen in their cryogenic machining works, Kaynak and Gharibi studied cryogenic turning of Ti-5Al-5V-3Cr-0.5Fe with both liquid nitrogen and liquid carbon dioxide and compared with dry machining [32]. At low cutting speed (below 120 m/min), tool wear did not depend on cutting condition (dry/liquid nitrogen/liquid carbon dioxide) and no significant different in wear mechanism was observed on the cutting inserts from three different machining conditions. At higher cutting speeds of above 120 m/min, cryogenic machining with liquid nitrogen outperformed cryogenic machining with liquid carbon dioxide and dry machining. Different wear mechanisms were detected on the cutting inserts from the three different machining conditions.

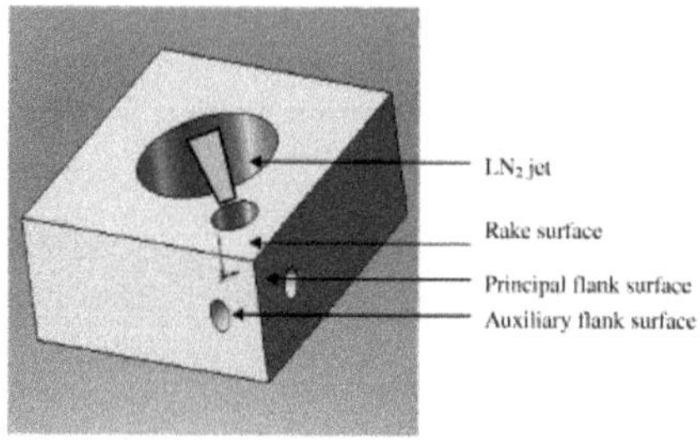

Figure 1. Modified cutting insert for cryogenic cooling method [30].

Inconel is a nickel-based alloy with poor machinability [33]. Low thermal conductivity of inconel alloy (about 11 W/m °C) caused higher cutting temperature and high tool wear rate [1] and additional effort was required to reduce the cutting temperature. Wang and Rajurkar reported that machining Inconel 718 with LN_2 cooling was able to delay the tool wear and improve the surface roughness of workpiece surface [1]. Similar results were reported by Pusavec et al. [34]. In addition to tool wear and surface roughness, a thicker compressive zone is noticed beneath the machining surface and smaller grain size is observed after cryogenic machining with liquid nitrogen. Beside this, the sub-surface layer under cryogenic machining is harder and thinner than sub-surface layer under dry or Minimum Quantity Lubrication (MQL) machining. In other words, cryogenic machining is able to improve the surface roughness, and increase the hardness of the machined workpiece. The majority of previous cryogenic machining works supplied cryogen at a constant supply pressure. To study the effect of supply pressure and the corresponding flowrate of the cryogen, Klocke et al. conducted cryogenic machining tests at supply pressures of 7–30 MPa on Inconel 718 and then measured maximum flank wear of the cutting inserts [35]. They reported that the higher the supply pressure and flowrate of cryogen, the lower the maximum flank wear at a cutting speed of 500 m/min. However, they obtained a negative result at cutting speed of 60 m/min, where higher supply pressure and flowrate produced higher maximum flank wear. Earlier researchers used cryogen only in their cryogenic turning experiments, but recently hybrid cooling which combined cryogen with cutting fluids has been proposed. Bagherzadeh and Budak used four different cooling strategies in turning titanium alloy Ti6Al4V and Inconel 718 [36]. They introduced a new method, CMOL, where CO_2 and vegetable oil were mixed in the form of frozen oil particles, before reaching the tool–work interface. The new method, CMOL, is better than cryogenic turning with carbon dioxide cooling only, in terms of tool wear and surface finish. Yildirim compared cryogenic, nanofluids, and hybrid cooling in turning Inconel 625 [37]. Six cooling techniques such as dry, pure MQL, nMQL, LN_2, and their hybrid were used in his work. Three types of nanofluids based on Al_2O_3, hBN, and hBN + Al_2O_3 were included in the vegetable cutting fluid in 0.5 vol% and 1.0 vol%. Hybrid cooling (cryogenic liquid nitrogen and 0.5 vol% hBN) outperformed cryogenic cooling only and vegetable oil with nanofluids only in terms of reducing cutting temperature, prolonging tool life and improving surface roughness. Tantalum has low thermal conductivity, low specific heat, high shear strength, high work-hardening capacity, and gummy consistency, which cause it to be difficult to machine. In order to study the roles of cryogenic machining in machining tantalum, Wang and Rajurkar conducted a preliminary study on cryogenic machining of tantalum. They reported that liquid nitrogen was able to reduce the cutting temperature and extend the tool life [1]. In their further research in cryogenic machining of tantalum with liquid nitrogen [22], they found that cryogenic machining reduced tool wear by 70%, improved surface roughness by 200% and reduced cutting forces by 60%.

In general, cryogen is used to reduce the maximum cutting temperature and reduce/delay tool wear in machining superalloys. The majority of the previous works show positive results at selected ranges (cutting speed, or pressure). However, no consensus is obtained on the precise conditions where cryogenic machining is beneficial. Some works reported cryogens work best at lower cutting speed, but some other suggested cryogens only contribute at high cutting speed. The amount of the cryogen used and the location of cooling (workpiece, tool, or both) are important factors in deciding the final results. Overcooling superalloy will produce a negative effect. The roles of liquid nitrogen, whether beneficial or detrimental, depend on workpiece materials, cutting speed, and supply pressure of the cryogen. Furthermore, hybrid cooling (cryogenic machining with lubricant) was evolved from cryogenic machining, and it shows encouraging outcomes in machining superalloys.

2.3. Cryogenic Turning of Ferrous Metals

Ferrous metals such as steel were also investigated in cryogenic turning research. Zurecki et al. [38] found that the spraying of cryogenic liquid nitrogen on the tool rake during hard turning of AISI 52100 steel decreased the thickness of white layer, maintained the hardness of steel, and improved residual stress distribution of the machined surface. Kumar and Choudhury [39] investigated the effect of cryogenic liquid nitrogen on tool wear and cutting forces during high-speed turning of stainless steel. They obtained positive results in cryogenic machining in terms of reducing tool wear and cutting force. However, they raised their concern in the high consumption of liquid nitrogen because this increases the total cost of machining. Stanford, et al. [40] investigated tool wear in turning carbon steel BS 970-080A15 using uncoated tungsten carbide–cobalt insert in six different cutting environments (dry cutting, flood coolant, compressed air blast, nitrogen gas at ambient temperature, nitrogen gas at −40 °C and liquid nitrogen at −196 °C). They concluded that liquid nitrogen machining can provide a similar effect as provided by the flood coolant, and thus, liquid nitrogen can replace flood coolants for machining steel. Dhar et al. [14,15] reported that reduction in tool wear produced better cutting surfaces under cryogenic liquid nitrogen. Furthermore, the reduction in cutting temperature was more obvious at a lower feed and lower cutting velocity. Sivaiah and Chakradhar compared cryogenic, MQL, wet, and dry machining conditions in turning 17-4 PH stainless steel and they found that cryogenic turning outperformed in terms of tool wear, surface finish, and chip morphology compared to other three conditions [41]. The improvement was mainly due to retainment of cutting tool edge under cryogenic condition. Sivaiah and Chakradhar furthered their works and identified the optimum cryogenic cutting condition with two optimisation techniques [42]. The two techniques, 'Taguchi incorporated Gray relational analysis' (TGRA) and 'Taguchi coupled Technique for Order Preference by Similarity to Ideal Solution' (T-TOPSIS) were used to optimise the surface roughness, tool wear, and material removal rate of cryogenic turning 14-4 PH stainless steel with tungsten coated carbide. Both techniques were able to optimise the cryogenic turning process but the *'Taguchi incorporated Gray relational analysis'* was a better technique.

In contrast, Yap et al. [43] reported a different result. They studied turning of carbon steel S45C under three conditions; dry, wet, and cryogenic. Their results showed that even through a cryogenic liquid nitrogen jet is able to reduce friction coefficient in turning carbon steel, it failed to produce good surface roughness of the machined surface. Their experimental results suggested that conventional machining with cutting fluid is still the best method to obtain good surface roughness of carbon steel at a machining speed of 226 m/min). While most of the reported works focused on cutting forces, friction coefficient, tool wear, and surface roughness, some researchers investigated the surface integrity and corrosion behaviour cryogenic machined steel. Bruschi et al. investigated the effect of cryogenic machining to the surface integrity and corrosion behaviour of AISI 316L stainless steel [44]. Cryogenic cooling altered the microstructure of machined part, especially the outer surface. Beside this, stainless steel machined under cryogenic condition showed better corrosion resistance behaviour. Similarly, phase transformation from austenite in AISI 347 to martensite was reported by Kirsch et al. [45]. Cryogenic carbon dioxide was used to reduce the temperature of the contact zone during machining AISI 347 and precool the workpiece. Martensite was detected in the surface of the workpiece after cryogenic turning. In additional, microhardness of the surface was also increased after cryogenic cooling. Cryogenic turning can be further explored as a surface hardening improvement integrated in manufacturing methods.

For ferrous metals, cryogenic machining with cryogens such as liquid nitrogen and liquid carbon dioxide can reduce the cutting temperature, alter the microstructure/phase, and harden the surface of the workpiece. Similar to cryogenic turning of superalloy, a correct selection of parameter such as cryogen pressure, machining speed, etc. is important. Cryogen is used to 'reduce' the cutting temperature. However, overcooling of ferrous metals brings negative effects. Overcooling of ferrous metals increases brittleness and reduces toughness in ferrous metal, and then causes higher tool wear and poorer surface finishes. Therefore a proper cooling strategy is prerequisite in cryo-machining of ferrous metals.

2.4. Cryogenic Machining of Viscoelastic Polymers and Elastomers

Viscoelastic polymers are hard to be machined precisely at ambient temperature because of their characteristics such as softness, high elasticity, and adhesion [46]. Kakinuma et al. [46] proposed to solve this problem by applying cryogenic machining. They milled polydimethylsiloxane (PDMS) inside a liquid nitrogen chamber (as shown in Figure 2) and obtained positive results. This is because cryogenic liquid nitrogen maintains the glassy state of PDMS during the milling process. Cryogenic milling of PDMS was also investigated by Song et al. [47]. Cryogenic liquid nitrogen improved the surface roughness of machined PDMS with machined temperature lower than −143 °C. In addition, the problems of adhesion and shrinkage in machining PDMS were also solved with correct amount of liquid nitrogen flow. A similar technique was applied by Dhokia et al., in machining elastomer [48]. With the technique of cryogenic liquid nitrogen machining, formation of adiabatic shear band was reduced in milling ethylene vinyl acetate (EVA) and Neoprene [49].

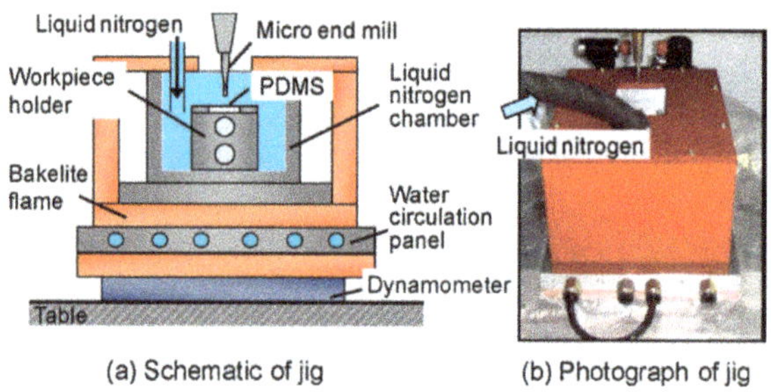

Figure 2. Jig for cryogenic micromachining [46].

This concept can be applied to cryogenic turning of viscoelastic polymers although less published works in cryogenic turning of viscoelastic polymers can be found in established databases. Putz et al. studied the effect of cryogenic liquid nitrogen on the cutting forces and mechanism of chip formation in turning elastomer [50]. They found that cryogenic machining can improve the machining accuracy of elastomer. This is mainly due to the elastomer's behaviour being changed from viscoelastic to energy-elastic with higher modulus of elasticity, under the influence of liquid nitrogen.

In general, cryogenic machining viscoelastic polymers and elastomers is not as popular as cryogenic machining of superalloys or ferrous steel. The role of cryogen in cryogenic machining of viscoelastic polymers and elastomers is different from the role of cryogen in cryogenic machining of superalloy or ferrous steel. Cryogen is used to change the mechanical properties of the polymer or elastomer during machining, i.e., to increase the modulus of elasticity so that dimensional accuracy can be improved.

3. Challenges of Cryogenic Machining in Industries

A large number of scientific studies have been conducted for more than 20 years, and most of them showed that cryogenic machining is generally better than conventional machining especially in machining superalloys. However, conventional industries still prefer the conventional machining technique. The advantages and disadvantage of cryogenic turning are listed in Table 1. As shown in Table 1, one of the major obstructions is the initial set-up cost. Industries are required to invest heavily on the cryogenic cooling system. Beside this, they are unable to reuse the cryogen as how they recirculate cutting fluids in conventional cutting. High consumption of cryogen during cryogenic machining, as reported by Hong [11] and Kumar and Choudhury [39], is another barrier in applying cryogenic machining in industries. To minimise the usage of liquid nitrogen, Hong designed and patented a nozzle to deliver the liquid nitrogen onto the workpiece or/and cutting tool [18,51]. The newly designed nozzle can direct the liquid nitrogen to the rake face, or the flank face, or both faces and inject less liquid nitrogen. A similar concept was also adapted by Ahmed and his research team [52,53]. They used a modified tool to apply cryogenic liquid nitrogen in turning AISI 4340 steel and SUS 304 stainless steel. With their modified tool, a smaller amount of liquid nitrogen was used. However, this special nozzle and modified tool are still not commonly available as commercial products.

Table 1. Advantages and disadvantages of cryogenic turning.

Advantages	Disadvantages
Performs better than conventional cutting fluid in high-speed machining of titanium alloys.	Requires high initial set-up cost of the cryogenic cooling system
Greener machining	Unable to reuse the fluid (cryogen)—higher production cost caused by consumption of cryogen
Able to machine viscoelastic polymers	Requires special knowledge

The third obstruction in applying cryogenic machining in industries is lack of clear guidelines for optimum combination of machining and cooling parameters for cryogenic machining. As mentioned previously, several factors determine whether cryogenic machining is beneficial or detrimental. The factors are cooling strategy (cooling the chip, cooling the cutting tool and cutting zone), workpiece materials, cutting speed, and supply pressure of the cryogen. Contradictory or different optimum factors/value were reported in literature. Currently, a general guideline to select the correct parameters is still unavailable for machine operators to refer to. The science behind cryogenic machining is a complicated field of knowledge and requires further investigation.

4. Conclusions

Cryogenic cooling is an alternative technique in machining. Cryogenic machining is applied on superalloys, ferrous metal, and viscoelastic polymers/elastomers. In general, cryogenic cooling during turning improved various performance indicators (surface roughness, tool life, tool wear, cutting forces, etc.) in machining superalloys (titanium alloys, inconel alloys, and tantalum alloys). This is due to cryogen being able to solve the major problem in machining superalloys, i.e., heat accumulated at the cutting zone due to poor thermal conductivities. Beside this, cryogenic machining also showed beneficial improvement in machining ferrous metal, if a correct setting is used. Correct application of cryogen can delay/reduce the tool wear in high-speed machining of ferrous steel and also modify the surface behaviour of the machined parts. On the other hand, cryogenic cooling altered hardness and adhesion properties of viscoelastic polymers, and improved the machinability of viscoelastic polymers. Cryogenic machining has several advantages compared to machining with conventional cooling. Nevertheless, additional set-up cost for a cryogenic cooling system is required. Beside this, the cost of cryogen is another factor that must be considered by industries in order to adopt this technique. A correct selection of machining and cooling techniques/parameters is another challenge for applying this technique in industries. Although cryogenic machining is more than 20 years old, this technique is not widely adopted by machine shops. A general guideline for selection of cryogenic machining parameters is crucial in order to promote cryogenic turning to machine shops.

Funding: This research received no external funding.

Acknowledgments: The author wishes to thank former mentors who led him to the research in cryogenic machining.

Conflicts of Interest: The author declares no conflict of interest.

References

1. Wang, Z.Y.; Rajurkar, K.P. Cryogenic machining of hard-to-cut materials. *Wear* **2000**, *239*, 168–175. [CrossRef]
2. El Baradie, M.A. Cutting fluids: Part II. Recycling and clean machining. *J. Mater. Process. Technol.* **1996**, *56*, 798–806. [CrossRef]
3. Kleber, M.; Föllmann, W.; Blaszkewicz, M. Assessing the genotoxicity of industrial cutting fluids under conditions of use. *Toxicol. Lett.* **2004**, *151*, 211–217. [CrossRef]
4. Shvedova, A.A.; Kisin, E.; Murray, A.; Goldsmith, T.; Reynolds, J.S.; Castranova, V.; Frazer, D.G.; Kommineni, C. Metal working fluids: Sub-chronic effects on pulmonary functions in B6C3F1 mice given vitamin E deficient and sufficient diets. *Toxicology* **2002**, *177*, 285–297. [CrossRef]
5. Suliman, S.M.A.; Abubakr, M.I.; Mirghani, E.F. Microbial contamination of cutting fluids and associated hazards. *Tribol. Int.* **1997**, *30*, 753–757. [CrossRef]
6. Shaw, M.C. *Metal Cutting Principles*; Oxford University Press: New York, NY, USA, 2005.
7. Dhar, N.R.; Paul, S.; Chattopadhyay, A.B. Machining of AISI 4140 steel under cryogenic cooling-tool wear, surface roughness and dimensional deviation. *J. Mater. Process. Technol.* **2002**, *123*, 483–489. [CrossRef]
8. Hilal, N.; Busca, G.; Talens-Alesson, F.; Atkin, B.P. Treatment of waste coolants by coagulation and membrane filtration. *Chem. Eng. Process.* **2004**, *43*, 811–821. [CrossRef]
9. Hong, S.Y.; Zhao, Z. Thermal aspects, material considerations and cooling strategies in cryogenic machining. *Clean Prod. Process.* **1999**, *1*, 107–116. [CrossRef]
10. Fillippi, A.D.; Ippolito, R. Face milling at 180 °C. *Ann. CIRP* **1970**, *19*, 399–405.
11. Hong, S.Y. Economical and ecological cryogenic machining. *J. Manuf. Sci. Eng. ASME* **2001**, *123*, 331–338. [CrossRef]
12. Uehara, K.; Kumagai, S. Characteristics of tool wear in cryogenic machining. *Ann CIRP* **1970**, *19*, 273–277.
13. Dhar, N.R.; Islam, S.; Kamruzzaman, M.; Paul, S. Wear behavior of uncoated carbide inserts under dry, wet and cryogenic cooling conditions in turning C-60 steel. *J. Brazilian Soc. Mech. Sci. Eng.* **2006**, *28*, 146–152. [CrossRef]
14. Dhar, N.R.; Kamruzzaman, M. Cutting temperature, tool wear, surface roughness and dimensional deviation in turning AISI-4037 steel under cryogenic condition. *Int. J. Mach. Tools Manuf.* **2007**, *47*, 754–759. [CrossRef]
15. Dhar, N.R.; Paul, S.; Chattopadhyay, A.B. The influence of cryogenic cooling on tool wear, dimensional accuracy and surface finish in turning AISI 1040 and E4340C steels. *Wear* **2001**, *249*, 932–942. [CrossRef]
16. Dhar, N.R.; Paul, S.; Chattopadhyay, A.B. Role of cryogenic cooling on cutting temperature in turning steel. *J. Manuf. Sci. Eng. ASME* **2002**, *124*, 146–154. [CrossRef]
17. Hong, S.Y.; Ding, Y.C. Cooling approaches and cutting temperatures in cryogenic machining of Ti-6Al-4V. *Int. J. Mach. Tools Manuf.* **2001**, *41*, 1417–1437. [CrossRef]
18. Hong, S.Y.; Markus, I.; Jeong, W. New cooling approach and tool life improvement in cryogenic machining of titanium alloy Ti-6Al-4V. *Int. J. Mach. Tools Manuf.* **2001**, *41*, 2245–2260. [CrossRef]
19. Venugopal, K.A.; Paul, S.; Chattopadhyay, A.B. Tool wear in cryogenic turning of Ti-6Al-4V alloy. *Cryogenics* **2007**, *47*, 12–18. [CrossRef]
20. Venugopal, K.A.; Paul, S.; Chattopadhyay, A.B. Growth of tool wear in turning of Ti-6Al-4V alloy under cryogenic cooling. *Wear* **2007**, *262*, 1071–1078. [CrossRef]
21. Wang, Z.Y.; Rajurkar, K.P. Wear of CBN tool in turning of silicon nitride with cryogenic cooling. *Int. J. Mach. Tools Manuf.* **1997**, *37*, 319–326. [CrossRef]
22. Wang, Z.Y.; Rajurkar, K.P.; Fan, J.; Petrescu, G. Cryogenic Machining of Tantalum. *J. Manuf. Process.* **2002**, *4*, 122–127. [CrossRef]
23. Wang, Z.Y.; Rajurkar, K.P.; Murugappan, M. Cryogenic PCBN turning of ceramic (Si3N4). *Wear* **1996**, *195*, 1–6. [CrossRef]
24. Paul, S.; Chattopadhyay, A.B. A study of effects of cryo-cooling in grinding. *Int. J. Mach. Tools Manuf.* **1995**, *35*, 109–117. [CrossRef]

25. Paul, S.; Chattopadhyay, A.B. Effects of cryogenic cooling by liquid nitrogen jet on forces, temperature and surface residual stresses in grinding steels. *Cryogenics* **1995**, *35*, 515–523. [CrossRef]
26. Paul, S.; Chattopadhyay, A.B. The effect of cryogenic cooling on grinding forces. *Int. J. Mach. Tools Manuf.* **1996**, *36*, 63–72. [CrossRef]
27. Paul, S.; Chattopadhyay, A.B. Determination and control of grinding zone temperature under cryogenic cooling. *Int. J. Mach. Tools Manuf.* **1996**, *36*, 491–501. [CrossRef]
28. Geddes, B.; Leon, H.; Huang, X. *Superalloys: Alloying and Performance*; ASM International: Geauga, OH, USA, 2010.
29. Hong, S.Y.; Ding, Y.C.; Jeong, J. Experimental evaluation of friction coefficient and liquid nitrogen lubrication effect in cryogenic machining. *Mach. Sci. Technol.* **2002**, *6*, 235–250. [CrossRef]
30. Dhananchezian, M.; Pradeep Kumar, M. Cryogenic turning of the Ti-6Al-4V alloy with modified cutting tool inserts. *Cryogenics* **2011**, *51*, 34–40. [CrossRef]
31. Yap, T.C.; El-Tayeb, N.S.M.; von Brevern, P. Cutting forces, friction coefficient and surface roughness in machining Ti-5Al-4V-0.6Mo-0.4Fe using carbide tool K313 under low pressure liquid nitrogen. *J. Brazilian Soc. Mech. Sci. Eng.* **2013**, *35*, 11–15. [CrossRef]
32. Kaynak, Y.; Gharibi, A. Cryogenic machining of titanium Ti-5553 alloy. *J. Manufact. Sci. Eng.* **2019**, *141*, 041012-1–041012-9. [CrossRef]
33. Ezugwu, E.O.; Bonney, J.; Fadare, D.A.; Sales, W.F. Machining of nickel-base, Inconel 718, alloy with ceramic tools under finishing conditions with various coolant supply pressures. *J. Mater. Process. Technol.* **2005**, *162–163*, 609–614. [CrossRef]
34. Pusavec, F.; Hamdi, H.; Kopac, J.; Jawahir, I.S. Surface integrity in cryogenic machining of nickel based alloy—Inconel 718. *J. Mater. Process. Technol.* **2011**, *211*, 773–783. [CrossRef]
35. Klocke, F.; Krämer, A.; Sangermann, H.; Lung, D. Thermo-Mechanical Tool Load during High Performance Cutting of Hard-to-Cut Materials. *Procedia CIRP* **2012**, *1*, 295–300. [CrossRef]
36. Bagherzadeh, A.; Budak, E. Investigation of machinability in turning of difficult-to-cut materials using a new cryogenic cooling approach. *Tribol. Int.* **2018**, *119*, 510–520. [CrossRef]
37. Yıldırım, Ç.V. Experimental comparison of the performance of nanofluids, cryogenic and hybrid cooling in turning of Inconel 625. *Tribol. Int.* **2019**, *137*, 366–378. [CrossRef]
38. Zurecki, Z.; Ghosh, R.; Frey, J.H. Investigation of White Layers Formed in Conventional and Cryogenic Hard Turning of Steels. *Proc. IMECE'03* **2003**, 1–10. [CrossRef]
39. Kumar, K.; Choudhury, S.K. Investigation of tool wear and cutting force in cryogenic machining using design of experiments. *J. Mater. Process. Technol.* **2008**, *203*, 95–101. [CrossRef]
40. Stanford, M.; Lister, P.M.; Morgan, C.; Kibble, K.A. Investigation into the use of gaseous and liquid nitrogen as a cutting fluid when turning BS 970-80A15 (En32b) plain carbon steel using WC-Co uncoated tooling. *J. Mater. Process. Technol.* **2009**, *209*, 961–972. [CrossRef]
41. Sivaiah, P.; Chakradhar, D. Effect of cryogenic coolant on turning performance characteristics during machining of 17-4 PH stainless steel: A comparison with MQL, wet, dry machining. *CIRP J. Manufact. Sci. Technol.* **2018**, *21*, 86–96. [CrossRef]
42. Sivaiah, P.; Chakradhar, D. Performance improvement of cryogenic turning process during machining of 17-4 PH stainless steel using multi objective optimization techniques. *Measurement* **2019**, *136*, 326–336. [CrossRef]
43. Yap, T.C.; Sivaraos, C.S.; Lim, C.S.; Leau, J.W. Surface roughness and cutting forces in cryogenic turning of carbon steel. *J. Eng. Sci. Technol.* **2015**, *10*, 911–920.
44. Bruschi, S.; Pezzato, L.; Ghiotti, A.; Dabalà, M.; Bertolini, R. Effectiveness of using low-temperature coolants in machining to enhance durability of AISI 316L stainless steel for reusable biomedical devices. *J. Manufact. Process.* **2019**, *39*, 295–304. [CrossRef]
45. Kirsch, B.; Hotz, H.; Müller, R.; Becker, S.; Boemke, A.; Smaga, M.; Beck, T.; Aurich, J.C. Generation of deformation-induced martensite when cryogenic turning various batches of the metastable austenitic steel AISI 347. *Product. Eng.* **2019**, *13*, 343–350. [CrossRef]
46. Kakinuma, Y.; Kidani, S.; Aoyama, T. Ultra-precision cryogenic machining of viscoelastic polymers. *CIRP Ann. Manuf. Technol.* **2012**, *61*, 79–82. [CrossRef]
47. Song, K.; Gang, M.G.; Jun, M.B.G.; Min, B.-K. Cryogenic machining of PDMS fluidic channel using shrinkage compensation and surface roughness control. *Int. J. Precis. Eng. Manuf.* **2017**, *18*, 1711. [CrossRef]

48. Dhokia, V.G.; Newman, S.T.; Crabtree, P.; Ansell, M.P. A process control system for cryogenic CNC elastomer machining. *Robot. Comput. Integr. Manuf.* **2011**, *27*, 779–784. [CrossRef]
49. Dhokia, V.G.; Newman, S.T.; Crabtree, P.; Ansell, M.P. Adiabatic shear band formation as a result of cryogenic CNC machining of elastomers. *Proc. Inst. Mech. Eng. Part B J. Eng. Manuf.* **2011**, *225*, 1482–1492. [CrossRef]
50. Putz, M.; Dix, M.; Neubert, M.; Schmidt, T. Mechanism of Cutting Elastomers with Cryogenic Cooling. *CIRP Annals* **2016**, *65*, 73–76. [CrossRef]
51. Hong, S.Y. Cryogenic Machining. U.S. Patent 5,901,623, 11 May 1999.
52. Ahmed, M.I.; Ismail, A.F.; Abakr, Y.A.; Amin, A. Effectiveness of cryogenic machining with modified tool holder. *J. Mater. Process. Technol.* **2007**, *185*, 91–96. [CrossRef]
53. Khan, A.A.; Ahmed, M.I. Improving tool life using cryogenic cooling. *J. Mater. Process. Technol.* **2008**, *196*, 149–154. [CrossRef]

© 2019 by the author. Licensee MDPI, Basel, Switzerland. This article is an open access article distributed under the terms and conditions of the Creative Commons Attribution (CC BY) license (http://creativecommons.org/licenses/by/4.0/).

technologies

Article

Wire Tool Electrode Behavior and Wear under Discharge Pulses

Sergey N. Grigoriev [1], Marina A. Volosova [1], Anna A. Okunkova [1,*], Sergey V. Fedorov [1], Khaled Hamdy [1,2], Pavel A. Podrabinnik [1], Petr M. Pivkin [1], Mikhail P. Kozochkin [1] and Artur N. Porvatov [1]

1 Department of High-Efficiency Processing Technologies, Moscow State University of Technology «STANKIN», Vadkovsky per. 1, 127055 Moscow, Russia; s.grigoriev@stankin.ru (S.N.G.); m.volosova@stankin.ru (M.A.V.); sv.fedorov@icloud.com (S.V.F.); eng_khaled2222@mu.edu.eg (K.H.); p.podrabinnik@stankin.ru (P.A.P.); p.pivkin@stankin.ru (P.M.P.); astra-mp@yandex.ru (M.P.K.); porvatov_artur@mail.ru (A.N.P.)

2 Production Engineering and Mechanical Design Department, Faculty of Engineering, Minia University, Minia 61519, Egypt

* Correspondence: a.okunkova@stankin.ru; Tel.: +7-909-913-1207

Received: 16 August 2020; Accepted: 16 September 2020; Published: 20 September 2020

Abstract: This work is devoted to researching the tool electrode behavior and wear under discharge pulses at electrical discharge machining. The experiments were conducted on the workpieces of 12Kh18N10T (AISI 321) chrome-nickel anti-corrosion steel and D16 (AA 2024) duralumin by a 0.25-mm-diameter CuZn35 brass tool in a deionized water medium. The developed diagnostic and monitoring mean based on acoustic emission registered the oscillations accompanying machining at 4–8 kHz. The obtained workpiece and non-profiled tool surfaces were investigated by optical and scanning electron microscopy. Calculated volumetric and mass removal rates showed the difference in the character of wear at roughing and finishing. It was shown that interaction between material components in anti-corrosion steel machining had an explosive character between Zn of brass and Ni of steel at a micron level and formed multiple craters of 30–100 μm. The secondary structure and topology of worn tool surfaces were caused by material sublimation, chemical interaction between material components at high heat (10,000 °C), explosive deposition of the secondary structure. Acoustic diagnostics adequately registered the character of interaction. The observed phenomena at the submicron level and microstructure of the obtained surfaces provide grounding on the nature of material interactions and electrical erosion wear fundamentals.

Keywords: erosion; tool wear; sublimation; $ZnNi_x$; explosive deposition

1. Introduction

The subject of electrical discharge machining (EDM) and wear of a wire tool electrode is not new, but the physical processes that occur during processing are still not sufficiently studied [1–4]. It is related to the absence of the possibility of visual control over the processes occurring in the discharge gap during EDM especially for large workpieces [5,6]. Plenty of studies are devoted to the processes related to the physical phenomena of erosion wear with various conclusions [7–11]. However, at the industrial level, there is no solution to avoid the negative consequences of the accident wire electrode breakage or dumping of the machined part into the working tank at the end of machining. It is especially actual in the case of splitting two co-dependent parts—die and punch for the injection mold, micro-gears [12,13].

An experienced operator often determines the control over processing and process conditions by the specific noise that occurred in the working area. It grows with an increase in the intensity of processing (roughing or finishing) and varies during the electrical discharge machining of materials with uneven structures—porous material, set of tubes, nanocomposites, or composites. The changes in the specific sound are especially noticeable during wire tool penetration into the workpiece and at the end of machining.

The electrical discharge machining of materials occurs in specific conditions between two electrodes. A bias increase is followed by ionizing the space between two electrodes at the moment of their approach. Dielectric breakdown by spark provokes a discharge channel that creates the conditions of low-temperature plasma with 10,000 °C that can be observed in particular conditions (forming intermetallics of Al_2Cu [14,15], $ZnNi_x$ [16], $CeNi_2$ [17], burning titanium at 1200 °C [18,19]) in the form of an instant growing gas bubble surrounded the discharge channel. Then, pulses interrupt to restore the breakdown conditions for the next pulse—the bias in the gap, erosion products' washing away from the working area. The occurred conditions are close to the conditions of lightning formation [20]. The temperature in the interelectrode gap achieves high value in a microsecond [21–23].

All monitoring means can be divided into optical and non-optical—electrical and vibroacoustic. The absence of visual contact with the working zone due to its tiny sizes hampers the application of any optical monitoring means. At the same time, the existed monitoring of the electrical parameters does not provide adequate data on the effectiveness of the discharge pulses since for the modern control systems all the produced pulses in the working zone are counted as working ones when a part of them can be spending on the destruction of erosion products [24,25]. It can be grounded by difficulties that met electrical discharge machining in processing materials with threshold conductivity, uneven structure, or microstructure, and inclined surfaces with a thickness of more than 100 mm.

The vibroacoustic monitoring method does not have this kind of disadvantage, as it counts only effective discharge addressed to the material to be processed on the destruction of the surface that was recorded with the help of the accelerometers placed at the working table of the machine [26,27]. It can be an effective means for not only tool behavior investigation and its influence on the quality of the machined surfaces but also an effective means for adaptive control of electrical discharge machining in real manufacturing conditions.

Simultaneously, research of the character of tool electrical erosion wear, sublimation phenomena [28–30], and the nature of the material destruction of the machined surfaces can give additional, comprehensive, and exhausting data.

This paper is aimed at the research of electrical discharge machining by the developed diagnostic means to obtain new data on the influence of wire tool behavior on the quality of the machined surfaces; wire tool wear at roughing and finishing, the nature of material destruction under discharge pulses, and sublimation phenomena.

The research is conducted for two materials:

- 12kH18N10T (AISI321) chrome-nickel anti-corrosion structural steel that is often used in injection mold manufacturing;
- D16 (AA 2024) duralumin used for aviation purposes.

The work's scientific novelty is in new data on electrical erosion wear of materials, sublimation phenomena, nature and mechanism of material destruction for two types of structural materials, dependencies between detected acoustic emission and machined surface quality, and classification of the eroded surfaces of the tool.

The tasks of the study are:

1. Applying the vibroacoustic means for research on dependencies between the tool behavior and surface quality;
2. Research of the tool and machined surface morphology and chemical composition;

3. Classification of the observed defects and traces of tool destruction at roughing and finishing of two material types;
4. Analyses of the chemical interactions between components that occurred in the discharge gap and conclusions on type material destruction and changes occurred at surface and subsurface layers.

2. Materials and Methods

2.1. Equipment

A four-axis wire electrical discharge machine, Seibu M500S, was used in the experiments for research of wire tool behavior and wear under pulses. The main characteristics of the machines are presented in Table 1.

Table 1. Main characteristics of wire electrical discharge machines used in experiments.

Characteristic	Value and Description
Max axis motions along the axes X×Y×Z, mm	500 × 350 × 310
Max angle of conical machining, degree	±10°
Max weight of workpiece, kg	800
Accuracy of positioning along the axes, μm	± 1 ÷ 2
Achievable roughness Ra, μm	0.4
Dielectric medium	Deionized water
Machine body	The frame is made of gray cast iron having good thermal and vibration compensating characteristics

The machines are located in a thermo-constant room to reduce ambient temperature's effect on the results of processing. Workpieces were immersed in a dielectric for 10 min before processing to avoid dimensional fluctuations related to the difference in temperatures between the environment and working fluid. The dielectric height was established at 1 ÷ 2 mm above the workpiece. The upper guide of the machine was placed at a minimum distance above 2 ÷ 5 mm from the dielectric [31–33].

The tool electrode is a brass wire with a diameter d_w of 0.25 mm made of CuZn35 (Cu—65%; Zn—35%) with a processing temperature of 260 °C and annealing temperature of 425–750 °C.

The choice of the electrode type was made since a brass tool of 0.25 mm in diameter is the most widespread for wire electrical discharge machining and suitable for the broad field of applications when the forced choice of any other electrode is due to a severe technological need and is associated with the need to purchase and reinstall expensive nozzles.

It should be noted that the positive polarization of the workpiece and negative polarization of the tool electrode is traditional for the electrical discharge machining. However, modern machine tools can switch the electrodes' orientation for some particular modes or even during machining uneven and hard-to-machine materials in automatic mode.

A CNC program was prepared manually; path offsets were not taken into account. The EDM-factors were chosen using recommendations mentioned in previously conducted works and developed by other scientific groups [25,34–36] (Table 2). The maximum working voltage V_o varied in the range of 40–70 V with a pitch of 10 V for characterization of the discharge pulses by for oscilloscope research and to provoke the conditions of wire breakage for microscopic research. At least five samples and cuts were produced for each group of parameters.

Table 2. Electrical discharge machining (EDM) factors during experimental work.

Factor	Value
Seibu M500S [1]	
Operational voltage in the interelectrode gap before the approach of the tool electrode to the workpiece, V_o	55, 60, 65
Auxiliary voltage that occurs at the moment of discharge between the tool electrode and workpiece, V_g	32
Strength of the working current in the interelectrode gap, I	8
Auxiliary current to increase the cutting efficiency when the circuit is turned back on, N_s	43
Time of disconnection of the current source, the percentage ratio of the time of the discharge pulse to the time of its absence, T_{off}	6
Time intermittent pause to ensure the stability of the processing process, A_d	305
Speed of the tool rewinding, W_s	35
Feed speed, S_g	5
Wire tension, W_t	30, 35, 40
Dielectric pressure in nozzles for flushing, F_l	245

[1] Provided in equivalent unit of the machine.

2.2. Materials

The chemical composition of 12Kh18N10T (AISI 321) austenite steel is presented in Table 3; the composition of D16 (AA2024, AlCuMg2) duralumin is in Table 4. The thickness of the samples was 20 ± 0.1 mm for both materials.

Table 3. Chemical composition of 12Kh18N10T steel (AISI 321) in wt%.

Element	Fe	Cr	Ni	Ti	Si	S	Mn	Cu	P	C
wt%	Balance	17–19	9–11	About 0.8	Up to 0.8	Up to 0.02	Up to 2.0	Up to 0.03	Up to 0.035	About 0.12

Table 4. Chemical composition of D16 alloy (AA2024, AlCuMg2) in wt%.

Element	Al	Cu	Mg	Mn	Fe	Si	Zn	Ni	Ti
wt%	90.8–94.7	3.8–4.9	1.2–1.8	0.3–0.9	Up to 0.5	Up to 0.5	Up to 0.3	Up to 0.1	Up to 0.1

The high chromium content of the proposed in the research steel ensures the metal's ability to passivate and causes strong corrosion resistance of steel. The addition of nickel converts steel to austenite class. This property is of exceptional importance, allowing to combine the machinability with an expanded set of performance properties. The addition of strong carbide-forming element titanium eliminates the tendency to intergranular corrosion. In turn, carbon forms a refractory titanium carbide and excludes a decrease in the concentration of chromium by chromium carbides formation. It should be noted that the field of chromium-nickel steel applications dominates in the modern rolled metal market [37–40].

Duralumin D16 is a structural alloy mainly used in the aviation and space industries. D16 is rarely used in its pure form since it has less strength and hardness in the non-quenched state. The alloy is classified as a durable thermo-hardened material [41,42].

A Fischer Sigmascope SMP10 instrument (Helmut Fischer GmbH, Sindelfingen, Germany) measured the specific electrical resistance ρ of the materials used in experiments (Table 5, Figure 1a). The device measures the material electric conductance in Siemens and the percentage of the control sample's electrical conductance produced from annealed bronze in the range of 1 ÷ 112%. All measured values were converted to $\frac{\Omega \cdot mm^2}{m}$. The melting/sublimation points of the materials T provided in Figure 1b [43–46].

Table 5. Specific electrical resistance ρ of some materials at +20 °C.

Material	Specific Electrical Resistance ρ [$\frac{\Omega \cdot mm^2}{m}$]
12Kh18N10T (AISI 321) steel	0.746
D16 (AA2024) alloy	0.028
Brass alloy, CuZn35	0.065

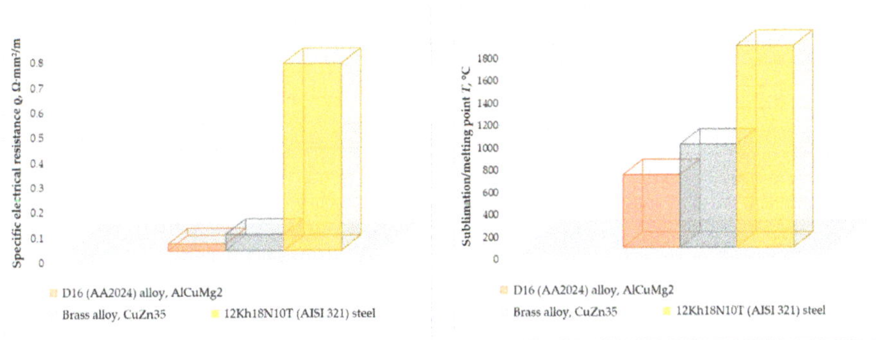

(**a**) (**b**)

Figure 1. Electrophysical properties of the materials used in experiments: (**a**) specific electrical resistance ρ at +20 °C; (**b**) sublimation/melting point T.

The stress-strain curves of chosen materials have the presence of elastic and plastic deformation zones [47,48]. The fracture formation schemes have a ductile phase that increases with the material's plasticity—from 12Kh18N10T (AISI 321) steel and brass to D16 (AA2024) alloy. The reduction area for these materials is ~38 ÷ 42 % for steels [47], ~52 ÷ 53 % for brass alloys [49], and ~75 ÷ 77 % for aluminum alloys [50]. Tensile strength σ_B (UTS) for these materials are 510–830 MPa for 12Kh18N10T (AISI 321) steel, 450 MPa for CuZn35 brass alloy, and 345–420 MPa for D16 (AA2024) alloy.

Reduction area S_{RA} is calculated by the following equation [47]:

$$S_{RA} = \frac{S_0 - S_{min}}{S_0} \cdot 100[\%], \quad (1)$$

where S_0 is an original transverse area, mm^2, and S_{min} is a minimal area of the final neck, mm^2.

2.3. Monitoring

A digital oscilloscope TDS2014B (Tektronix, Berkshire, UK) produced the characterization of the discharge pulses.

The vibroacoustic monitoring was provided by piezoelectric accelerometers mounted on the elastic system of the machine [23,25–27,51,52] (Figure 2). The data received by the accelerometers signals were forwarded to preamplifiers, amplifiers VShV003 (JSC Izmeritel, Taganrog, Russia), and an analog-to-digital converter E440 (L-card, Saint-Petersburg, Russia), and recorded with a personal computer (PC). Data were recorded at 1 min, 30 s, and 5 s before the end of processing. Spectral analysis was performed at frequencies 4–8 kHz. The signal was preliminarily cleaned from low-frequency noise using a high-frequency filter. The signal's effective amplitude was determined. The square of this amplitude is proportional to the signal that arises in the machine's elastic system under disturbing influences of the discharge pulses. The cutoff frequency of filters is 2 kHz.

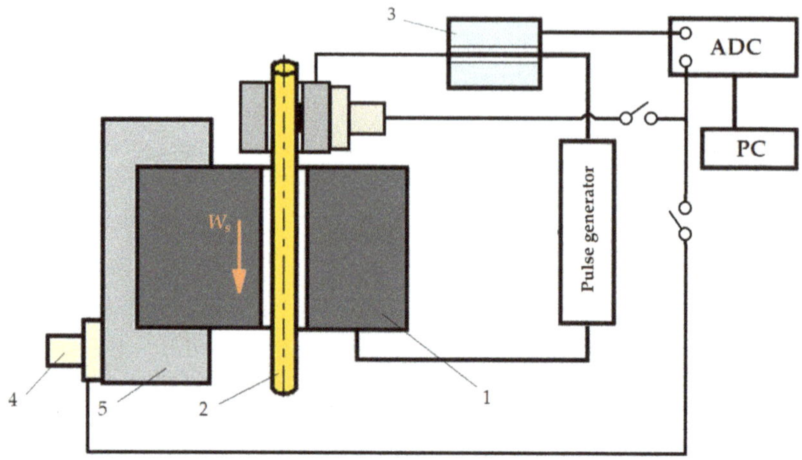

Figure 2. Scheme of monitoring sensor setup at electrical discharge machine: (**1**) is a workpiece; (**2**) is a wire tool; (**3**) is a current sensor; (**4**) is accelerometers; (**5**) is a workpiece fastening system; ADC is the analog-digital converter; PC is a personal computer.

2.4. Physical Relationship of EDM Factors and Vibration Amplitude

Typically, electrical discharge machining has a very narrow range of working factors for machining every material type. Nevertheless, up to 16 factors can be varied during machining [6]. The papers related to EDM research concentrate on some of them as we have done regarding the research subject. It was decided to vary two of the most important factors—operational voltage, which influences the density of discharge pulses distribution, and wire tension, which influences system stiffness and consequently wire oscillation amplitude. The detailed force diagram is presented in [27].

The amplitude of the wire can be presented by summarized force of working impulses in the system's action ΣF_{imp} and stiffness k_n:

$$A_n = \frac{\sum F_{imp}}{k_n}. \tag{2}$$

At the same time, the stiffness of the system is determined by its mass:

$$k_n = 4\pi^2 \frac{m_n}{T^2} \tag{3}$$

where T is a period of self-oscillations; thus, the signal amplitude has an inverse relationship with the weight.

Regarding operational voltage, it has a physical dependence on the signal's amplitude, since it influences the density of the discharge pulses and, consequently, the summarized force factor. The electrical impulse itself is a short-term burst of electrical breakdown voltage and working current that can be presented as follows:

$$\sum F_{imp} = I \cdot V_0. \tag{4}$$

The energy of pulses then will be:

$$\sum E_{imp} = I \cdot A_n. \tag{5}$$

The wire tension has dependence that is even more evident—W_t influences the system stiffness:

$$k_n = \frac{F_e}{\Delta l}, \tag{6}$$

where Fe is a restoring force that is opposite and equal to the applied wire tension W_t and Δl is a change in the wire length. Thus:

$$k_n = \frac{|W_t|}{\Delta l}. \tag{7}$$

Moreover, the height of the workpiece also influences the stiffness of the system and wire amplitude:

$$k_n = \frac{E \cdot S_0}{l_n}, \tag{8}$$

where l_n is a wire length, E is Young's modulus, and S_0 is a wire area.

2.5. Characterization of the Samples, Wear Rate, and Discharge Gap

An EL104 (Mettler Toledo, Columbus, OH, USA) laboratory balance with a measurement range of 0.0001 ÷ 120 g weighed the obtained samples with an error of 0.0001 g.

The samples' surface roughness was controlled by a high-precision profilometer, Hommel Tester T8000 (Jenoptik GmbH, Villingen-Schwenningen, Germany) with a resolution of 1 ÷ 1000 nm and a measurement error of 2%.

An Olympus BX51M instrument (Ryf AG, Grenchen, Switzerland) provided the optical microscopy; the discharge gap was measured optically.

A VEGA 3 LMH instrument (Tescan Brno s.r.o., Brno, The Czech Republic) with magnification up to 1,000,000× provided the scanning electron microscopy and spectrometry of the sample.

The cross-sections were prepared according to the standard probe techniques by an ATM sample equipment—Opal 410, Jade 700, and Saphir 300 (ATM, Haan, the Netherlands). Epoxy resin with quartz sand provided pouring of the samples as a filler was used.

The worn area of the tool can be calculated by the equation of the circle segment area (S_w):

$$S_w - \frac{1}{2} r_w^2 \left(\frac{\pi \cdot \alpha}{180^{\circ}} - \sin \alpha \right) \left[mm^2 \right], \tag{9}$$

where r_w is a wire radius, mm and α is an angle of the segment, degree. The volumetric (R_v) and mass wear rates (R_m) are calculated by the following equations [53,54]:

$$R_v = \frac{\Delta V}{t} \left[mm^3 \cdot s^{-1} \right], \tag{10}$$

$$R_m = \frac{\Delta m}{t} \left[g \cdot s^{-1} \right], \tag{11}$$

where ΔV is volumetric wear, mm^3, Δm is a worn mass, g, and t is the wire length wear time; for a thickness of 20 mm, $t = 0.343 \pm 0.005$ s.

The discharge gap is calculated by the following equation:

$$\Delta_{DB} = \frac{l_s - d_w}{2} [mm], \tag{12}$$

where l_s is the measured width of the slot, mm, and d_w is the wire diameter, mm.

3. Results

3.1. Electrical Discharge Pulses

Electrical discharge machining occurs at the value of V_g, approximately half of the value of operational voltage V_0. Idle pulses of a particular amplitude and frequency were detected with the value of the interelectrode gap Δ (distance between two electrodes) more than the value of the gap of dielectric breakdown Δ_{DB} ($\Delta > \Delta_{DB}$), Figure 3a. The idle pulse repetition frequency is $f = 10$ kHz, and the amplitude depends entirely on the factor of the voltage in the interelectrode gap (V_0) applied to the electrodes when the electrodes are at a distance Δ.

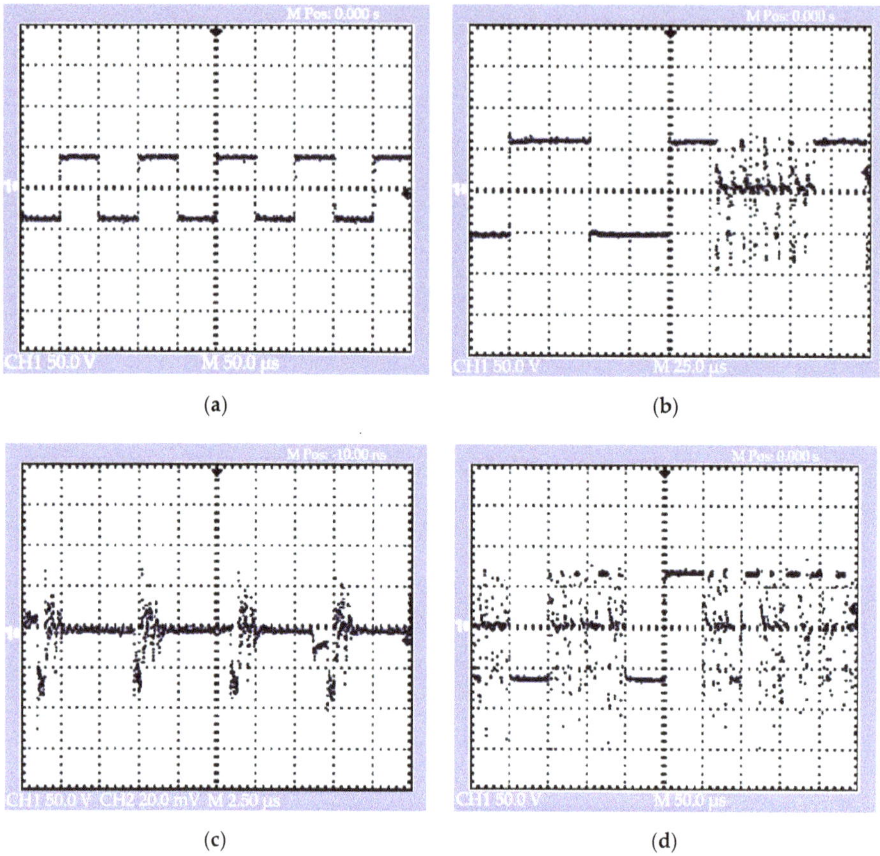

Figure 3. Electrical discharge machining pulses detected by the oscilloscope: (**a**) idle pules at a distance $\Delta > \Delta_{DB}$; (**b**) interrupted introduction of operational pulses at the distance $\Delta \geq \Delta_{DB}$; (**c**) operational pulses during machining $\Delta = \Delta_{DB}$; (**d**) interrupted operational pulses during unstable wire feed and variation with the factors, $\Delta \approx \Delta_{DB}$.

An infrequent dielectric breakdown of the gap occurs with a decrease in distance $\Delta \geq \Delta_{DB}$ by a single series of pointed operational pulses (Figure 3b). Regular operational pulses that are similar to damped harmonic oscillations followed the tool penetration (Figure 3c). The working pulses frequency is quite high and is equal to tenths of MHz and detected at a level of about 0.2 MHz. Furthermore, the idle and operational pulses start to alternate with different frequencies and amplitudes that depend on factors present in the stable mode (Figure 3d).

3.2. Wire Electrode Oscilations

It was determined that the vibroacoustic signal has a periodic character and decreases gradually during the first 2–3 s after tool penetration. It continues to decline during the next 15–20 s of machining (Figure 4a). At the same time, it increases approximately 20 s before the end of the machining. At 5–7 s before the end of machining, the signal interrupts repeatedly. The character of the signal at the beginning and the end of machining has a parabolic character. It corresponds to the final bridge destruction under the cutoff sample's weight (Figure 4b) that was detected for the samples of both materials with a weight up to 1.8 g.

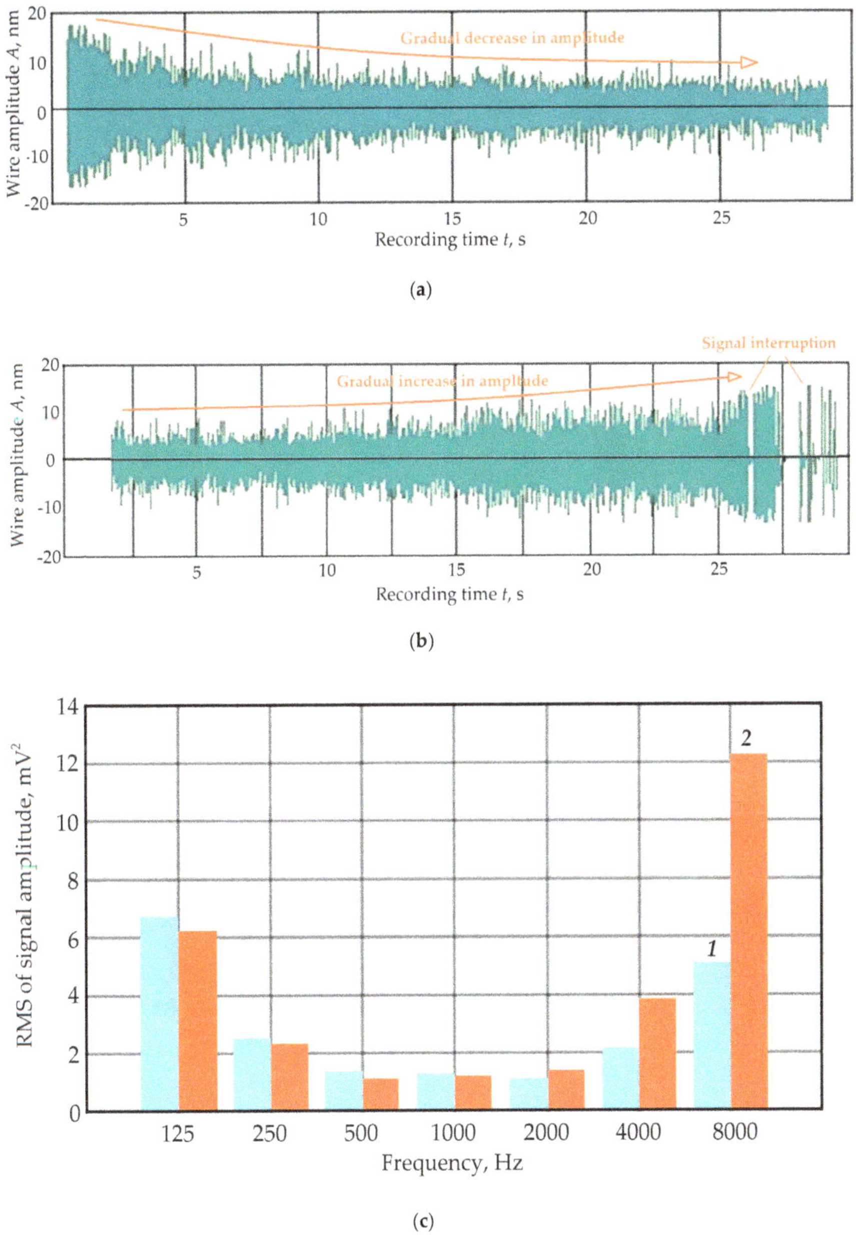

Figure 4. Recorded signal and its root mean-square value during electrical discharge machining of 12Kh18N10T (AISI 321) samples: (**a**) at 30 s of the tool penetration into the workpiece; (**b**) at 30 s before the end of the processing; (**c**) octave spectra of the root-mean-square value of the signal amplitude of 24.5 g sample, V_0 = 65 c.u., W_t = 35 c.u., where (**1**) at 30 s before the end of the operation, (**2**) at 5 s.

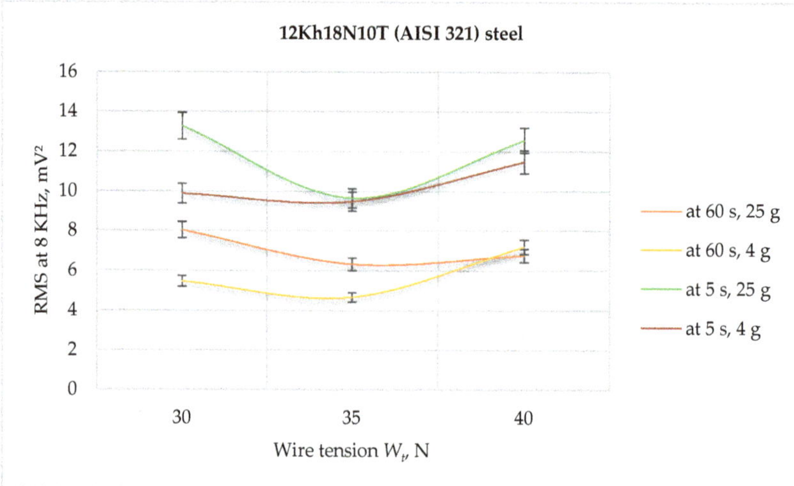

(**b**)

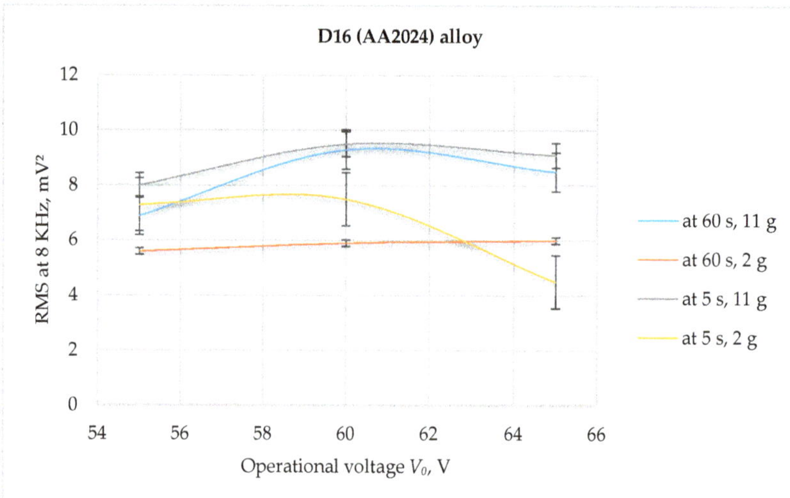

(**c**)

Figure 6. *Cont.*

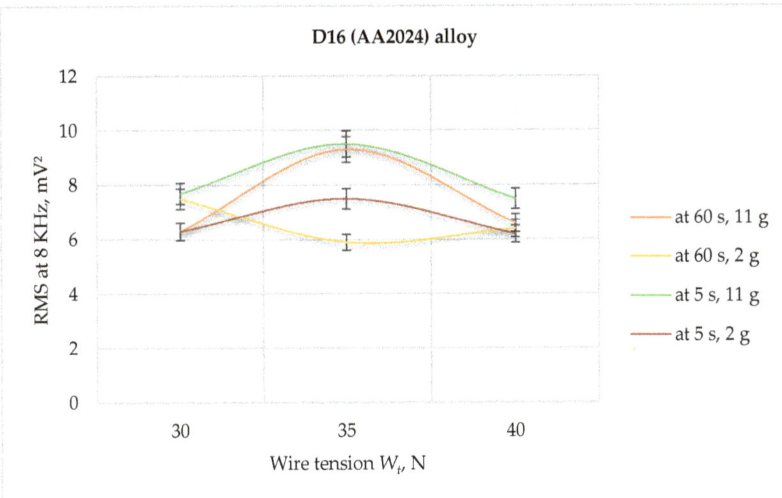

(d)

Figure 6. RMS of the received electrical discharge machining oscillation signal at 8 kHz for two types of materials: (**a**) 12Kh18N10T (AISI 321), in dependence on operational voltage V_0; (**b**) 12Kh18N10T (AISI 321), in dependence on wire tension W_t; (**c**) D16 (AA2024), in dependence on operational voltage V_0; (**d**) D16 (AA2024), in dependence on wire tension W_t.

3.3. Morphology of the Samples

Roughness profile *Ra* and recorded signal at 30 s before the end of machining are presented in Figure 7. As can be seen, the density of the signal amplitude is higher at 12Kh18N10T (AISI 321) steel machining (Figure 7a,b), by 20%, compared to D16 (AA2024) alloy machining (Figure 7c,d); approximately 30 μm^{-1} and ~24 μm^{-1}, correspondingly. The three-dimensional (3D) graphs of the EDM factors' influence on the average roughness *Ra* are presented in Figure 8, where minimal value is associated with the stable machining process and the lowest RMS values of the signal's amplitude.

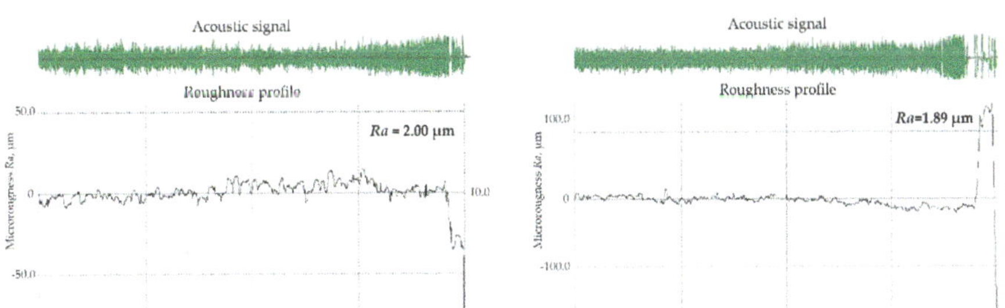

(**a**) (**b**)

Figure 7. *Cont.*

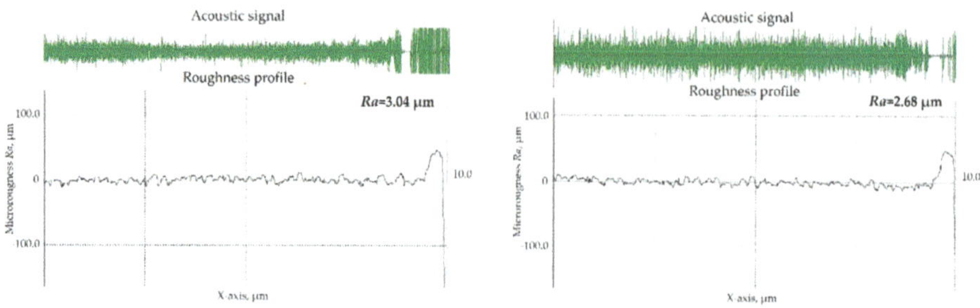

(c) (d)

Figure 7. Roughness R_a and recorded signal amplitude: (**a**) 12Kh18N10T (AISI 321) steel of 24.5 g, R_a of 2.00 μm, V_0 = 55 V; W_t = 35 N; (**b**) 12Kh18N10T (AISI 321) steel of 4.3 g, R_a of 1.89 μm, V_0 = 55 V; W_t = 35 N; (**c**) D16 (AA2024) alloy of 10.7 g, R_a of 3.04 μm, V_0 = 60 V; W_t = 35 N; (**d**) D16 (AA2024) alloy of 1.8 g, R_a of 2.68 μm, V_0 = 60 V; W_t = 40 N.

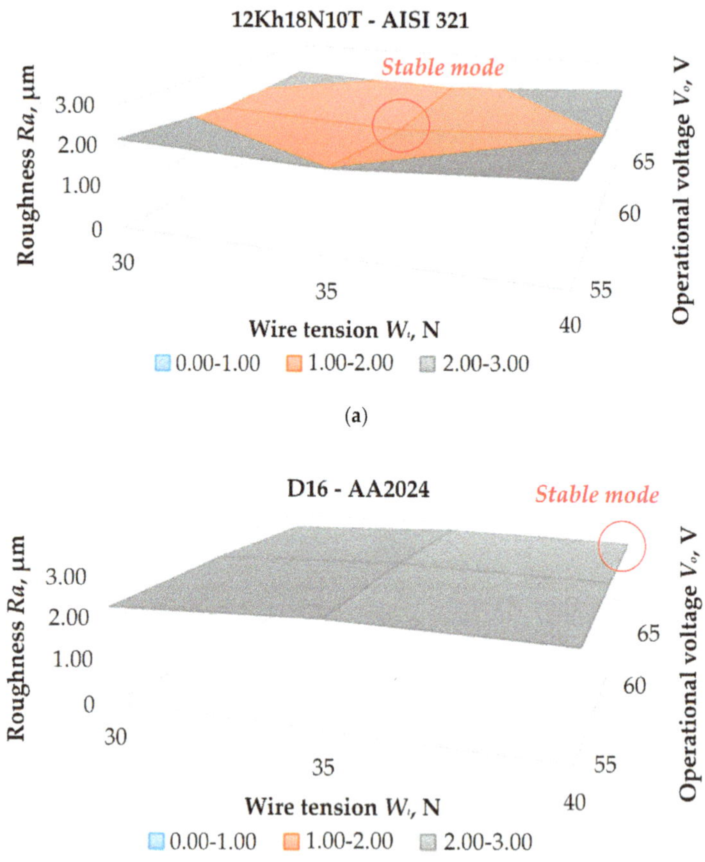

(a) (b)

Figure 8. Three-dimensional (3D) graphs of the surface roughness R_a dependences on operational voltage V_0 and wire tension W_t: (**a**) 12Kh18N10T (AISI 321) steel; (**b**) D16 (AA2024) alloy.

3.4. Discharge Gap

Figure 9 shows the measured offset Δ^*_{DB} of the path in the dependence of EDM factors for two types of materials. The offset Δ^*_{DB} includes the wire radius r_w of 0.125 mm. The optically measured effective discharge gap Δ_{DB} is in the range of 45 ÷ 53 μm for 12Kh18N10T (AISI 321) steel and in the range of 71 ÷ 78 μm for D16 (AA2024) alloy. The minimal values—170 μm for anti-corrosion steel and 196 μm for aluminum alloy—are associated with the stable machining process and corresponds to the lowest RMS values of the signal's amplitude.

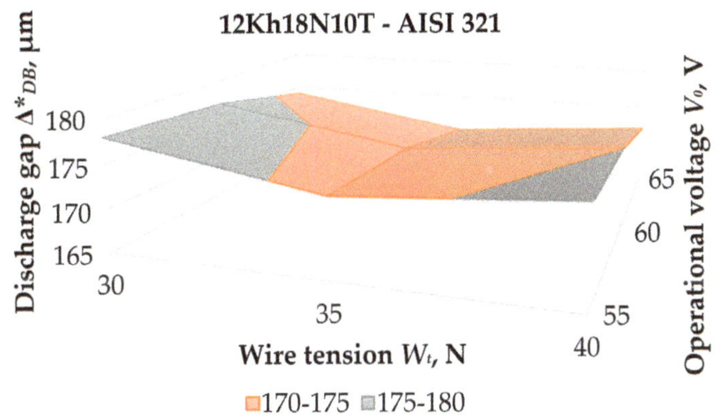

(**a**)

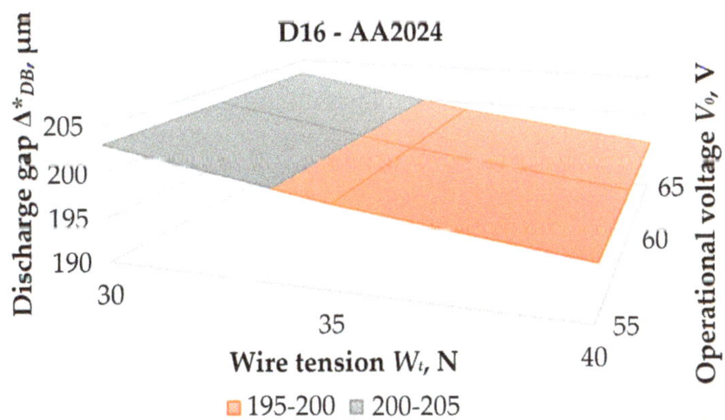

(**b**)

Figure 9. 3D graphs of the offsets Δ^*_{DB} dependencies on operational voltage V_0 and wire tension W_t: (**a**) 12Kh18N10T (AISI 321) steel; (**b**) D16 (AA2024) alloy.

3.5. Tool Wear

The tool electrode's rupture point (Figure 10) shows cup neck formation before destruction that corresponds to the ductile properties of the brass with the reduction area:

$$S_{RA} = \frac{0.049 - 0.003}{0.049} \cdot 100 = 93.8[\%], \tag{13}$$

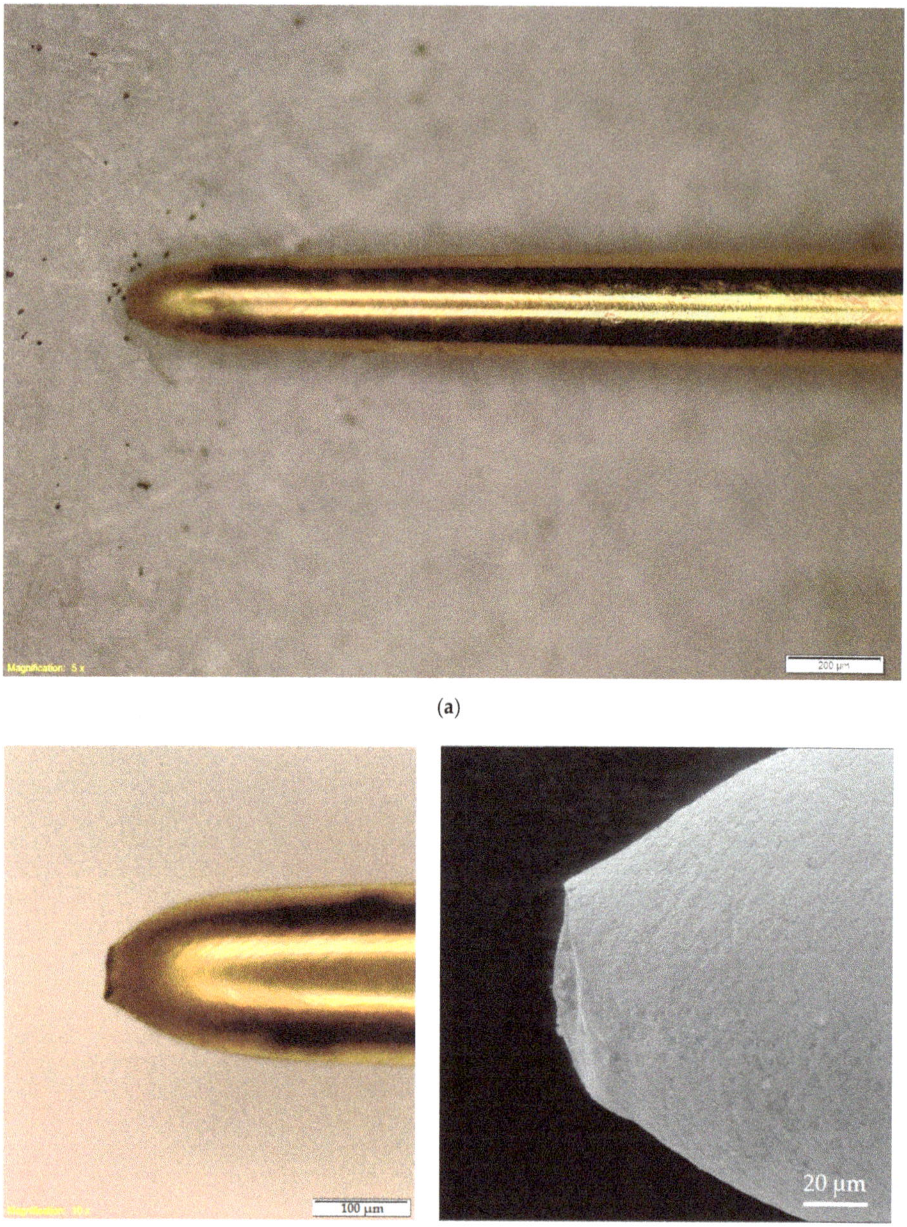

(**a**)

(**b**) (**c**)

Figure 10. Wire tool: (**a**) microphotograph, 5×; (**b**) microphotograph, 10×; (**c**) SEM-image, 2.0k×.

Figure 11 presents the electrode wear after roughing and finishing at the electrical discharge machining of 12Kh18N10T (AISI 321) steel and D16 (AA 2024) duralumin.

Figure 11a shows the conjugation of two interdependent surfaces with two types of wear—lateral at left and front at right. The wear has a different character. The front wear surface has the appearance of the typical eroded surface—sublimated and heat-affected material coated by the secondary structure of electrode components with pores and cracks. Moreover, the surface is covered by craters of solidified secondary material—usually consisting of the metallic material of the first order, solid solutions, and complex compounds of the second order (mostly oxides in the case of machining in deionized water). The craters have an explosive character that is not observed at lateral surfaces. The line of two wear surfaces conjugation is pronounced.

Figure 11b shows lateral wear at finishing. The formed surface has visible edges; the conjoined surface's left side has no presence of wear when the right side is also blank but with clear traces of secondary structure explosive deposition at the blank surface. The lateral wear surface showed two types of material destruction—the classical eroded surface of material sublimation with secondary structure deposition and mechanical wear traces.

Figure 11c shows the conjugation of two surfaces—of lateral wear and blank surface at roughing. The left side of the image—blank surface—has pronounced traces—drops, copious splashes—of explosive character of interaction occurred in the discharge gap at lateral wear. A significant volume of uneven sublimated material coated by the secondary structure with pores and cracks presents the surface with lateral wear at the right side of the image.

The front wear surface at roughing (Figure 11d) has secondary structure pellet formation that coat the sublimated surface. The secondary sublimated surface shows typical nanoframe structure formation more easy to melt material components sublimate from the secondary structure's coating (pellets) and are adsorbed by the refractory matrix.

Figure 11e shows the conjugation of two surfaces—front and lateral wear at roughing. The left side of the image—front wear surface—has a coating of secondary structure.

Figure 11f shows the lateral wear's surface at finishing when the obtained surface has traces of two types of wear—thermal and mechanical, which can be easily identified.

The cross-sections of the electrodes at roughing and finishing are presented in Figure 11g,h. Both the cross-sections showed quite intensive wear with affluent loose of the electrode material. The worn area, volume, and mass of the tool, volumetric, and mass wear rates were calculated using Equations (8)–(10) (Tables 6 and 7).

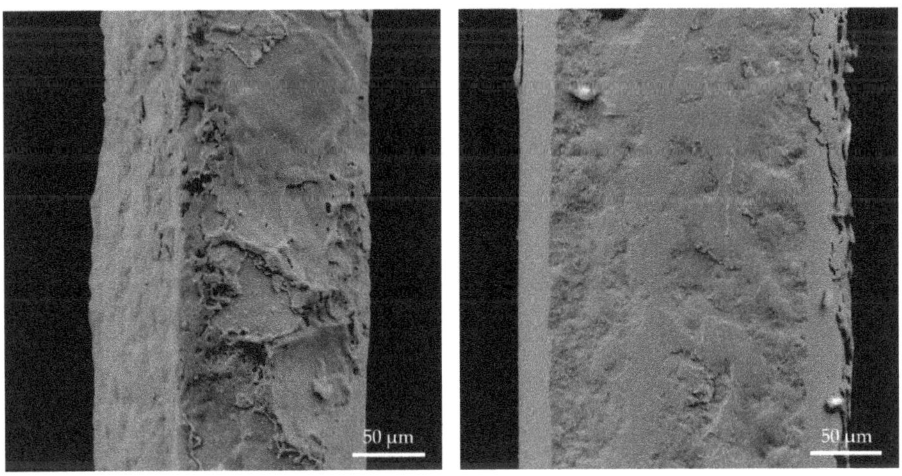

(a) (b)

Figure 11. Cont.

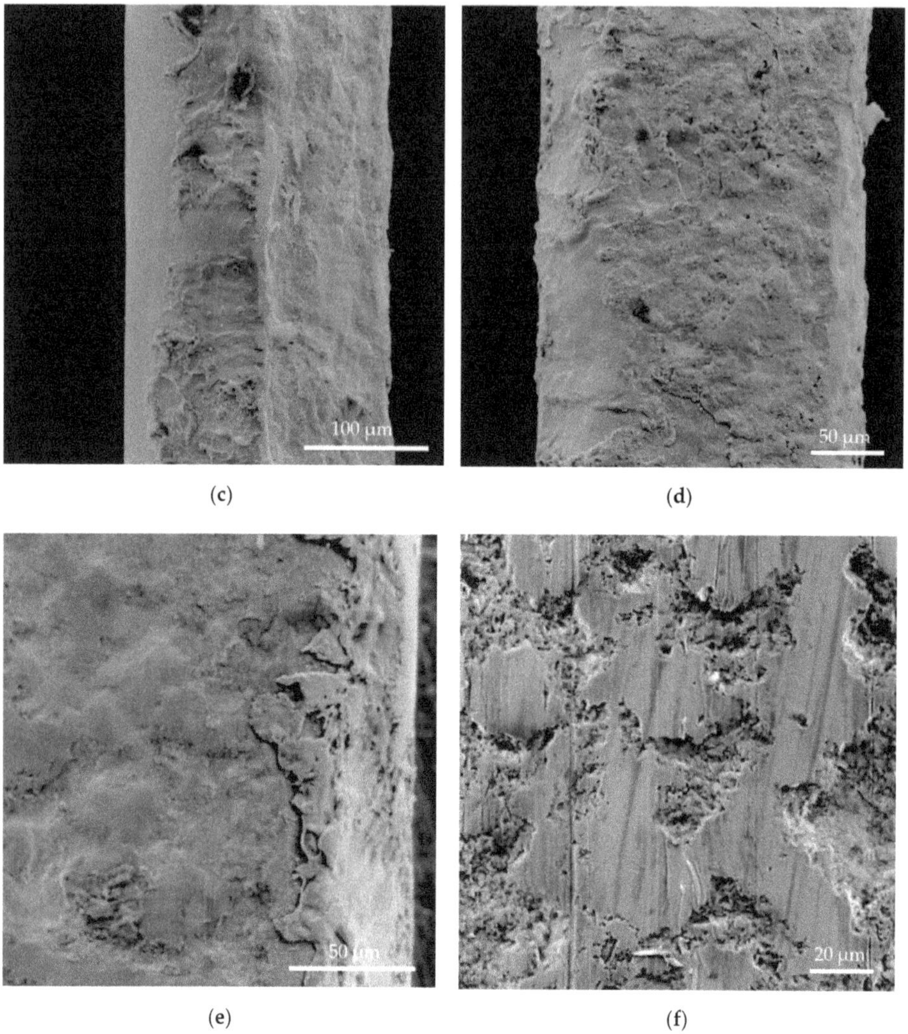

Figure 11. *Cont.*

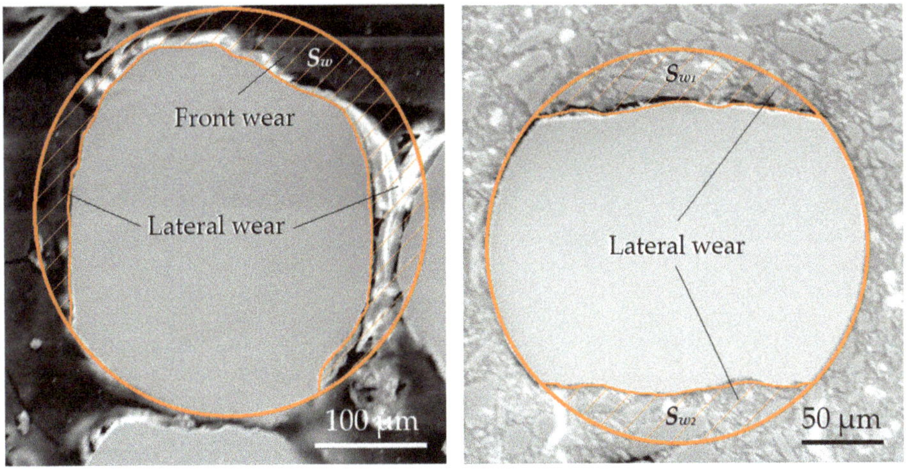

(g) (h)

Figure 11. SEM-images of wire tool electrode sample: (**a**) conjugation of front and lateral wear at roughing of 12Kh18N10T (AISI 321) steel, 916×; (**b**) lateral wear at finishing of 12Kh18N10T (AISI 321) steel, 909×; (**c**) conjugation of lateral wear and blank surface at roughing of 12Kh18N10T (AISI 321) steel, 780×; (**d**) front wear at roughing of D16 (AA2024) duralumin, 909×; (**e**) conjugation of front and lateral wear at roughing of D16 (AA2024) duralumin, 1.61k×; (**f**) lateral wear at finishing of D16 (AA2024) duralumin, 2.04k×; (**g**) cross-section after steel roughing, 696×; (**h**) cross-section after steel finishing, 1.02k×.

Table 6. Volumetric wear rate of electrical discharge machining of 12Kh18N10T (AISI 321) steel.

Type of Machining	Worn Surfaces	Measuring Error	Worn Area S_w	Summarized Worn Area S_s		Volumetric Wear ΔV	Volumetric Wear Rate [2] R_v	Error
		[µm]	[mm^2]	[mm^2]	[%] [1]	[mm^3]	[mm$^3 \cdot$s^{-1}]	[µm$^3 \cdot$µs^{-1}]
Roughing	Front	± 1 ÷ 2	0.0014 ± 0.00015	0.021 ± 0.0022	51 ± 10.5	0.42 + 0.0002	1.22 ± 0.04	± 20 ÷ 40
	Lateral		0.0048 ± 0.0010					
Finishing	Lateral		0.0045 ± 0.00005	0.009 ± 0.0001	18 ± 5.56	0.18 ± 0.00001	0.52 ± 0.002	

[1] Calculated to the entire cross-sectional area; [2] rewinding rate of 3.5 m/min.

Table 7. Mass wear rate of electrical discharge machining of 12Kh18N10T (AISI 321) steel.

Type of Machining	Measuring Error	Worn Mass [1] Δm	Mass Wear Rate [2] R_m	Error
	[g]	[g]	[g·s^{-1}]	[g·µs^{-1}]
Roughing	±0.0001	3.3×10^{-3} ± 0.00005	9.6×10^{-3} ± 0.01	± 0.01 ÷ 0.02
Finishing		1.4×10^{-3} ± 0.00004	4.0×10^{-3} ± 0.008	

[1] Density of 7.9×10^3 kg/m^3 or 0.0079 g/mm^3 at +20 °C; [2] rewinding rate of 3.5 m/min.

3.6. Chemical Content

The chemical content of the tool electrode's cross-section at roughing of 12Kh18N10T (AISI 321) steel is presented in Figure 12.

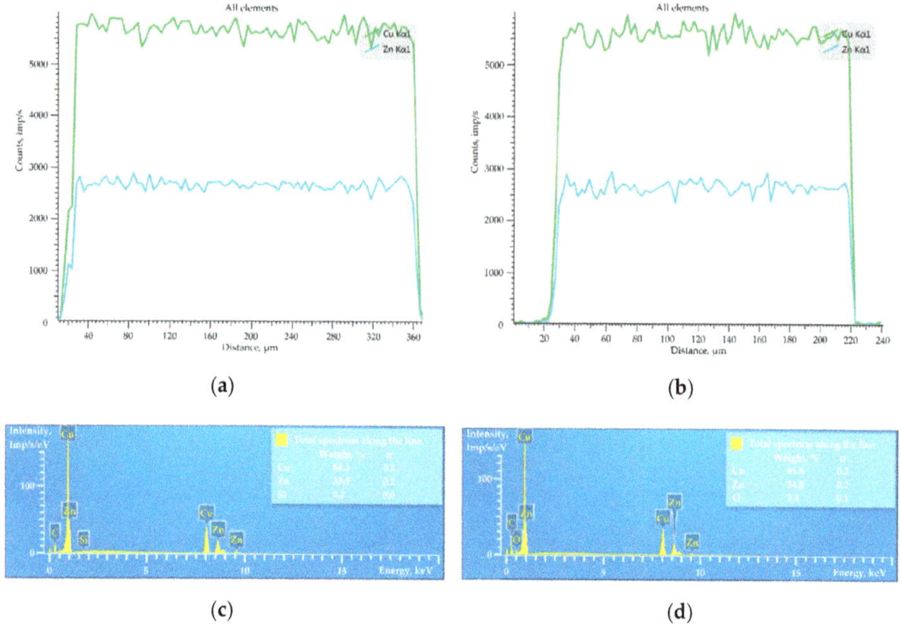

Figure 12. Chemical analyses of the worn tool electrode cross-section after machining: (**a**) chemical elements along the line at front wear; (**b**) chemical elements along the line at lateral wear; (**c**) EDX spectrum at front wear; (**d**) EDX spectrum at lateral wear.

4. Discussion

4.1. Discharge Pulses and Oscillations Control

Currently, a large number of EDM factors determine the machining mode, the value of which adaptively adjusts by the CNC system during processing. It is called an adaptive pulse-width modulation based on electrical response (Figure 3). At the same time, the value of interelectrode gap, the stability of processing, and consequently the quality of the obtained surfaces depend on the homogeneity of workpiece structure and microstructure, effectiveness of the erosion debris washout by the working fluid, the workpiece thickness, and electrophysical and electrochemical properties of the materials in the working zone. In this connection, the discharge pulses have a more chaotic, probabilistic nature, depending on many factors.

The vibroacoustic signal reflects the changes in the weight and structure of the workpiece, the main discharging factors that correlated with the force diagram in the working zone [27,55] that influence the amplitude of the signal in the wide range of spectra [23,25,26,51,52].

The recorded signal arises during processing and increases by 1.5 times from the initial level at the end of processing (Figure 4). The signal interruptions can be observed at 5 s before the end of processing, which is associated with the direct contact between the workpiece and tool electrode that occurs during changes in the cutting-off sample position in the working space in relation to the rest of the workpiece. It leads to the consequent clamping of the tool electrode to the workpiece by moving the sample and short circuits.

Optimum electrical discharge machining factors have the least value compared to the closest values (Figure 6a–c, except for the duralumin of 2 g). The EDM factors for the stable electrical discharge machining are V_0 = 60 V, W_t = 35 N for steel and V_0 = 55 V, W_t = 30 N for duralumin. At the same time, the sensibility of the system grows with its stiffness (Figure 6b,d) and decreases with the weight of the sample (Figure 6c,d).

An increase in RMS of the signal amplitude for stable factors at 8 kHz at 5 s before the end of processing was 40 ÷ 55% for steel and 12.5 ÷ 25% for aluminum alloy compared with data recorded at 60 s (Figure 5b,c and Figure 6).

The developed system showed its controversial response for the samples of 2 g of aluminum and adequate data for the samples' weight more than 4 g for the steel and 10 g of aluminum. The samples' weight varies by more than six times, but the RMS of the recorded signal demonstrates similar trends.

The observed behavior of the signal (Figures 5 and 6) can be correlated with the particular features of elastic and plastic deformation during the ductile failure according to the stress-strain curves and scheme of fracture formation (Figure 2) as aluminum shows better ductility during destruction that actually associated with a stretch of the interatomic bonds [56–60]. It correlates with the recorded signal when more brittle material—steel shows an adequate signal response even for the samples of 4 g when data received for ductile aluminum alloy are less significant but can also be registered for monitoring and control of samples of more than 2 g. That all make a basis for the development of multi-parameter control systems and switch to the next technological paradigm [61–67].

The minimum value of the measured roughness R_a of the samples (Figure 8) correlates with stable machining signals. The same tendency is observed for the measured discharge gap (Figure 9).

4.2. Wire Breakage and Tool Wear

The optical and scanning electron microscopy (Figure 10) showed that the character of rupture had mechanical nature corresponding to cup neck formation and stress-strain curve of middle ductile material destruction—brass alloy [47–49]. The observed area has the topology of the wire breakage that occurs in most of the cases during electrical discharge machining with unstable factors, in case the surface inclination or uneven structure of the workpiece need to be processed. There is no presence of thermal defects except an ashy shade at the formed cup. Additionally, there is no evidence of the rupture's external origin that can occur during the wire cut.

The excess in bias during wire blockage between the workpiece and cutting-off sample that did not allow adequate debris removal probably caused this rupture, since the current and pulses factors were constant. An increase in bias gave a denser distribution of discharges, while an increase in current gave more expressed discharges [28–30] (Figure 13a).

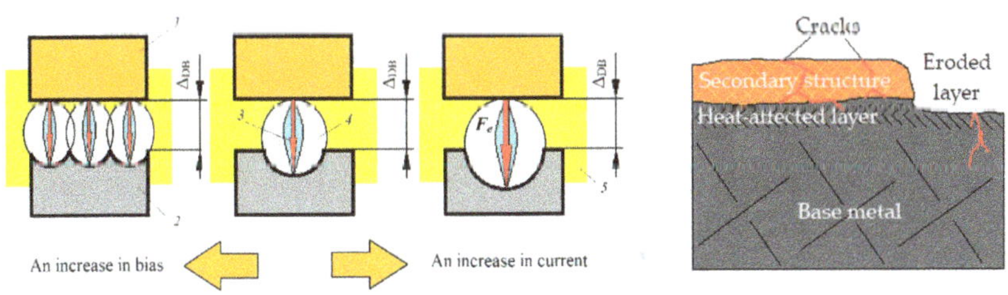

(a) (b)

Figure 13. Electrical discharge machining principles: (**a**) dependencies of the discharge pulse character on bias and current, where (**1**) is a tool electrode, (**2**) is a workpiece, (**3**) is a discharge channel, (**4**) is a plasma cloud, (**5**) is dielectric medium, and F_e is discharge force; (**b**) submicron structure of erosion wear.

The calculated enlarged value of the reduction area (S_{RA} = 93.8%) is obviously caused not only by mechanical rupture but a mechanical rupture in the softened state [68,69] due to the heat of the discharge gap that was definitely above 600–650 °C (dark red color) since the brass's surface around the formed cup neck is covered by the ashy shade of zinc oxide (Figure 10a) [70–72].

The formed craters that have different from the typical erosion morphology are 30–100 µm (Figure 11a). The explosive droplets reached a distance of ~100 µm from the wear edge on the backside surface (Figure 11c). The submicron structure of these droplets (Figure 11e) is different from the typical eroded surfaces as on the lateral surface at roughing and at finishing (Figure 11a,b,e,f).

The observation area of the wire tool presented in the SEM-microphotographs showed:

- Thermal traces (Figure 11a,e), which partly have a relation to the erosion that occurred between electrodes;
- Mechanical traces (Figure 11b,f), which has no relation to the processing;
- A topology that is more correlated to the electrical erosion of the materials under discharge pulses (Figure 11c,d).

The samples with typical erosion wear traces (Figure 11c,d) correlate to non-oxide (oxygen unsaturated) structures [73,74]—secondary submicrostructures of the complex compounds (of second order) adsorbed by the eroded surface of the base material—of the first order (Figure 14b) [75–77], which probably contain metastable and insoluble solid solution in the form of adherent and brittle thin film and heat-affected sublayer [78–81].

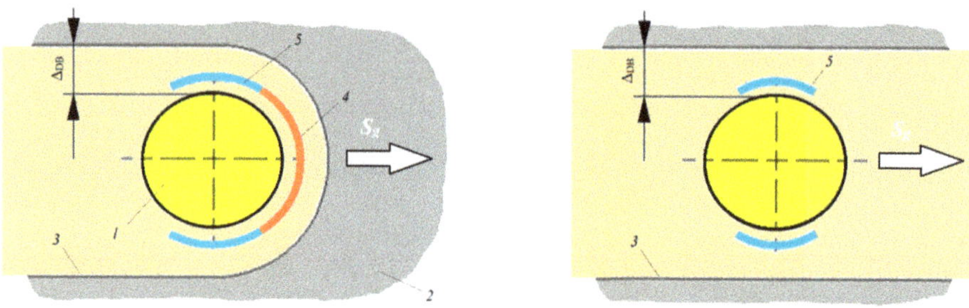

(**a**) (**b**)

Figure 14. Non-profiled tool electrode wear: (**a**) at roughing; (**b**) at finishing, where (**1**) is a tool electrode, (**2**) is a workpiece, (**3**) is machined surface, (**4**) is front wear, (**5**) is lateral wear, and S_g is a guidance feed of wire.

The mechanical traces on the surface of the wire at lateral wear correspond to the mechanical destruction of the tool during rewinding (Figure 11b,f) that occurred after electrical erosion (secondary wear). Probably, wire tool pinch rollers or diamond nozzle destroyed the morphology of the lateral wear mechanically.

The thermal traces were very pronounced at roughening (front wear), which are different from the lateral wear morphology that was especially apparent at the conjunction of the front and lateral wear surfaces (Figure 11a,e), and the lateral and blank surfaces (Figure 11c) have a different origin, related to the chemical composition of the tool and workpiece.

This is due to the picture of the erosion process observed visually and based on the interaction of the components of the electrodes—CuZn35 brass alloy and 12Kh18N10T (AISI 321) steel. The nickel and zinc reaction at a temperature of 1000 °C has an explosive character and results in the formation of intermetallic $ZnNi_x$ (x = 0, 5, 10, 15, 20 wt%) [82–86]. It can be easily observed by the formation of non-periodic orange flashes in the discharge gap with the release of abundant black sediment during the processing of anti-corrosion austenite (nickel-containing) steels [25,34,87–89]. Visually, the density of the flashes is less than the density of discharges and occurs with a lower frequency. However, the flashes cannot be detected with a higher workpiece, especially with a height of more than 100 mm that often occurred at machining in tool and mold production, when the visual access to the working zone is absent. Thus, the signal was adequately registered by the developed vibroacoustic diagnostic mean—RMS of the amplitude signal was higher by 12.5% and more intense by 20% for 12Kh18N10T (AISI 321) steel than for D16 (AA 2024) alloy for the thickness of the sample of 20 mm.

As can be seen, the deposition of the secondary submicron structure of the sublimated electrodes' components and working medium in the case of anti-corrosion steel processing occurred explosively (with craters of 30 ÷ 100 μm). The presence of the explosive character of reaction between metals accompanies the electrical erosion wear can be seen in Figure 11c on the blank surface when a clear edge limits the area of deep EDM wear on the lateral surface during roughing. However, the explosive nature of the secondary phase deposition overcomes the wear edge and is visible from the electrode's blank side.

The front wear at roughing has a more pronounced topology that correlates with the non-oxidized erosion wear an explosive reaction between wire and workpiece components (Figure 14), where the deposed film of secondary structure coat eroded base metal surface. The lateral wear at roughing has a less pronounced topology that corresponds to the typical wear that occurred under discharge pulses. It correlated to the degree of the involvement of the sides of the electrode in the formation of the slot when the front surface has the presence of secondary wear of the formed films: the front side is more involved in the formation of the slot, and the side surfaces are involved in the erosion process only partly by secondary "polishing" formed surfaces [24,90,91]. The lateral wear at finishing has a similar character. However, wire tool pinch rollers destroyed the morphology of the lateral wear mechanically.

The electrode's cross-section shows the intensity of the two types of wear (Figure 11g,h). In the considered sample, the front wear does not predominate the lateral one at roughing, and distributes quite even at the periphery of the tool (Figure 11g). The conjugation of the worn surfaces was pronounced for all of the samples.

Analysis of chemical elements along the line and EDX spectrum of the wire tool at roughing and finishing (Figure 12) showed mostly chemical elements except for chemical elements of the brass wire in balance—61.8 ÷ 64.3% of Cu and 34.8 ÷ 35.5% of Zn. However, less than 3.4% of oxygen is proof of semiconductive and amphoteric zinc oxide formation, which usually occurs during brass heating (Figure 10a) [92,93], when copper (II) oxide decomposes in the presence of hydrogen [94,95]:

$$2ZnO + O_2 \rightarrow 2ZnO, \tag{14}$$

$$CuO + H_2 \rightarrow Cu + H_2O. \tag{15}$$

Both of the oxides do not interact with water. Zinc oxide gets yellow with heating and sublimates at 1800 °C. It should be noted that that oxygen was present quantitatively more in the samples after finishing and at later wear of roughing, while it was not possible to quantify it along the line in some cases at front wear after roughing. A small amount of carbon that was not quantitatively evaluated (less than 0.2%) is associated with normal atmospheric contamination.

5. Conclusions

5.1. Monitoring System and Tool Behavior

A comprehensive study of the tool electrode's wear process during electrical discharge machining was accomplished by the developed monitoring system based on oscillation detecting. That gives detailed data on the character of electrode tool wear and stability of workpiece machining in the high-frequency acoustic band of 8 kHz.

The optimum electrical discharge machining factors are detected by monitoring the vibroacoustic signal—RMS value of the amplitude at 8 kHz for steels and more ductile duralumin with a weight of more than 2 g. The stable electrical discharge machining are V_0 = 60 V, W_t = 35 N for steel and V_0 = 55 V, W_t = 30 N for duralumin. An increase in RMS of the signal amplitude at 5 s before the end of processing was 40 ÷ 55% for steel and 12.5 ÷ 25% for aluminum alloy compared with data recorded at 60 s. The proposed approach can be used to develop a multiparameter controlling system of EDM-equipment to carry out the modern CNC-systems at a principally new level.

5.2. *Wire Tool Topology and Wear Rate*

Classification of the obtained surface topology of the tool electrode determines two types of wear under discharge pulses related to the thermal nature: material sublimation and chemical interaction between components of the working zone when mechanical destruction of the finishing electrode sample has a different origin.

Volumetric wear rate R_v was 1.22 ± 0.04 mm^3·s^{-1} at roughing and 0.52 ± 0.002 mm^3·s^{-1} at finishing; mass wear rate R_m—$9.6 \times 10^{-3} \pm 0.01$ g·s^{-1} and $4.0 \times 10^{-3} \pm 0.008$ g·s^{-1}, respectively. 41 ÷ 62% of the tool subjected wear under discharge impulses at roughing during electrical discharge machining of anti-corrosion steel when summarized lateral wear exceed front wear by 29.17%. 12 ÷ 24% of the tool sublimates under lateral wear at finishing.

The study showed that the processing of the materials with inadequate process parameters or the not proper combination of tool and workpiece materials causes more intensive wear of the tool correlated with the chemical interaction of the electrodes and dielectric medium components. This leads to the micro explosive character of processing with formation intermetallic ZnNix (x = 0, 5, 10, 15, 20 wt%), with Zn of the brass and nickel of austenite steel that was also registered the mean of vibroacoustic diagnostic. The crater diameter was of 30 ÷ 100 μm; RMS of the amplitude signal was higher by 12.5% and more intense by 20% for 12Kh18N10T (AISI 321) steel than for D16 (AA 2024) alloy.

5.3. *Further Procpects amd Paractical Significance of the Work*

As was shown, the amplitude is up to 55% higher for steel and up to 25% higher for duralumin at convenient machining than 5 s before the end of processing that always stays critical for precision cutting, especially in the conditions of tool production—profiled cutters, hot channels, and injection molds. The obtained data were for the thickness of 20 mm when it stays one of the most often used thickness for EDM workpieces in tool production. The developed system proved its reliability for the samples up to 2 g when the standard sample weight for discharge gap and machining mode verifying is 15.6 g for steels and 5.4 g for aluminum for a sample of 10 × 10 mm in the plan with a thickness of 20 mm.

The tool wear under electrical discharge pulses has a complex character related to the thermal type of wear with a heat-affected sublayer, and the upper layer consisted of a secondary structure formed from the components of electrodes with the traces of chemical reactions at a heat of 10,000 °C. Thus, electrical discharge machining wear forms in the following stages:

- Sublimation of the electrode surfaces under discharge;
- Chemical interaction of the sublimated electrode components in the presence of high heat;
- Explosive deposition of the formed secondary structure of first and second order material;
- Re-sublimation of the secondary structure.

The explosive character of interaction between Zn and Ni should be considered while designing experiments and electrical discharge machining of chrome-nickel anti-corrosion steels. For high-precision and nano-works, machining of nickel-containing steels should be provided by a tool with no Zn in its content—copper, steel, or tungsten wire have a few disadvantages due to the softness of copper, the relatively low electrical conductivity of steels, and heat-resistance of tungsten. However, it is a promising direction for further research.

The obtained knowledge has a fundamental character and can be used as a recommendation for the industrial applications on the choice of the electrode tool material and searching the optimum EDM-factors; in this context, not only structural requirements are addressed for the working and auxiliary surfaces of the final product, but also functionality in the exploitation conditions.

Author Contributions: Conceptualization, S.N.G.; methodology, M.P.K.; software, K.H.; validation, P.M.P., A.N.P.; formal analysis, S.V.F.; investigation, M.P.K.; resources, P.M.P. and S.V.F.; data curation, P.A.P. and K.H.; writing—original draft preparation, A.N.P.; writing—review and editing, A.A.O.; visualization, P.A.P. and A.A.O.; supervision, M.A.V.; project administration, M.A.V.; funding acquisition, S.N.G. All authors have read and agreed to the published version of the manuscript.

Funding: This project has received funding from the Ministry of Education and Science of the Russian Federation within the framework of the state task for scientific research, under Grant Agreement No. 0707-2020-0025.

Acknowledgments: The research was done at the Department of High-Efficiency Processing Technologies of MSTU Stankin.

Conflicts of Interest: The authors declare no conflict of interest. The funders had no role in the design of the study; in the collection, analyses, or interpretation of data; in the writing of the manuscript, or in the decision to publish the results.

Nomenclature

Symbol	Description	Unit
V_o	Operational voltage	V
W_t	Wire tension	N
I	Strength of the working current	A
f	Frequency of discharge pulses	s^{-1}
Δ	Distance between electrodes	µm
Δ_{DB}	Effective discharge gap	µm
Δ^*_{DB}	Offset of the path	µm
d_w	Wire diameter	mm
r_w	Wire radius	mm
S_{RA}	Reduction area	mm^2
S_0	Original transverse area	mm^2
S_{min}	Minimal area of the final neck	mm^2
S_w	Circle segment area	mm^2
α	Segment angle	degree
A_n	Wire amplitude of nth vibration, n is a positive integer (1, 2, 3 ,...)	mm
A_n'	Registered signal amplitude	mV
RMS	Root-mean-square mean of signal amplitude	mV^2
ΣF_{imp}	Summarized force of working impulses	N
ΣE_{imp}	Summarized energy of working impulses	J
k_n	Stiffness (coefficient of elasticity)	$N{\cdot}mm^{-1}$
m_n	Mass of system	g
l_n	Wire length	mm
Δl	Change in the wire length	mm
T	Period of self-oscillations	s
F_e	Restoring force (opposite and equal to W_t)	N
E	Young's modulus	Pa
R_v	Volumetric wear rates	$mm^3{\cdot}s^{-1}$
R_m	Mass wear rates	$g{\cdot}s^{-1}$
ΔV	Volumetric wear	mm^3
Δm	Worn mass	g
t	Wire length wear time	s
l_s	Slot width	mm

References

1. Faisal, N.; Kumar, K. Optimization of Machine Process Parameters in EDM for EN 31 Using Evolutionary Optimization Techniques. *Technologies* **2018**, *6*, 54. [CrossRef]
2. Wu, Y.-Y.; Huang, T.-W.; Sheu, D.-Y. Desktop Micro-EDM System for High-Aspect Ratio Micro-Hole Drilling in Tungsten Cemented Carbide by Cut-Side Micro-Tool. *Micromachines* **2020**, *11*, 675. [CrossRef] [PubMed]

3. Borchers, F.; Clausen, B.; Eckert, S.; Ehle, L.; Epp, J.; Harst, S.; Hettig, M.; Klink, A.; Kohls, E.; Meyer, H.; et al. Comparison of Different Manufacturing Processes of AISI 4140 Steel with Regard to Surface Modification and Its Influencing Depth. *Metals* **2020**, *10*, 895. [CrossRef]
4. Tran, T.-H.; Nguyen, M.-C.; Luu, A.-T.; Do, T.-V.; Le, T.-Q.; Vu, T.-T.; Tran, N.-G.; Do, T.-T.; Vu, N.-P. Electrical Discharge Machining with SiC Powder-Mixed Dielectric: An Effective Application in the Machining Process of Hardened 90CrSi Steel. *Machines* **2020**, *8*, 36. [CrossRef]
5. Chai, H.; Phung, B.; Mitchell, S. Application of UHF Sensors in Power System Equipment for Partial Discharge Detection: A Review. *Sensors* **2019**, *19*, 1029. [CrossRef] [PubMed]
6. Porvatov, A.N.; Kozochkin, M.P.; Fedorov, S.V.; Okunkova, A.A. About possibility of vibroacoustic diagnostics of electrical discharge machining and characterization of defects. *Mech. Ind.* **2015**, *16*, 707. [CrossRef]
7. Gu, L.; Zhu, Y.; Zhang, F.; Farhadi, A.; Zhao, W.S. Mechanism analysation and parameter optimisation of electro discharge machining of titanium-zirconium-molybdenum alloy. *J. Manuf. Process.* **2018**, *32*, 773–781. [CrossRef]
8. Scherjau, D.; Meyer, G.; Rosc, J.; Mai, T.; Gschirr, A.; Wimmer, A. Erosion processes of electrodes—Experiments and modeling. *Wear* **2019**, *428*, 85–92. [CrossRef]
9. Singh, M.A.; Sarma, D.K.; Hanzel, O.; Sedlacek, J.; Sajgalik, P. Surface characteristics and erosion phenomena in WEDM of alumina composites. *Mater. Manuf. Process.* **2018**, *33*, 1815–1821. [CrossRef]
10. Zhang, Y.; Guo, S.; Zhang, Z.; Hao, H.; Wenyuan, L.; Guojun, Z.; Yu, H. Simulation and experimental investigations of complex thermal deformation behavior of wire electrical discharge machining of the thin-walled component of Inconel 718. *J. Mater. Process. Technol.* **2019**, *270*, 306–322. [CrossRef]
11. Roy, T.; Balasubramaniam, R. Influence of ion-rich plasma discharge channel on unusually high discharging points in reverse micro electrical discharge machining. *Int. J. Adv. Manuf. Technol.* **2020**, *106*, 4467–4475. [CrossRef]
12. Yu, Z.; Li, D.; Yang, J.; Zeng, Z.J.; Yang, X.L.; Li, J.Z. Fabrication of micro punching mold for micro complex shape part by micro EDM. *Int. J. Adv. Manuf. Technol.* **2019**, *100*, 743–749. [CrossRef]
13. Chen, X.; Wang, Z.; Xu, J.; Wang, Y.K.; Li, J.W.; Liu, H.Z. Sustainable production of micro gears combining micro reciprocated wire electrical discharge machining and precision forging. *J. Clean. Prod.* **2018**, *188*, 1–11. [CrossRef]
14. Korznikova, G.F.; Nazarov, K.S.; Khisamov, R.K.H.; Sergeev, S.N.; Shayachmetov, R.U.; Khalikova, G.R.; Baimova, J.A.; Glezer, A.M.; Mulyukov, R.R. Intermetallic growth kinetics and microstructure evolution in Al-Cu-Al metal-matrix composite processed by high pressure torsion. *Mater. Lett.* **2019**, *253*, 412–415. [CrossRef]
15. Liu, M.; Jin, Y.; Pan, J.; Leygraf, C. Co-Adsorption of H_2O, OH, and Cl on Aluminum and Intermetallic Surfaces and Its Effects on the Work Function Studied by DFT Calculations. *Molecules* **2019**, *24*, 4284. [CrossRef]
16. Rekha, M.Y.; Srivastava, C. Microstructural Evolution and Corrosion Behavior of ZnNi-Graphene Oxide Composite Coatings. *Metall. Mater. Trans. A* **2019**, *50*, 5896–5913. [CrossRef]
17. Endo, N.; Ito, S.; Tomishige, K.; Kameoka, S.; Tsai, A.P.; Hirata, T.; Nishimura, C. CO hydrogenation over a hydrogen-induced amorphization of intermetallic compound CeNi2. *Catal. Today* **2011**, *164*, 293–296. [CrossRef]
18. Kochetov, N.A.; Seplyarskii, B.S. Effect of Initial Temperature and Mechanical Activation on Synthesis in a Ti plus Al System. *Combust. Explo Shock* **2020**, *56*, 308–316. [CrossRef]
19. Miloserdov, P.A.; Gorshkov, V.A.; Sachkova, N.V.; Khomenko, N.Y.; Miloserdova, O.M. Synthesis of Composite Materials in the System Cr-Ti-B by the Self-Propagating High-Temperature Synthesis from Mixtures $CaCrO_4/TiO_2/Al/B$. *Russ. J. Appl. Chem.* **2020**, *93*, 362–368. [CrossRef]
20. Vasilyev, N.; Borisov, E.N.; Novikov, B.V.; Akopyan, I.K.; Labzovskaya, M.E. Random lasing in ZnO self-organized nanoparticles produced by laser induced breakdown. *J. Lumin.* **2019**, *215*, 116668. [CrossRef]
21. Yue, X.; Yang, X.; Tian, J.; He, Z.F.; Fan, Y.Q. Thermal, mechanical and chemical material removal mechanism of carbon fiber reinforced polymers in electrical discharge machining. *Int. J. Mach. Tool Manu.* **2018**, *133*, 4–17. [CrossRef]
22. Volosova, M.A.; Okunkova, A.A.; Fedorov, S.V.; Hamdy, K.; Mikhailova, M.A. Electrical Discharge Machining Non-Conductive Ceramics: Combination of Materials. *Technologies* **2020**, *8*, 32. [CrossRef]

23. Grigoriev, S.N.; Kozochkin, M.P.; Porvatov, A.N.; Volosova, M.A.; Okunkova, A.A. Electrical discharge machining of ceramic nanocomposites: Sublimation phenomena and adaptive control. *Heliyon* **2019**, *5*, e02629. [CrossRef]
24. Shlykov, E.; Ablyaz, T. Complex Analysis of the Process of Electrical Discharge Machining of Bimetallic Steel-Copper Material. *Obrab. Metallov-Metal Work. Mater. Sci.* **2020**, *22*, 16–26.
25. Grigoriev, S.N.; Kozochkin, M.P.; Kropotkina, E.Y.; Okunkova, A.A. Study of wire tool-electrode behavior during electrical discharge machining by vibroacoustic monitoring. *Mech. Ind.* **2016**, *17*, 717. [CrossRef]
26. Dwaraka, R.; Arunachalam, N. Investigation on the effect of EDM process variables and environments on acoustic emission signals. *Mach. Sci. Technol.* **2020**, *24*, 638–662. [CrossRef]
27. Melnik, Y.A.; Kozochkin, M.P.; Porvatov, A.N.; Okunkova, A.A. On adaptive control for electrical discharge machining using vibroacoustic emission. *Technologies* **2018**, *6*, 96. [CrossRef]
28. Lazarenko, B.R.; Mikhailov, V.V.; Gitlevich, A.E.; Verkhoturov, A.D.; Anfimov, I.S. Distribution of elements in surface layers during electric spark alloying. (Raspredelenie Elementov V Poverkhnostnykh Sloyakh Pri Elektroiskrovom Legirovanii). *Surf. Eng. Appl. Electrochem. (Elektron. Obrab. Mater.)* **1977**, *3*, 28–33.
29. Lazarenko, B.R.; Duradzhi, V.N.; Bryantsev, I.V. Effect of Incorporating an additional inductance on the characteristics of anode and cathode processes. (O Vliyanii Vklyucheniya Dopolnitel'noi Induktivnosti Na Kharakteristiki Anodnogo I Katodnogo Protsessov). *Surf. Eng. Appl. Electrochem. (Elektron. Obrab. Mater.)* **1979**, *5*, 8–13.
30. Lazarenko, B.R.; Lazarenko, N.I. Electric spark machining of metals in water and electrolytes. (Elektroiskrovaya Obrabotka Metallov V Vode I Elektrolitakh). *Surf. Eng. Appl. Electrochem. (Elektron. Obrab. Mater.)* **1980**, *1*, 5–8.
31. Volosova, M.A.; Okunkova, A.A.; Povolotskiy, D.E.; Podrabinnik, P.A. Study of electrical discharge machining for the parts of nuclear industry usage. *Mech. Ind.* **2015**, *16*, 706. [CrossRef]
32. Gavrin, V.N.; Kozlova, Y.P.; Veretenkin, E.P.; Logachev, A.V.; Logacheva, A.I.; Lednev, I.S.; Okunkova, A.A. Reactor target from metal chromium for "pure" high-intensive artificial neutrino source. *Phys. Part. Nucl. Lett.* **2016**, *13*, 267–273. [CrossRef]
33. Steuer, P.; Weber, O.; Baehre, D. Structuring of wear-affected copper electrodes for electrical discharge machining using Pulse Electrochemical Machining. *Int. J. Refract. Met. H* **2015**, *52*, 85–89. [CrossRef]
34. Ay, M.; Etyemez, A. Optimization of the effects of wire EDM parameters on tolerances. *Emerg. Mater. Res.* **2020**, *9*, 527–531. [CrossRef]
35. Markopoulos, A.P.; Papazoglou, E.-L.; Karmiris-Obratański, P. Experimental Study on the Influence of Machining Conditions on the Quality of Electrical Discharge Machined Surfaces of aluminum alloy Al5052. *Machines* **2020**, *8*, 12. [CrossRef]
36. Yun, J.D.; Go, C.; Wang, D.H.; Ahn, Y.C. Electrical discharge machining of aluminum oxide matrix composites containing titanium carbide as a conductive second phase. In *Processing and Fabrication of Advanced Materials VI, Vols 1 & 2, Proceedings of International Symposium on Processing and Fabrication of Advanced Materials, Singapore, 24–26 November 1997*; Khor, K.A., Srivatsan, T.S., Moore, J.J., Eds.; IOM Communications Ltd.: London, UK, 1998; pp. 1773–1781.
37. Yarovchuk, A.V.; Maksimkin, O.P.; Tsay, K.V. Effect of Low-Cycle Thermocycling Treatment on Corrosion and Mechanical Properties of Corrosion-Resistant Steel 12Kh18N10T Irradiated with Neutrons. *Met. Sci. Heat Treat.* **2017**, *59*, 446–453. [CrossRef]
38. Zhelezkov, O.S.; Galiakhmetov, T.S.; Malakanov, S.A. Shaping the Hexahedral Head of a 12Kh18N10T Stainless Steel Bolt. *Steel Transl.* **2017**, *47*, 824–826. [CrossRef]
39. Gnedenkov, S.V.; Sinebryukhov, S.L.; Egorkin, V.S.; Vyaliy, I.E.; Imshinetskiy, I.M.; Kostina, M.V.; Muradyan, S.O.; Sergienko, V.I. Corrosion stability of austenitic steels 05Kh22AG15N8M2F and 12Kh18N10T in chloride-containing media. *Prot. Met. Phys. Chem. Surf.* **2017**, *53*, 910–915. [CrossRef]
40. Denisov, E.A.; Kompaniets, T.N.; Yukhimchuk, A.A.; Boitsov, I.E.; Malkov, I.L. Hydrogen and helium in nickel and 12Kh18N10T steel. *Tech. Phys.* **2013**, *58*, 779–786. [CrossRef]
41. Krymskiy, S.V.; Ilyasov, R.R.; Avtokratova, E.V.; Sitdikov, O.S.; Markushev, M.V. Intergranular Corrosion of Cryorolled and Aged D16 Aluminum Alloy. *Prot Met.* **2017**, *53*, 1091–1099. [CrossRef]
42. Markushev, M.; Krymskiy, S.; Ilyasov, R.; Avtokratova, E.; Khazgalieva, A.; Sitdikov, O. Influence of Zr on intergranular corrosion of cast and cryorolled D16 aluminum alloy. *Lett. Mater.* **2017**, *7*, 447–451. [CrossRef]

43. Ponomarev, A.N. An investigation of diffusion in droplets and of evaporation of volatile components into vacuum. *High. Temp.* **2005**, *43*, 930–936. [CrossRef]
44. Nikitina, E.V.; Kazakovtseva, N.A. High Temperature Corrosion of 12Kh18N10T Steel in Molten Lithium and Potassium Chlorides with a Cerium Trichloride Addition. *Russ. Metall.* **2018**, *8*, 747–749. [CrossRef]
45. Elshina, L.A.; Elshina, V.A. Synthesis of a Nanocrystalline alpha-Al2O3 Powder in Molten Halides in the Temperature Range 700–800 degrees C. *Russ. Metall.* **2020**, *2*, 138–141. [CrossRef]
46. Pang, W.K.; Low, I.M. Understanding and improving the thermal stability of layered ternary carbides in ceramic matrix composites. In *Advances in Ceramic Matrix Composites*; Low, I.M., Ed.; Woodhead Publishing Series in Composites Science and Engineering; Woodhead Publ Ltd.: Cambridge, UK, 2014; Volume 45, pp. 340–368. [CrossRef]
47. Lekhov, O.S.; Chukov, I.Y.; Karpova, N.M.; Peskov, A.V. Investigation of Cyclic Deformation of Continuously Cast Stainless-Steel Slabs. *Steel USSR* **1990**, *20*, 279–280.
48. Armstrong, R.W. Dislocation Mechanics Pile-Up and Thermal Activation Roles in Metal Plasticity and Fracturing. *Metals* **2019**, *9*, 154. [CrossRef]
49. Maksimov, E.A. Test of Stress-Strain State and Flatness Parameter under Bending of the Corrugated Part of a Flat Bar in the Course of Sheet Rolling of Flat Bars and Bands of Brass Alloys. *Russ. J. Non-Ferrous Met.* **2010**, *51*, 74–78. [CrossRef]
50. Zhu, J.-H.; Young, B. Tests and Design of Aluminum Flexural Members. In *Proceeding of the 10th International Symposium on Structural Engineering for Young Experts, Changsha, China, 19–21 Octber 2008, Vols I and II*; Xu, B., Xiao, Y., Ru, J.P., Ren, W.X., Eds.; Science Press Beijing: Beijing, China, 2008; pp. 1769–1774.
51. Grigor'ev, S.N.; Kozochkin, M.P.; Fedorov, S.V.; Porvatov, A.N.; Okun'kova, A.A. Study of electroerosion processing by vibroacoustic diagnostic methods. *Meas. Tech.* **2015**, *58*, 878–884. [CrossRef]
52. Goswami, K.; Samuel, G.L. Monitoring of material-removal mechanism in micro-electrical discharge machining by pulse classification and acoustic emission signals. *Proc. Inst. Mech. Eng. B-J. Eng.* **2020**. [CrossRef]
53. Shilov, M.A.; Smirnova, A.I.; Gvozdev, A.A.; Rozhkova, N.N.; Dyachkova, T.P.; Burkov, A.A.; Stolbov, D.N.; Savilov, S.V.; Usol'tseva, N.V. Rheology of Plastic Lubricants with Additives of Carbon Nanostructures of Various Type. *J. Frict. Wear* **2019**, *40*, 546–554. [CrossRef]
54. Patzer, G.; Woydt, M.; Shah, R.; Miller, C.; Iaccarino, P. Test Modes for Establishing the Tribological Profile under Slip-Rolling. *Lubricants* **2020**, *8*, 59. [CrossRef]
55. Haroon, M.; Adams, D.E. Development of component-level damage evolution models for mechanical prognosis. *J. Appl. Mech.-T ASME* **2008**, *75*, 021017. [CrossRef]
56. Montanari, R.; Varone, A. Synergic Role of Self-Interstitials and Vacancies in Indium Melting. *Metals* **2015**, *5*, 1061–1072. [CrossRef]
57. Sun, H.; Singh, C.V. Temperature dependence of grain boundary excess free volume. *Scr. Mater.* **2020**, *178*, 71–76. [CrossRef]
58. Metel, A.; Bolbukov, V.; Volosova, M.; Grigoriev, S.; Melnik, Y. Source of metal atoms and fast gas molecules for coating deposition on complex shaped dielectric products. *Surf. Coat. Technol.* **2013**, *225*, 34–39. [CrossRef]
59. Metel, A.S.; Grigoriev, S.N.; Melnik, Y.A.; Bolbukov, V.P. Broad beam sources of fast molecules with segmented cold cathodes and emissive grids. *Instrum. Exp. Tech.* **2012**, *55*, 122–130. [CrossRef]
60. Kuzin, V.V.; Grigoriev, S.N.; Fedorov, M.Y. Role of the thermal factor in the wear mechanism of ceramic tools. Part 2: Microlevel. *J. Frict. Wear* **2015**, *36*, 40–44. [CrossRef]
61. Sevastianov, S.V. An introduction to multi-parameter complexity analysis of discrete problems. *Eur. J. Oper. Res.* **2005**, *165*, 387–397. [CrossRef]
62. Perez, C. Technological revolutions and techno-economic paradigms. *Camb. J. Econ.* **2010**, *34*, 185–202. [CrossRef]
63. Bobovich, B.B. Glass-fiber reinforced plastics—Construction materials of the sixth technological paradigm? *Glas. Ceram.* **2019**, *76*, 38–41. [CrossRef]
64. Wonglimpiyarat, J. Towards the sixth Kondratieff cycle of nano revolution. *Int. J. Nanotechnol. Mol. Comput.* **2011**, *3*, 87–100. [CrossRef]
65. Grigoriev, S.; Melnik, Y.; Metel, A. Broad fast neutral molecule beam sources for industrial-scale beam-assisted deposition. *Surf. Coat. Technol.* **2002**, *156*, 44–49. [CrossRef]

66. Kozochkin, M.P.; Porvatov, A.N. Estimation of Uncertainty in Solving Multi-Parameter Diagnostic Problems. *Meas. Tech.* **2015**, *58*, 173–178. [CrossRef]
67. Grigoriev, S.N.; Masterenko, D.A.; Teleshevskii, V.I.; Emelyanov, P.N. Contemporary state and outlook for development of metrological assurance in the machine-building industry. *Meas. Tech.* **2013**, *55*, 1311. [CrossRef]
68. Alsagabi, S.; Shrestha, T.; Charit, I. High temperature tensile deformation behavior of Grade 92 steel. *J. Nucl. Mater.* **2014**, *453*, 151–157. [CrossRef]
69. Maziasz, P.J.; Mckamey, C.G. Microstructural Characterization of Precipitates Formed During High-Temperature Testing and Processing Of Iron Aluminide Alloys. *Mater. Sci. Eng. A* **1992**, *152*, 322–334. [CrossRef]
70. Ryczek, K.; Koziel, M.; Wiercigroch, E.; Malek, K.; Jarosz, M.; Sulka, G.D.; Zaraska, L. Fast fabrication of nanostructured semiconducting oxides by anodic oxidation of brass. *Mater. Sci. Semicond. Process.* **2020**, *113*, 105035. [CrossRef]
71. Panova, T.V.; Kovivchak, V.S. Formation of Oxide Layers on the Surface of Copper and its Alloys Modified by a High-Power Ion Beam. *J. Surf. Investig.* **2019**, *13*, 1098–1102. [CrossRef]
72. Murzin, S.P.; Kryuchkov, A.N. Formation of ZnO/CuO heterostructure caused by laser-induced vibration action. *Procedia Eng.* **2017**, *176*, 546–551. [CrossRef]
73. Chorley, R.W.; Lednor, P.W. Synthetic Routes to High Surface-Area Nonoxide Materials. *Adv. Mater.* **1991**, *3*, 474–485. [CrossRef]
74. Grigoriev, S.N.; Sobol, O.V.; Beresnev, V.M.; Serdyuk, I.V.; Pogrebnyak, A.D.; Kolesnikov, D.A.; Nemchenko, U.S. Tribological characteristics of (TiZrHfVNbTa)N coatings applied using the vacuum arc deposition method. *J. Frict. Wear* **2014**, *35*, 359–364. [CrossRef]
75. Volosova, M.A.; Grigor'ev, S.N.; Kuzin, V.V. Effect of titanium nitride coating on stress structural inhomogeneity in oxide-carbide ceramic. Part 4. Action of heat flow. *Refract. Ind. Ceram.* **2015**, *56*, 91–96. [CrossRef]
76. Kuzin, V.V.; Grigor'ev, S.N.; Volosova, M.A. Effect of a TiC Coating on the Stress-Strain State of a Plate of a High-Density Nitride Ceramic Under Nonsteady Thermoelastic Conditions. *Refract. Ind. Ceram.* **2014**, *54*, 376–380. [CrossRef]
77. Volosova, M.A.; Grigoriev, S.N.; Ostrikov, E.A. Use of laser ablation for formation of discontinuous (discrete) wear-resistant coatings formed on solid carbide cutting tool by electron beam alloying and vacuum-arc deposition. *Mech. Ind.* **2016**, *17*, 720. [CrossRef]
78. Vereschaka, A.A.; Grigoriev, S.N.; Vereschaka, A.S.; Popov, A.Y.; Batako, A.D. Nano-scale multilayered composite coatings for cutting tools operating under heavy cutting conditions. *Proc. CIRP* **2014**, *14*, 239–244. [CrossRef]
79. Vereschaka, A.A.; Volosova, M.A.; Grigoriev, S.N.; Vereschaka, A.S. Development of wear resistant complex for high-speed steel tool when using process of combined cathodic vacuum arc deposition. *Proc. CIRP* **2013**, *9*, 8–12. [CrossRef]
80. Gusarov, A.V.; Grigoriev, S.N.; Volosova, M.A.; Melnik, Y.A.; Laskin, A.; Kotoban, D.V.; Okunkova, A.A. On productivity of laser additive manufacturing. *J. Mater. Process. Technol.* **2018**, *261*, 213–232. [CrossRef]
81. Abdulhadi, H.A.; Ahmad, S.N.A.S.; Ismail, I.; Ishak, M.; Mohammed, G.R. Thermally-Induced Crack Evaluation in H13 Tool Steel. *Metals* **2017**, *7*, 475. [CrossRef]
82. Liu, Z.G.; Guo, J.T.; Hu, Z.Q. Mechanical Alloying Of The Ni-Al(M) (M Ti, Fe) System. *Mat. Sci. Eng. A-Struct.* **1995**, *192*, 577–582. [CrossRef]
83. Talawar, M.B.; Agrawal, A.P.; Asthana, S.N. Energetic co-ordination compounds: Synthesis, characterization and thermolysis studies on bis-(5-nitro-2H-tetrazolato-N-2)tetraammine cobalt(III) perchlorate (BNCP) and its new transition metal (Ni/Cu/Zn) perchlorate analogues. *J. Hazard. Mater.* **2005**, *120*, 25–35. [CrossRef]
84. Friedrich, M.; Teschner, D.; Knop-Gericke, A.; Armbruster, M. Surface and Subsurface Dynamics of the Intermetallic Compound ZnNi in Methanol Steam Reforming. *J. Phys. Chem. C* **2012**, *116*, 14930–14935. [CrossRef]
85. Grigoriev, S.; Metel, A. Plasma- and Beam-Assisted Deposition Methods. In *Nato Science Series, Series Ii: Mathematics, Physics and Chemistry*; Nanostructured Thin Films and Nanodispersion Strengthened Coatings. NATO Science Series II: Mathematics, Physics and Chemistry; Voevodin, A.A., Shtansky, D.V., Levashov, E.A., Moore, J.J., Eds.; Springer: Dordrecht, The Netherlands, 2004; Volume 155, pp. 147–154. [CrossRef]

86. Grigoriev, S.N.; Melnik, Y.A.; Metel, A.S.; Panin, V.V. Broad beam source of fast atoms produced as a result of charge exchange collisions of ions accelerated between two plasmas. *Instrum. Exp. Tech.* **2009**, *52*, 602–608. [CrossRef]
87. Mouralova, K.; Benes, L.; Bednar, J.; Zahradnicek, R.; Prokes, T.; Fries, J. Analysis of Machinability and Crack Occurrence of Steels 1.2363 and 1.2343ESR Machined by Die-Sinking EDM. *Coatings* **2020**, *10*, 406. [CrossRef]
88. Metel, A.S.; Grigoriev, S.N.; Melnik, Y.A.; Prudnikov, V.V. Glow discharge with electrostatic confinement of electrons in a chamber bombarded by fast electrons. *Plasma Phys. Rep.* **2011**, *37*, 628–637. [CrossRef]
89. Sobol', O.V.; Andreev, A.A.; Grigoriev, S.N.; Gorban', V.F.; Volosova, M.A.; Aleshin, S.V.; Stolbovoy, V.A. Physical characteristics, structure and stress state of vacuum-arc tin coating, deposition on the substrate when applying high-voltage pulse during the deposition. *Probl. Atom. Sci. Technol.* **2011**, *4*, 174–177.
90. Yuan, H.; Song, J. An Improved Calculation Model for the Prediction of the Wear of Coated Electrical Contacts. *Technologies* **2019**, *7*, 77. [CrossRef]
91. Phipon, R.; Shivakoti, I.; Sharma, A. Sustainable processing of Inconel 718 super alloy in electrical discharge machining process. *World J. Eng.* **2020**. [CrossRef]
92. Brzezińska, M.; García-Muñoz, P.; Ruppert, A.M.; Keller, N. Photoactive ZnO Materials for Solar Light-Induced Cu_xO-ZnO Catalyst Preparation. *Materials* **2018**, *11*, 2260. [CrossRef]
93. Lee, H.; Zhang, X.; Hwang, J.; Park, J. Morphological Influence of Solution-Processed Zinc Oxide Films on Electrical Characteristics of Thin-Film Transistors. *Materials* **2016**, *9*, 851. [CrossRef]
94. Silva, N.; Ramírez, S.; Díaz, I.; Garcia, A.; Hassan, N. Easy, Quick, and Reproducible Sonochemical Synthesis of CuO Nanoparticles. *Materials* **2019**, *12*, 804. [CrossRef]
95. Zheng, W.; Chen, Y.; Peng, X.; Zhong, K.; Lin, Y.; Huang, Z. The Phase Evolution and Physical Properties of Binary Copper Oxide Thin Films Prepared by Reactive Magnetron Sputtering. *Materials* **2018**, *11*, 1253. [CrossRef] [PubMed]

© 2020 by the authors. Licensee MDPI, Basel, Switzerland. This article is an open access article distributed under the terms and conditions of the Creative Commons Attribution (CC BY) license (http://creativecommons.org/licenses/by/4.0/).

technologies

MDPI

Article

On Adaptive Control for Electrical Discharge Machining Using Vibroacoustic Emission

Yury A. Melnik, Mikhail P. Kozochkin, Artur N. Porvatov and Anna A. Okunkova *

Department of High-efficiency Machining Technologies, Moscow State University of Technology STANKIN, Vadkovskiy per. 3A, 127055 Moscow, Russia; yu.melnik@stankin.ru (Y.A.M.); m.kozochkin@stankin.ru (M.P.K.); vto@stankin.ru (A.N.P.)

* Correspondence: a.okunkova@stankin.ru; Tel.: +7-909-913-12-07

Received: 16 October 2018; Accepted: 23 October 2018; Published: 24 October 2018

Abstract: The article is related to the research of the parameters of vibroacoustic emission for development of the monitoring and adaptive control system for electrical discharge machining. The classical control system based on a response of electrical parameters does not give an adequate data in the cases of a new class of materials processing as conductive ceramics reinforced by conductive nano additives and carbon nanotubes and whiskers. The idle pulses, which are working on the destruction of the erosion products in the gap, count as working pulses. The application of the monitoring and control tools based on vibroacoustic emission gives adequate data about conditions in the working zone. The developed system is available to count only impulses involved in working on the destruction of the workpiece. The experiments were conducted on the samples of materials with a low melting point as austenitic steel and aluminum alloy, and hard alloys. The records of vibroacoustic signals were analyzed for detection of the monitoring and adaptive control criteria.

Keywords: electrical discharge machining; vibroacoustic emission; adaptive control; monitoring; discharge gap; erosion products

1. Introduction

Today it is impossible to imagine the production of high-accuracy parts with complex volumetric geometry and internal cooling system without electrical discharge machining (EDM) [1–3]. The application of the technology was significantly extended by adding from 1 to 3 independent rotating axes [4–6].

Hence, practice shows that up to now EDM has uncontrollable problems related to decreasing of productivity and quality of machined surfaces during conventional processing of easy-to-work materials [7–9]. Mainly it is critical in the case of machining of narrow slots array. E.g., 20 ÷ 24 slots with the sizes in the plan 1.5 × 8 mm, and height of 15 ÷ 25 mm for production of the mold die with push-type ejectors for the part as a body of motor vehicle backlight [10]. In this case, a wire tool can be stuck between a workpiece and a cut-out industrial waste (or part), that leads to the undesirable issues related to process instability, short circuit, wire break, sometimes even to diamond nozzle damage [11–14].

Nowadays, the use of ceramic products in the industry is continuously growing due to its excellent exploitation properties [15,16]. Plenty of scientists aim their study on the development of a new class of materials as conductive ceramic nanocomposites [17–19]. This kind of material should provide mechanical properties up to the level of continent ceramics or even exceed them, but also be suitable for electrical discharge machining. By other words, it should be conductive [20]. The conductivity should be assured by conductive nano-sized additives as TiC, ZrC, $Nb_{(2)}C$ [21–23], when SiCw or carbon in the form of nanotubes or whiskers are responsible for supporting and improving mechanical properties [24,25]. It makes the questions related to ensuring of the uninterrupted electrical discharge machining using in-situ monitoring critical and significant to adapt the electrical parameters of machining and to prevent all undesirable effects during processing.

The tools of diagnostics based on vibroacoustic emission are known for application to the conventional mechanical machining for providing better reliability of the machine tools [26,27]. The vibroacoustic diagnostic tools aim to detect the moment of beating between a tool and a workpiece [28–30]. It allows controlling the operating element movements, preventing accidents. It was proved that based on the vibroacoustic signal it is possible to develop a new class of multiparametric diagnostic and adaptive control system [31]. The significant advantage of vibroacoustic diagnostics is the simplicity of the accelerometers' installation on the elastic system of the machine.

The scientific novelty of the article is in development and application diagnostic system based on vibroacoustic emission to the electrical discharge machining methods, which are known by an impossibility for the visual monitoring tools. The tasks of the current study are:

(1) Research a fundamental possibility of vibroacoustic diagnostics of wire electrical discharge machining;
(2) Adapt the existed method of in-situ vibroacoustic monitoring for EDM, make the comparison of the receive spectra of the signal with the electrical parameters of the machine;
(3) Analyze the obtained spectra and make understandable the criteria of deciding for development of the adaptive control system based on vibroacoustic emission;
(4) Demonstrate an opportunity of the in-situ adaptive control of EDM by specific examples of the machining.

2. Materials and Methods

The experiments were carried out at an industrial 4-axis EDM machine with CNC-controller Seibu M500S (Seibu, Fukuoka, Japan). A 0.25 mm-diameter brass wire of CuZn35 was used as a tool. Deionized water was used as a dielectric medium. Electrical discharge machining was produced with the immersion of the workpiece.

The accelerometers (Figure 1) of the developed vibroacoustic monitoring systems were installed at the upper guide (5) of the machine and on the system of the fastening of the workpiece (3). It was found that during electrical discharge machining the upper accelerometer (7) gave a more informative and symmetric signal than the accelerometer from the lower guide (8). It can be explain by presence of the disturbance influence from the working dielectric pump and driving gears of the machine. Further, only the signal from the upper guide was taken into account.

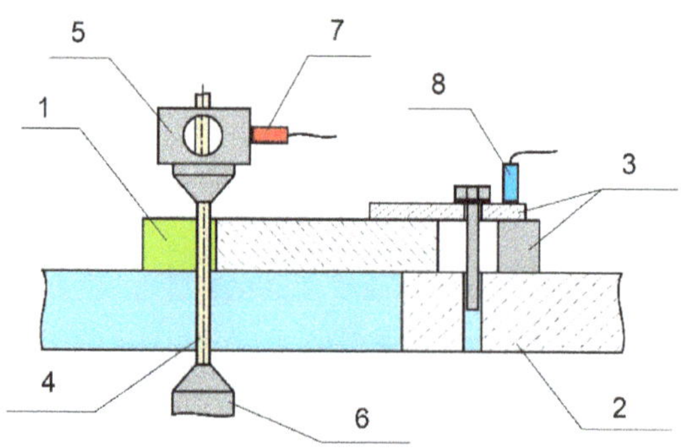

Figure 1. The positions of accelerometers in the working area of EDM machine, where (1) is a workpiece, (2) is a worktable, (3) is a fastening system, (4) is a wire electrode, (5) is an upper guide, (6) is a lower guide, (7) is an upper accelerometer, (8) is a lower accelerometer.

A 200 × 20 × 16 mm-blank of austenitic stainless steel AISI 321 [32,33] and a 200 × 16 × 16 mm-blank of aluminum alloy AISI 2024 [34–36] samples (Table 1) were used at the first stage of the experiments for developing of the system of diagnostics and monitoring based on vibroacoustic emission for the needs of electrical discharge machining. The blanks were fastened on the worktable of EDM machine tool.

Table 1. The chemical composition of AISI 321 steel and AISI 2024 alloy, %.

Material	Fe	Cr	C	Ni	Mn	Ti	P	Cu	Si	Zn	Mg	Al
AISI 321	44.9–59.1	17–19	~12	9–11	~2	~0.8	~0.035	~0.03	~0.02	-	-	-
AISI 2024	~0.5	-	-	~0.1	0.3-0.9	~0.1	-	3.8-4.9	~0.5	~0.3	1.2–1.8	90.8–94.7

The processing was done with the immersion of the blanks in a dielectric for 10 min before machining; it was done according to the EDM standards for high precision processing to exclude the thermal shrinkage of the materials [37]. The level of the dielectric was 1–2 mm above the blanks. The nozzle of the upper guide was installed as close as possible to the level of dielectric [38].

2- and 10 mm-width samples were cut off from the blanks. During the processing, offset of the pass on the value of the discharge gap and the radius of the wire was ignored. The EDM parameters of processing were calculated following the standard machine tool recommendations for the proposed materials. The parameter of wire tension and voltage were varied during processing in the range of ±5 machine measurement units from the recommended values of the mentioned parameters.

The developed system of diagnostic and monitoring was approbated on EDM machine CUT1000 (GF AgieCharmilles SA, Losone, Switzerland). For this purpose, 100 × 12 × 12 mm-blanks of hard alloys M05 [39,40] and P10 [41,42] by ISO (Table 2) were processed according to the prepared manually CNC-program. The standard EDM parameters for SKD-61 alloy [43] were chosen for machining. The frequencies of pulses changed during processing to create conditions for instability and provocation of a series of short circuits and wire breaks.

Table 2. The chemical composition of M05 and P10 hard alloys, %.

Material	WC	TaC	TiC	Co
M05	92	2	-	6
P10	79	-	15	6

The developed in-situ monitoring system based on vibroacoustic emission for electrical discharge machining is presented in Figure 2. The signals received from the accelerometers depend on the nature and intensity of the vibrations during processing, which depends on the parameters of electrical discharge machining. The signals were directed to an ADC and recorded. It was decided to take attention on the signal at the moment wire tool penetration into the blank, and at the end of processing: on 1st minute, 30th and 5th seconds before the end of CNC-program. Spectral analysis was performed in the range of frequencies from 1 to 32 kHz.

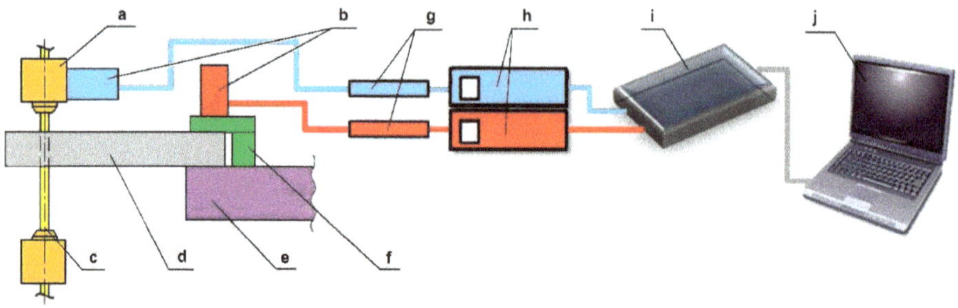

Figure 2. Principal scheme of the system of vibroacoustic monitoring of wire electrical discharge machining: (**a**) is a guide, (**b**) is accelerometers, (**c**) is a wire-tool, (**d**) is a workpiece, (**e**) is a working table, (**f**) is a fastening system, (**g**) is preamplifiers, (**h**) is amplifiers model VShV003 (OOO Izmeritel, Taganrog, Russia), (**i**) is an ADC E440 (L-card, Saint-Petersburg, Russia), (**j**) is a signal-recording device.

The pulsed solid-state laser of diode pumping U15 (RMI, Moscow, Russia) was used for laser ablation of AISI 321 samples (160 × 100 × 5 mm) for better understanding the possibilities of in-situ monitoring and adaptive control during pulsed machining processes.

The characterization of electrical discharge impulses was conducted with the use of the digital oscillograph Tektronix TDS2014B (Tektronix Inc., Beaverton, OR, USA).

The characterization of the machined surfaces and the specific wear of the tool was provided by an SEM VEGA 3 LMH (Tescan, Brno, The Czech Republic) and by an optical microscope Olympus BX51M (RYF AG, Grenchen, Switzerland). For the study, the surfaces of workpieces and a wire, and the changes of chemical composition in the sublayer were controlled before and after processing [10,33].

3. Results

3.1. Research A Fundamental Possibility of Vibroacoustic Diagnostics of Wire Electrical Discharge Machining

3.1.1. Mathematical Approach and Evaluation of Wire Amplitude Under Discharge Impulses

The scientists of MSTU Stankin under the supervision of Prof M.P. Kozochkin developed the in-situ monitoring system based on vibroacoustic emission [44,45] and adopted it for the needs of electrical discharge machining (Figures 1 and 2). The preliminary study showed that the oscillations of the wire in the slot during processing might be a reason of instability and turn into the self-oscillation process, which results in the defects of the machined surfaces [46–48].

A diagram of applied forces on the wire during electrical discharge machining presented in Figure 3.

If not taken into attention to the mass of the wire sections h_1 and h_2 and mass losses on electroerosion wear [49,50], an amplitude of oscillation may be presented as a complex amplitude of the harmonic signal [51–53]:

$$\overline{A}_n = \overline{A}_0 \cdot e^{\overline{\beta}\tau}, \tag{1}$$

where A_0 is an amplitude in the direction of the wire feed, which is always limited by a discharge gap Δ and much less than Δ, $A_n \leq A_0$ for stable EDM processing; $\overline{\beta}$ is the damping coefficient presented as a complex number in the form $(a + bi)$; τ is a period of oscillation T.

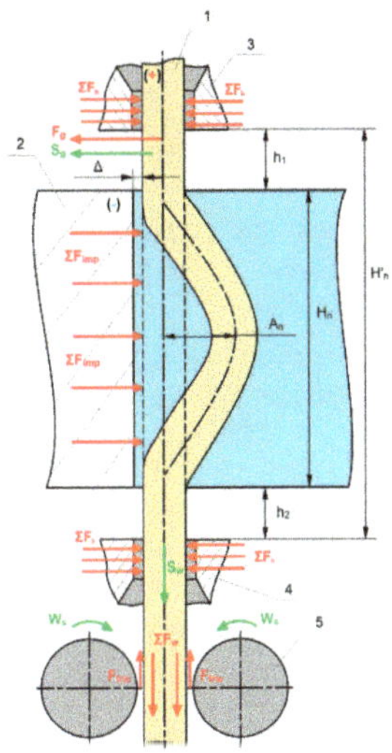

Figure 3. A diagram of applied forces during electrical discharge machining: (1) is a wire tool, (2) is a workpiece, (3) is an upper nozzle, (4) is a lower nozzle, (5) is rollers.

A_n is an amplitude of the wire in the inverse direction to the wire feed; Δ is a discharge gap; H_n is a thickness of workpiece; $H_n{}'$ is a distance between nozzles; F_g is total forces of wire guiding; ΣF_{imp} is total forces of discharge impulses; ΣF_h is total forces of wire holding; ΣF_w is total forces of wire tension; F_{fric} is a friction forces; S_g is a wire feed speed; S_w is a wire rewinding speed; W_s is a torsional moment of rewinding rollers.

$$\overline{\beta} = q - \mu, \tag{2}$$

where $\overline{q}$ is an index of excited oscillations; μ is a coefficient of dielectric medium resistance. In this case, $\overline{q}$ is

$$\overline{q} = \sqrt{\frac{\sqrt{\overline{h}} - \overline{k}}{2 \cdot m_n}}, \tag{3}$$

where $\overline{h}$ is a ratio between H_n and H_0, H_0 is a value of the workpiece thickness for stable electrical discharge machining 80 ÷ 100 mm for the diameter of wire d_w = 0.25 mm and dielectric based on deionized water; m_n is a wire mass; $\overline{k}$ is a ratio between K_n and K_0, K_n is an index of stiffness and can be presented as:

$$K_n = \Sigma F_{imp} / A_n. \tag{4}$$

Thus, $\overline{\beta}$ depends on the ratio of workpiece thickness and the ratio of the system stiffness, the EDM parameters and the viscosity of the working medium; it can vary in the limits |1|.

$\overline{q}$ is a criterion of self-oscillation:

$$\overline{q} = \frac{\lambda}{T}, \tag{5}$$

where λ is the logarithmic decrement of the damping ratio of the self-oscillatory and resonant process, describing the decrease in the amplitude of the oscillation process and equal to the natural logarithm of the ratio of two successive amplitudes of the oscillating quantity A_n to the same side:

$$\lambda = ln(A_n / A_{n+1}), \tag{6}$$

The period of self-oscillations T depends on the wire mass and the system stiffness, which depends as well on the wire tension:

$$T = 2\pi\sqrt{\frac{m_n}{K_n}} \quad (7)$$

If $q < \mu$, then self- and resonant oscillations do not arise. If $q > \mu$, then the self-oscillations are formed more intense and higher as q excess over μ. μ is higher for the oil medium and lower for the water-based dielectric. Thus, electrical discharge machining in the oil medium gives a priori higher accuracy of the machined surface.

Simplified calculations based on the measurement of the frequency of forced impulses in the working area gives a possibility to evaluate the wire amplitude during the processing with varied wire tension (Table 3).

Table 3. Results of simplified mathematics for evaluation of wire amplitude under discharge impulses.

Parameter	Value
Wire Mass	
A wire radius r_w, m	0.00125
A height of the workpiece H_n, m	0.016
A distance between the nozzles H_n', m	0.020
Brass density ζ [1], kg/m^3	8580
A volume of the wire Q_w, m^3	9.8×10^{-8}
Wire mass m_n, kg	8.4×10^{-4}
Force of Medium Resistance	
Water density ζ [1], kg/m^3	997
Resistance area P_n $(=\pi \cdot r_w \cdot H_n')$ [1], m^2	7.8×10^{-5}
Wire feed speed Sg, m/s	1.7×10^{-5}
A force of medium resistance F_σ [2], N	2.2×10^{-11}
Force of Impulses	
A force of impulses F_{imp} [3], N [54,55]	4.8×10^{-3}
An Amplitude of Wire under Forced Impulses	
A frequency of forced impulses f_{frc} [4], Hz	0.2×10^6
System stiffness K_w, N/m	1.1×10^6
An amplitude of wire under forced impulses A_{frc}, m	4.36×10^{-9}

[1] Given for reference; [2] Further ignored due to its small value; [3] Taken as an approximate value for evaluation of the amplitude for the first pass of electrical discharge machining of the same height workpiece; [4] Measured in absolute values during experiments.

3.1.2. Experiments on the Fundamental Possibility of Vibroacoustic Diagnostics

During the first stage of the experiments, it was established that miserable changes in the weight of the part influence on the vibroacoustic signal. As an example, a 24.5 g-weight of 10 mm-length AISI 321 sample shows a significant increase of the vibroacoustic signal amplitude at 5 s before the end of processing. For a 4.28 g-weight of 10 mm-length AISI 2024 sample, it was recorded at 2 s before the end of processing. It was noticed in the full frequency range of the signal. As well, it was characterized by instability in the low frequencies of the signal. Thus, it was chosen to use the frequency range higher than 4 kHz for vibroacoustic monitoring and analyzing.

Figure 4 presents a high-frequency spectrum of vibroacoustic signals recorded at 60 and 5 s before the part separation. The octave spectra are shown in the inset of the graph. The effective amplitude in the octave-frequency band is more stable than high-frequency spectra of the vibroacoustic signals due to averaging. The effective amplitude with approaching of the end of processing (the moment of separation of the part) increases approximately in 2 times for the octave-frequency band 4 kHz and in 1.5 times for the octave-frequency band 8 kHz. So as it can be seen, the occurred moment can be evaluated timely and distantly. The changes can be presented by visual demonstration, which may be understandable for any EDM operator.

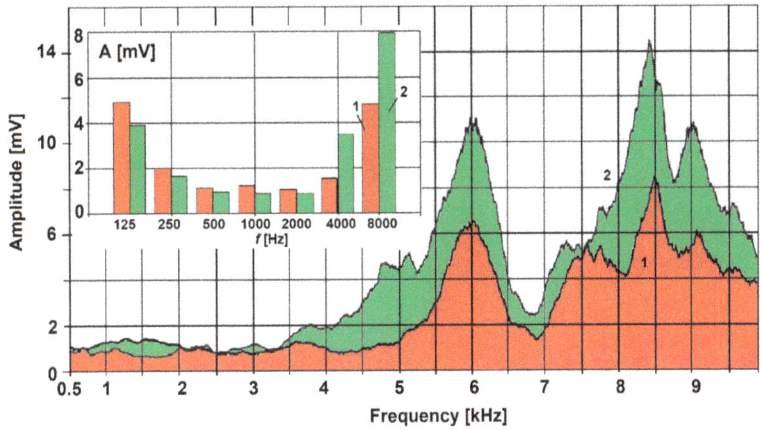

Figure 4. High-frequency spectra of a vibroacoustic signal for electrical discharge machining of AISI 321: (1) is for electrical discharge machining at 60 s before the end of the operation; (2) is for electrical discharge machining at 5 s before the end; the inset shows the octave spectra.

Figure 5 presents two spectra of vibroacoustic signals, which were obtained during the electrical discharge machining of the AISI 2024 sample. Spectrum (1) shows the moment of wire penetration, and spectrum (2) shows the moment when the discharge gap is already formed. The changes in root-mean-square value (RMS) of the vibroacoustic signal at the duration of the moment of wire penetration into the workpiece are shown on the inset of the graph.

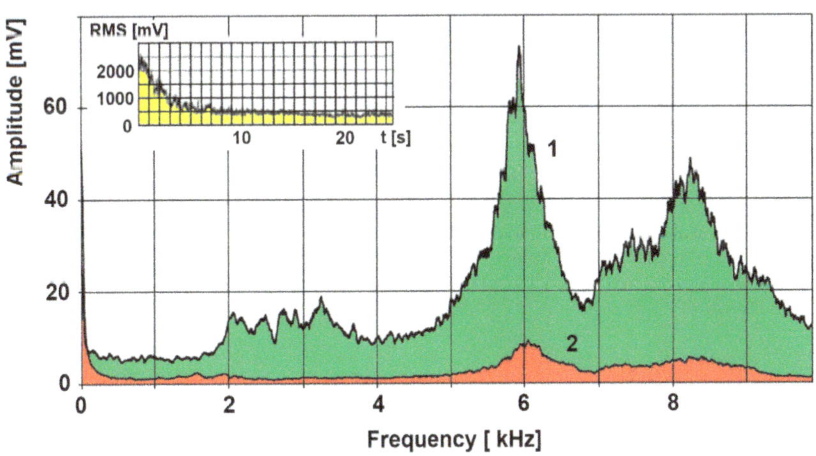

Figure 5. High-frequency spectra of vibroacoustic signals for AISI 2024: (1) is at the moment of wire penetration; (2) at the moment of the discharge gap formation; RMS-t diagram inset presents the spectra of the signal at the moment of wire penetration.

The experimental research of the influence of the wire tool tension F_w on vibroacoustic signal showed that decrease of the hardness of elements of the technological system [1,56,57] influences negatively on the growth of the vibration. The growth of vibration was detected with the decline tension force. The detected vibrations were also associated with a decrease of flushing ability of the erosion products from the working zone, which influent negatively on the quality of the machined surfaces.

3.2. Adaption of the In-Situ Monitoring Method for the Needs of Electrical Discharge Machining

The optimization of any machining technology means to ensure its maximum productivity, efficiency, and quality of the machined surfaces up to the required level. It demands the in-time regulation of EDM parameters as the discharge gap Δ, the concentration of erosion product in the gap, the temperature of the working fluid, and its flushing rate. The gap Δ is the primary parameter, which is responsible for determining the quality of the final product [58–60] A small increase in Δ may change the conditions in the working area and interrupt the discharge. A decrease in Δ impairs the yield of erosion products, reduces the productivity, increase accumulation of slag, and provoke short circuits. Electrical discharge machining cannot be effective without automatic regulation of the gap [61,62]. For regulation of the optimal value of the gap, the rate of particle formation M_p in the discharge gap should be equal to the rate of particles leaving the gap M_{ex}. The rate M_p is a function of the concentration of the particles γ:

$$M_p = f(\gamma). \tag{8}$$

When M_p and M_{ex} are unequal, the change in concentration of particles is

$$\Delta\gamma = \Delta M \cdot \Delta t / Q, \tag{9}$$

where Q is the volume of the discharge gap, and

$$\Delta M = M_p - M_{ex}. \tag{10}$$

For stable electrical discharge machining $\Delta\gamma = 0$.

However, it is complicated to ensure constant γ due to the many factors acting in the working area. Any fluctuations should be timely eliminated by control signals, which change the parameters of electrical discharge machining. The analysis indicates that the maximum rate M_p decrease as M_{ex} decreases. It occurs due to the deterioration of the erosion products' evacuation as wire penetrates in the workpiece, and the number of working pulses reduces [58].

Figure 6 presents the dependences of the machining rate M_p, the number of pulses n on the discharge gap Δ, where n_w is for working pulses, n_{id} is for idling pulses, n_{sc} is for short-circuit pulses. Analysis indicates that the gap corresponding to the maximum rate M_{max} is higher than the gap corresponding to the maximum rate of working pulses as an excess of erosion products at the maximum rate of working pulses and short-circuiting pulses destabilize processing [61].

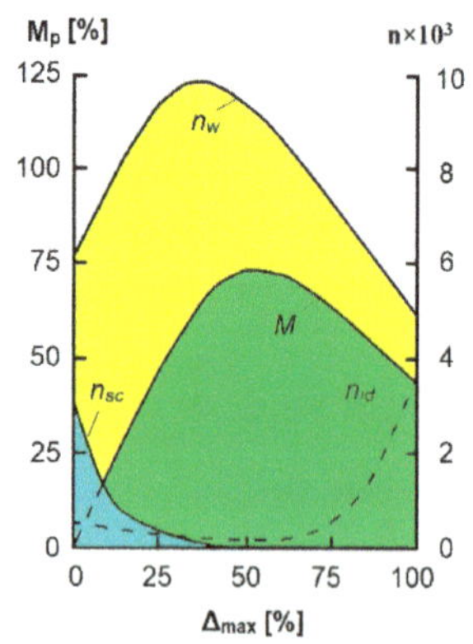

Figure 6. Dependence of the rate M_p and the number of pulses n on the discharge gap Δ.

Thus, the in-situ monitoring systems for electrical discharge machining aim to maintain the efficiency of pulse utilization, which is the ratio of the number of working pulses (n_w) to the total number n of pulses:

$$\psi = \frac{n_w}{n} = 0.7 \div 0.9. \tag{11}$$

The rate ψ can be informative for in-situ monitoring and adaptive control, but its use is complicated by the inertia of the required measurement instruments.

Hence, there are difficulties in assessing the efficiency of electrical discharge machining concerning ψ [33,44]. The energy of the individual pulses is not entirely consumed in the processing of material sublimation when the dielectric medium is contaminated with erosion product. In this case, a part of the energy is consumed in the destruction of the erosion products. Since the number of working pulses is assessed from the total number of the discharge impulses, it results in an imprecise assessment of the electrical discharge machining efficiency. The precise estimate can be obtained by relating the efficiency to the ratio of the useful energy consumption for sublimation of the material and the total incoming energy of the working impulses in the discharge gap. Normally, the energy of the discharge impulses is proportional to the effective discharge current (I_e) of independent generators in operation. The use of a current sensor can estimate this factor.

3.3. Analysis of the Obtained Data, Search for the Criteria for the Development of the Adaptive Control System Based on Vibroacoustic Emission

The results of vibroacoustic monitoring of another non-contact precise machining were considered for better-understanding impulse character of electrical discharge machining and possibility to improve its productivity. Laser ablation is close by its nature to the processes of material removal during electrical discharge machining as in both of the cases there are no mechanical contacts, the material removal occurs under thermal influences, initiated by pulses of concentrated energy fluxes. The material removal during electrical discharge machining is related to the complex processes of chaotic material sublimation under discharge impulses [63] as after laser ablation the surface presents the organized wells [64]. The nature of the initial processes is different.

As it was previously determined [61], the vibroacoustic signals coming in the time of processing demonstrate steady increase as the laser power increases and the volume of removed metal increases (Figure 7)

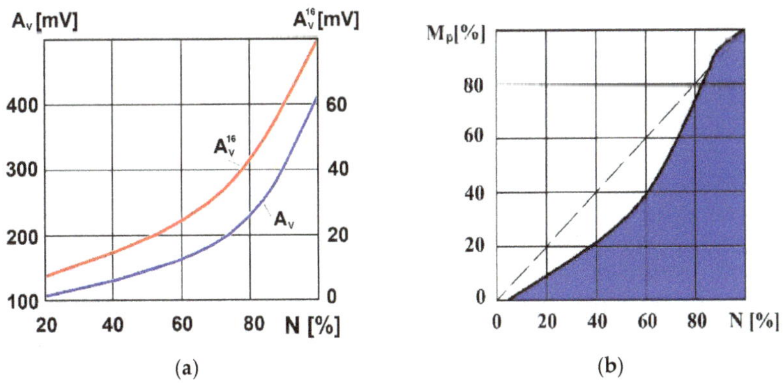

Figure 7. Dependencies of the effective amplitude of the vibroacoustic signal (**a**) and productivity (**b**) from laser power N. A_v is a vibroacoustic signal amplitude in wide frequency range; $A_v{}^{16}$ is an amplitudes of the signal of 16 kHz octave.

The presented in Figure 7b dependence has a monotonous character. It does not demonstrate any significant material removal at low values of laser power, that is related to lack of thermal energy for initiation of material ablation. It can be approximated with linear dependence or exponential function with specified accuracy for the industrial needs.

Figure 7a shows the correlation between vibroacoustic signal amplitude A_{va} and productivity of the pulses M_p after data processing. It can be approximated by an empirical function:

$$A_{va} = \zeta \cdot M_p{}^{\Lambda}, \tag{12}$$

where $\Lambda = 0.72 \div 0.76$, $\zeta = const$.

It is highly possible that the vibroacoustic signal of electrical discharge machining has the same dependences on processing parameters and instead of (8) is easier to use:

$$A_{va} = f(\Delta, \gamma). \tag{13}$$

There is a proved possibility of realization of the in-situ monitoring and adaptive control measures based on vibroacoustic emission for the needs of electrical discharge machining. The parameter of vibroacoustic signals can be observed without any difficulties in the distinction of current pulses productivity.

The dynamic system of electrical discharge machining is linear with dynamic characteristics depending on the state of the dielectric medium. The amplitude spectrum of $U(f)$ signal received by the accelerometer is defined according to the dynamic system linear properties [44,61]

$$U(f) = B_1(f) \cdot B_2(f) \cdot \Theta(f), \tag{14}$$

where $B_1(f)$ is the frequency response function (FRF) of the operational environment with changes of accumulation of the erosion products, the distance between electrodes, the temperature in the discharge gap (the temperature in the discharge channel is about 7000 K, the temperature of dielectric medium is about 293 K); $B_2(f)$ is FRF of a dynamic influence of the discharge impulses on the elastic system; $\Theta(f)$ is an amplitude spectrum, f is the frequency.

The energy of vibroacoustic signals changes in the dependence on the concentrated energy flux intensity on the machining surface. A part of energy flow is spent on repeated sublimation of erosion products in the discharge gap. As a result, it crates chaotic and overlapped wells on machined, it chanes as well the conditions in the discharge gap. The formed overlaps cause localization of the following discharge. Thus, the impulses created by phase transfer of the processed metal are longer in time.

Therefore, the concentration of erosion products' γ increase as short impulses decrease and long impulses increase. The transformations result in changes of dynamic characteristic $B_1(f)$ including changes in FRF of $B(f)$:

$$B(f) = B_1(f) \cdot B_2(f) \tag{15}$$

The current values $B(f)$ during electrical discharge machining can be evaluated by controlling discharging current and vibroacoustic signal. The experiment showed that $B_2(f)$ might be presented as constant in the short period of observation. Thus, the changes in $B(f)$ are related to the changes in dielectric medium conditions $B_1(f)$.

3.4. Demonstration of the Opportunity for The In-situ Adaptive Control of Edm by Specific Examples

The developed in-situ monitoring technique was approbated on the high-precision wire electrical discharge machine CUT1000 (GF AgieCharmilles SA, Losone, Switzerland) during processing of the samples made of hard alloys, which can be actual for the modern machine building industry: M05 and P10, ISO. The vibroacoustic signals and discharge current were recorded during experiments. FRF of $B(f)$ in the observation channel was fixed at the moment of wire approach to a workpiece and the moment wire break. Figure 8 presents an FRF of $B(f)$ in the channel of observation. The vibroacoustic signals were recorded in [mV], FRF values were non-dimensional.

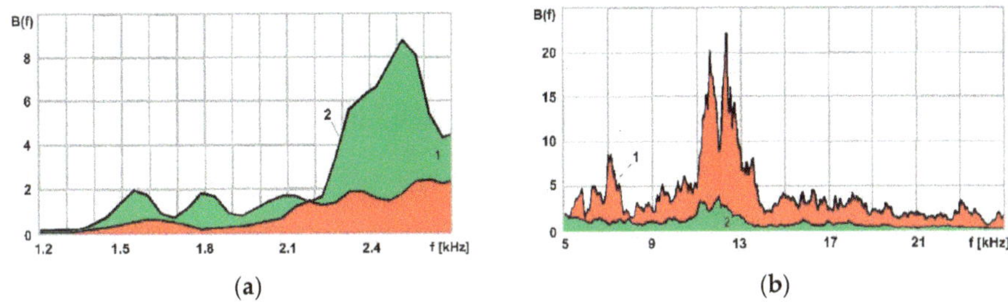

Figure 8. The record of FRF during electrical discharge machining of M05 hard alloy: (**a**) is in the low-frequency range; (**b**) is in the high-frequency range; (1) is at the moment of wire approach; (2) is before the wire break.

During the period of the initial stage of wire approach (1), the dielectric is clean of erosion products, the wire approaches to the flat surface of the workpiece, the erosion products easily flush from the discharge gap. The highest value of amplitude fixed at the frequency range 11 ÷ 13 kHz.

The wire break occurs at 12 s of electrical discharge machining (2). There were observed a decrease in high-frequency components and an increase in low-frequency components at the frequency range 2.4 ÷ 2.6 kHz. The flushing of erosion products was straitened before the wire break. The short impulses decreased, and long impulses increased for forming of the groove.

It was noticed that FRF changes in the full frequency range are not convenient due to possible disturbance from the work of the equipment. One or two frequency ranges where FRF changes can be easily monitored should be chosen for the development of the adaptive control method.

Figure 9 presents an example of RMS vibroacoustic signal change in various octave bands: from the wire approach to breaking.

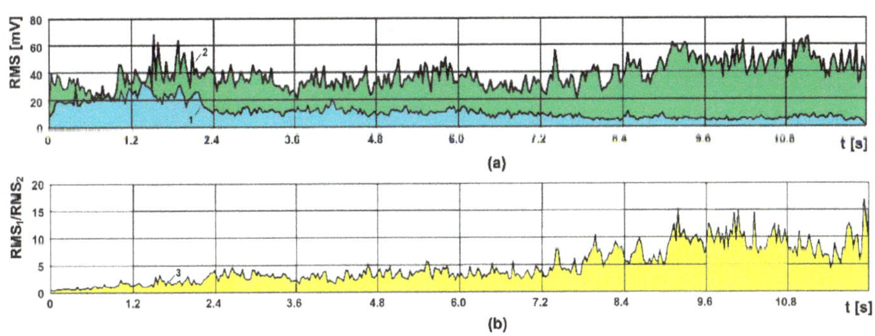

Figure 9. RMS of vibroacoustic signals during 12 s of electrical discharge machining of M05 hard alloy: (**a**) RMS of vibroacoustic signal; (**b**) proportional ratio of RMS1/RMS2, where (1) is an octave band of 2 kHz; (2) is an octave band of 32 kHz; (3) is the result of interconnection.

The power of discharging current has not any significant changes during changes of RMS of the vibroacoustic signal; hence, the components of the vibroacoustic signal demonstrates the tendency of changes in the conditions of the discharge gap. A gradual increase in RMS of low-frequency vibroacoustic signal (1) and a significant decrease in the RMS of the high-frequency signal (2) was recorded during machining. Graph 3 shows the change in the ratio of RMS of a low-frequency signal to RMS of a high-frequency signal. The ratio gives a more informative picture of a decrease in the conditions of electrical discharge machining. The stable phase of processing was observed until 7 s and then an increase in 10 times of RMS of a vibroacoustic signal from 3 ÷ 4 s was observed; it achieves briefly 15-fold value.

As the control system of the machine reacted only to indirect electrical parameters, in particular, to a coefficient of impulse use, the critical increase of the concentration of the erosion products was left without proper attention. It resulted in overheating of electrodes and consequent wire break.

The exceptional use of data in the electric format of impulses control system results in consideration of the impulses directed to the sublimation of the erosion products in the discharge gap as effective. At the same time, the vibroacoustic signals from such impulses cannot be created and detected at all, or they have a narrow range of frequency spectra.

Figure 10 shows the examples of vibroacoustic signal impulses of various shapes at various spark gaps. It can be seen the discharges current impulses in the form of vibroacoustic signals of less than 0.1 ms in length (a). The length of separate vibroacoustic signals' fragments at the high concentration of erosion products increases up to 5–6 times (b). The long fragments (impulses) provide an increase of low-frequency elements of the spectra of a vibroacoustic signal.

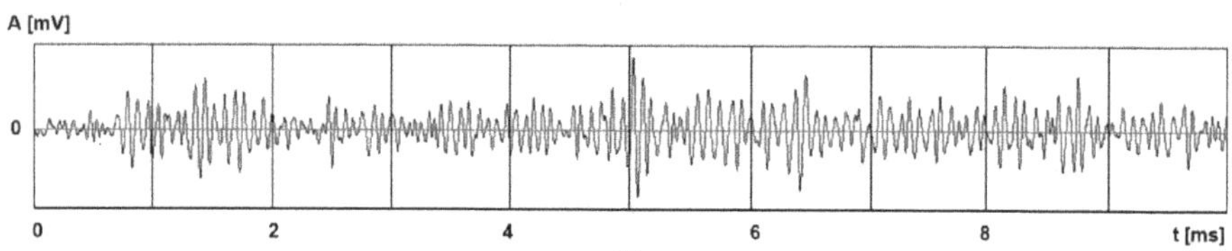

(a)

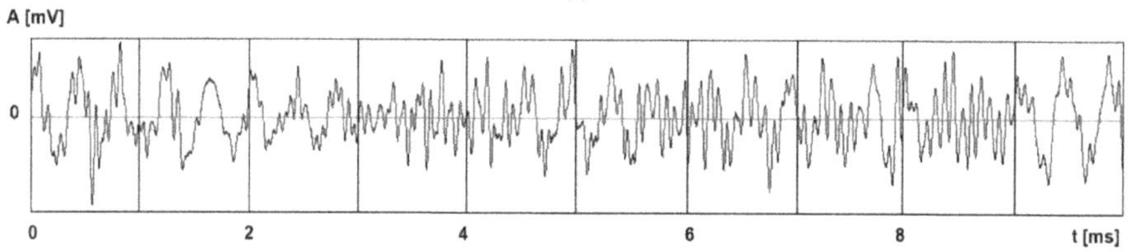

(b)

Figure 10. Vibroacoustic signals at 10 ms period during electrical discharge machining of M05 hard alloy: (**a**) is at the moment of wire approach; (**b**) is before wire break.

4. Discussion

In the begging, it was decided to focus attention at the moment of separation of the part from the workpiece, which gives changes in the specific noise accompanied electrical discharge machining. By this vibroacoustic emission, it is possible to recognize a specific noise related to each stage of plastic deformation and brittle fracture of the part separation.

It was shown that an amplitude of the vibroacoustic signal increases gradually with the approach of the moment of the separation (Figure 4). Then the series of peaks of the diagram related to the possible wire stuck in the discharge gap and consequent short-circuiting. It proves the final bridge is weakening due to rapid mass loss under electroerosion pulses. Then the part starts rotated motion relatively lower contact point between a part and a workpiece when the upper part of the bridge is separated. The wire tool stays clasped between a separated part and the rest of the workpiece. The series of the short circuits occur, and the vertical traces of the wire can be observed on the samples of the material with the electrical resistance more than 7.9×10^{-8} Ω·m.

During the experiments, it was noticed that the intensity of vibroacoustic signals increases as the symmetry of the wire tool position in the gap is disturbed. It is related to the prevalence of the discharge current pulses acting on one of the sides of the wire tool. This kind of phenomena was also detected at the moment of wire tool penetration into a workpiece when there is no stable discharge gap between the electrodes. At the moment when the discharge gap is established, and a wire tool penetrated the workpiece and formed a discharge groove, the acting pulses balance each other and the amplitude of the vibroacoustic signal decreases.

It can be detected from the obtained vibroacoustic diagram (Figure 5) of the spectra that at the 5 ÷ 7 s the RMS value of the spectra increases four times in frequency range of 8 kHz. It is noticeable even for alloys of transition metals with the relatively low-melting-point (more than 231.9 °C and less than 950 °C) and low electrical resistance (2.8×10^{-8} Ω·m). The described changes of the vibroacoustic signal are thoroughly enough for developing in-situ monitoring and adaptive control system for taking measures timely to prevent short-circuiting and consequent defects of the machined surfaces.

The absolute excess of the vibroacoustic amplitude during conventional machining after penetration of the tool in the workpiece can be explained by an excess of the erosion products (solidified particles of the eroded material) in the discharge gap. The excess of the erosion products usually related to the insufficient flushing in the working zone due to the relatively high height of the workpiece (more than 70 ÷ 100 mm for the wire tool 0.25 mm) or relatively low pressure of dielectric in the nozzles (code "1" instead of "2" for the first tool pass, rough cutting). It may also be related to the relatively small discharge gap (less than 0.005 mm) for the materials with the high electrical resistance (more than 1.12×10^{-6} Ω·m) due to the difficulties in controlling it by the servo drives of the machine tool.

Thus, it may be possible that:

(1) For the lower value of wire tension, the adequate flushing of the erosion products is hampered by bending of a wire tool during processing and higher wire amplitude due to the low circular frequency of the vibrations, when the circular frequency of the forced oscillations under electrical impulses is above the circular frequency of the self-oscillation;
(2) For the higher value of wire tension, the flushing of the erosion products is adequate, but the stiffness of the system is higher, then wire amplitude is lower, at the same time, the circular frequency of the vibrations grows, when the circular frequency of the forced oscillations under electrical impulses is below the circular frequency of the self-oscillation;
(3) For the value of wire tension associated with stable EDM processing, the circular frequency of forced oscillation may be compensated by the frequency of self-oscillation.

Therefore, it was shown that the non-contact electrical discharge machining generates the vibroacoustic signal in the full frequency range that is quite similar by its character to the character of vibroacoustic emission during the convenient mechanical machining. It was proved that the signal might be registered by accelerometers placed on the elastic system of the machine tool wirelessly and at a distance from a working zone to exclude the influence on accuracy positioning of drivers and the parameters of machining. It was determined that the vibroacoustic signal is a result of wire tool disturbance in the discharge gap by discharge current pulses, cavitation processes and intensity of flushing in the working medium, wire tool contacts with the workpiece or erosion products.

The positions of the accelerometers were chosen based on the idea that the accelerometers should be placed as close as possible to the working zone. However, the installation of the lower accelerometer on the lower guide was inconvenient as it is moving under a workpiece and the wires of the accelerometers can hamper the processing. Therefore, it was more suitable to place it closer on the fastening system of the workpiece for monitoring of workpiece vibration. The upper accelerometer was placed on the upper guide for monitoring of wire vibration, and it was most informative and convenient for development of the adaptive control system.

It was found during the study, the upper accelerometer records the vibrations which are associated with the total forces of the working discharge impulses, which are directed in the destruction of the workpiece ($\Sigma F_{imp.w}$), but not in the destruction of the erosion products in the discharge gap ($\Sigma F_{imp.idle}$). In this case, the total forces of initiated impulses, which can be detected by monitoring only electrical parameters, is:

$$\sum F_{imp} = \sum F_{imp.w} + \sum F_{imp.idle}. \tag{16}$$

It should be noted that not all working impulses carry out the same and useful work. Some of them spent a part of the energy or even all its energy on the destruction of the erosion products [65–67].

The experiments showed that the performance of the EDM (volume in mm^3) is related to the power of the vibroacoustic signal by a linear dependence (or close to linear) (Figure 7). Besides, it follows that the control of the share of working pulses at high frequencies is associated with great difficulties in the field of circuit engineering and with large errors as it is known from published sources [55,58,59]. It is the advantage of the vibroacoustic signal over the method of controlling the pulse utilization rate (ψ).

In other words, during discharge gap breakdown all initiated pulses are considered to be working pulses by measuring electrical parameters, which is not correct because a part of the pulses is spent on the destruction of the erosion products in the discharge gap. It is suitable for the cases when there is a low part of the erosion products in the discharge gap. In the case of vibroacoustic monitoring, only the pulses directed on the destruction of another electrode are taken into account as working impulses.

Moreover, the measurement of the vibroacoustic signal in [mV] is suitable as the amplitude of the vibroacoustic signal is proportional to the signal in [mV] at the output of the measuring channel. In principle, the measurements in [mV] can be converted to the measurement of vibration acceleration [m/s^2], but it is unnecessary work called "calibration." For the adaptive control system development, it is enough to have measurement units, which are proportional to amplitude.

The widespread use of vibroacoustic spectra for diagnostics and monitoring is complicated as it is a closed system with a nonlinear dependence of the vibroacoustic signal on the impulse load [68,69]. The dynamic system resembles better a linear model, where the dynamic relation between the load source and the workpiece is permanent.

In other words, the model of the dynamic system is significantly simplified, and its use in the in-situ monitoring and regulation of machining by high-energy fluxes is simplified [70,71].

5. Conclusions

The application of the developed in-situ monitoring technique using vibroacoustic spectra of the signal in the working zone can provide a current solution for the specific issues related to the insufficient productivity of electrical discharge machining and low quality of the machined surfaces due to technological issues of processing (e.g., wire breakage).

The developed method is based on the effect of the vibroacoustic emission during processing. Physical phenomena of vibroacoustic emission explained the fundamental possibility of this method. It was shown the interconnections between the electrical parameters of machining and vibroacoustic spectra, and the possibility to timely analyze the received vibroacoustic spectra data for development of the system of the adaptive control.

For further development, it is necessary to conduct more intensive research with the high electrical resistant materials to show the principle possibilities to adapt the electrical parameters of the processing for the needs of processing of new classes materials as nanocomposites based on ceramics and made with the use of carbon tubes and whiskers. It was shown that the developed method is suitable for operative positioning of wire tool-electrode or evaluation of wire tool-electrode bending during processing. That can be extremely important for the production of the parts with a complex linear configuration for the needs of the aircraft industry.

The study of processing of workpieces with high energy impulses shows that their effectiveness and parameters of vibroacoustic signals depend on the power of impulses supplied with monotonous analogic dependencies. It allows monitoring of the current effectiveness of flushing in the working zone and optimizes the value of the discharge gap.

The unstable electrical discharge processing is also associated with multiple contacts of electrodes and short-circuiting during processing that cannot be identified timely by modern CNC-system, but this problem can be solved with the adequate data received by monitoring system based on vibroacoustic emission. Analysis of vibroacoustic signals can provide a modern solution for resolving the manufacturing issues, which is not covered by existing methods of electrical parameters control and helps in forming a multi-parameter and multifunctional diagnostic system.

6. Patents

1. Kozochkin, M.P.; Grigoriev, S.N.; Porvatov, A.N., Okunkova, A.A. The method of controlling the electrical discharge machining of parts on an automated cutting machine with a system of CNC; RU 2598022
2. Kozochkin, M.P.; Khoteenkov, K.E.; Porvatov, A.N., Grigoriev, S.N. The method of EDM cutting of products; RU 2638607
3. Grigoriev, S.N.; Kozochkin, M.P.; Okunkova, A.A. The method of positioning the wire electrode on the EDM cutting machines; RU 2572678

Author Contributions: Conceptualization: A.A.O. and M.P.K.; methodology: M.P.K. and A.N.P.; software, A.N.P.; validation: M.P.K., A.N.P., and Y.A.M.; formal analysis: M.P.K.; investigation: A.N.P.; resources: Y.A.M.; data curation: A.A.O.; writing—original draft preparation: M.P.K. and A.N.P.; writing—review and editing: M.P.K. and A.A.O.; visualization: A.A.O. and A.N.P.; supervision: Y.A.M.; project administration: Y.A.M. and M.P.K.; funding acquisition: A.A.O. and Y.A.M.

Funding: This research was funded by the Ministry of Education and Science of Russian Federation, grant number No. 9.7453.2017/6.7.

Acknowledgments: The research was done at the Laboratories of the Department of High-efficiency Machining Technologies of MSTU Stankin.

Conflicts of Interest: The authors declare no conflict of interest.

References

1. Conde, A.; Arriandiaga, A.; Sanchez, J.A.; Portillo, E.; Plaza, S.; Cabanes, I. High-accuracy wire electrical discharge machining using artificial neural networks and optimization techniques. *Robot. Cim-Int. Manuf.* **2018**, *49*, 24–38. [CrossRef]
2. Maradia, U.; Filisetti, E.; Boccadoro, M.; Rotenb, M.; Dutoit, J.-M.; Hengsberger, S. Increasing the Injection Moulding Productivity through EDM Surface Modulation. *Proc. CIRP* **2018**, *68*, 58–63. [CrossRef]
3. Sahu, A.K.; Chatterjee, S.; Nayak, P.K.; Mahapatra, S.S. Study on effect of tool electrodes on surface finish during electrical discharge machining of Nitinol. *IOP Conf. Ser. Mater. Sci. Eng.* **2018**, *338*, 012033. [CrossRef]
4. Trung, K.H. Development of a WEDM system with high machining efficiency. In Proceedings of the 2017 International Conference on System Science and Engineering (ICSSE), Ho Chi Minh City, Vietnam, 21–23 July 2017; pp. 397–399. [CrossRef]
5. Ho, K.H.; Newman, S.T.; Rahimifard, S.; Allen, R.D. State of the Art in Wire Electrical Discharge Machining (WEDM). *Int. J. Mach. Tool Manuf.* **2004**, *44*, 1247–1259. [CrossRef]
6. Han, F.; Chen, L.; Yu, D., Zhou, X. Basic study on pulse generator for micro EDM. *Int. J. Adv. Manuf. Technol.* **2007**, *33*, 474–479. [CrossRef]
7. Faisal, N.; Kumar, K. Optimization of Machine Process Parameters in EDM for EN 31 Using Evolutionary Optimization Techniques. *Technologies* **2018**, *6*, 54. [CrossRef]
8. Gavrin, V.N.; Kozlova, Y.P.; Veretenkin, E.P.; Logachev, A.V.; Logacheva, A.I.; Lednev, I.S.; Okunkova, A.A. Reactor target from metal chromium for "pure" high-intensive artificial neutrino source. *Phys. Part. Nucl. Lett.* **2016**, *13*, 267–273. [CrossRef]
9. Prathipati, R.P.; Devuri, V.; Cheepu, M.; Gudimetla, K.; Kiran, R.U. Machining of AISI D2 Tool Steel with Multiple Hole Electrodes by EDM Process. *IOP Conf. Ser. Mater. Sci. Eng.* **2018**, *330*, 012067. [CrossRef]
10. Volosova, M.A.; Okunkova, A.A.; Povolotskiy, D.E.; Podrabinnik, P.A. Study of electrical discharge machining for the parts of nuclear industry usage. *Mech. Ind.* **2015**, *16*, 706. [CrossRef]
11. Ozerkan, H.B. Simultaneous machining and surface alloying of AISI 1040 steel by electrical discharge machining with boron oxide powders. *J. Mech. Sci. Technol.* **2018**, *32*, 4357–4364. [CrossRef]
12. Bhattacharya, S.; Abraham, G.J.; Mishra, A.; Kain, V.; Dey, G.K. Corrosion Behavior of Wire Electrical Discharge Machined Surfaces of P91 Steel. *J. Mater. Eng. Perform.* **2018**, *27*, 4561–4570. [CrossRef]
13. Nguyen, H.P.; Pham, V.D.; Ngo, N.V. Application of TOPSIS to Taguchi method for multi-characteristic optimization of electrical discharge machining with titanium powder mixed into dielectric fluid. *Int. J. Adv. Manuf. Technol.* **2018**, *98*, 1179. [CrossRef]

14. Uhlmann, E.; Oberschmidt, D.; Bolz, R. Application of Micro Structured, Boron Doped CVD-diamond as μEDM Tool Electrodes. *Proc. CIRP* **2018**, *68*, 649–653. [CrossRef]
15. Jayalakshmi, S.; Singh, R.A.; Gupta, M. Metallic Glasses as Potential Reinforcements in Al and Mg Matrices: A Review. *Technologies* **2018**, *6*, 40. [CrossRef]
16. Veronesi, P.; Leonelli, C.; Bondioli, F. Energy Efficiency in the Microwave-Assisted Solid-State Synthesis of Cobalt Aluminate Pigment. *Technologies* **2017**, *5*, 42. [CrossRef]
17. Fleischer, J.; Teti, R.; Lanza, G.; Mativenga, P.; Möhring, H.C.; Caggiano, A. Composite materials parts manufacturing. *CIRP Ann.* **2018**, *67*, 603–626. [CrossRef]
18. Nakamura, M.; Takeo, K.; Osada, T.; Ozaki, S. Finite Element Analysis of Self-Healing and Damage Processes in Alumina/SiC Composite Ceramics. *Technologies* **2017**, *5*, 40. [CrossRef]
19. Aimable, A.; Doubi, H.G.; Stuer, M.; Stuer, M.; Zhao, Z.; Bowen, P. Synthesis and Sintering of ZnO Nanopowders. *Technologies* **2017**, *5*, 28. [CrossRef]
20. Song, C.; Cheng, L.; Liu, Y.; Zhao, M.; Ye, F. Microstructure and electromagnetic wave absorption properties of RGO-SiBCN composites via PDC technology. *Ceram. Int.* **2018**, *44*, 18759–18769. [CrossRef]
21. Zhang, C.; Shi, Z.; Mao, F.; Yang, C.; Yang, J.; Zhu, X.; Zuo, H. Polymer composites with balanced dielectric constant and loss via constructing trilayer architecture. *J. Mater. Sci.* **2018**, *53*, 13230. [CrossRef]
22. Yan, Z.; Sun, H.; Chen, X.; Fu, X.; Chen, C.; Cheng, F.; Chen, J. Rapid low-temperature synthesis of perovskite/carbon nanocomposites as superior electrocatalysts for oxygen reduction in Zn-air batteries. *Nano Res.* **2018**, *11*, 3282. [CrossRef]
23. Mishra, M.K.; Moharana, S.; Mahaling, R.N. Fabrication of SiC-Treated P(VDF-HFP)-(BFO-SO3H) Composite Films for High Performance Energy Storage Device (HPESD) Applications. *Phys. Status Solidi A* **2018**, *215*, 1701057. [CrossRef]
24. Jang, J.-S.; Lee, G.Z.; Kim, H.; Hong, S.Y.; Ci, L.; Nam, J.D.; Suhr, J. High-damping and conducting epoxy nanocomposite using both zinc oxide particles and carbon nanofibers. *J. Mater.* **2018**, *4*, 187–193. [CrossRef]
25. Sato, K.; Tominaga, Y.; Hotta, Y.; Shibuya, H.; Sugie, M.; Saruyama, T. Cellulose nanofiber/nanodiamond composite films: Thermal conductivity enhancement achieved by a tuned nanostructure. *Adv. Powder Technol.* **2018**, *29*, 972–976. [CrossRef]
26. Grigoriev, S.N.; Volosova, M.A.; Gurin, V.D.; Seleznev, A.E. Wear of replaceable indexable inserts made of mixed cutting ceramics CC650 as a function of force parameters of steel ShKh15 face milling. *J. Frict. Wear* **2015**, *36*, 521–527. [CrossRef]
27. Grigoriev, S.N.; Gurin, V.D.; Volosova, M.A.; Cherkasova, N.Y. Development of residual cutting tool life prediction algorithm by processing on CNC machine tool. *Materialwiss. Werkst.* **2013**, *44*, 790–796. [CrossRef]
28. Grigoriev, S.N.; Sinopalnikov, V.A.; Tereshin, M.V.; Gurin, V.D. Control of parameters of the cutting process on the basis of diagnostics of the machine tool and workpiece. *Meas. Tech.* **2012**, *55*, 555–558. [CrossRef]
29. Volosova, M.; Grigoriev, S.; Metel, A.; Shein, A. The Role of Thin-Film Vacuum-Plasma Coatings and Their Influence on the Efficiency of Ceramic Cutting Inserts. *Coatings* **2018**, *8*, 287. [CrossRef]
30. Grigoriev, S.N.; Volosova, M.A.; Gurin, V.D.; Seleznyov, A.Y. Investigation of force parameters acting on a single cutting insert made of ceramics in face milling of hardened steel. *Mech. Ind.* **2015**, *16*, 702. [CrossRef]
31. He, Y.; Chen, X.; Liu, Z.; Qin, Y. Piezoelectric self-sensing actuator for active vibration control of motorized spindle based on adaptive signal separation. *Smart Mater. Struct.* **2018**, *27*, 065011. [CrossRef]
32. Schneider, F.; Mummert, K. Chemical aspects of stress corrosion cracking of AISI 321 stainless steel in hot aqueous chloride solutions. *ChemInform* **1997**, *28*. [CrossRef]
33. Grigoriev, S.N.; Kozochkin, M.P.; Kropotkina, E.Y.; Okunkova, A.A. Study of wire tool-electrode behavior during electrical discharge machining by vibroacoustic monitoring. *Mech. Ind.* **2016**, *17*, 717. [CrossRef]
34. Kustas, F.; Molian, P.; Kumar, A.S.; Besser, M.; Sordelet, D. Laser crystallization of amorphous sputter-deposited quasicrystalline coatings. *Surf. Coat. Technol.* **2004**, *188*, 274–280. [CrossRef]
35. Gautam, A.; Ajit, K.P.; Sarkar, P.K. A nonlinear CDM based damage growth law for ductile materials. *Mater. Res. Express* **2018**, *5*, 026518. [CrossRef]
36. Weidner, A.; Lippmann, T.; Biermann, H. Crack initiation in the very high cycle fatigue regime of nitrided 42CrMo4 steel. *J. Mater. Res.* **2017**, *32*, 4305–4316. [CrossRef]
37. Takada, N.; Hayashi, Y.; Goto, M.; Sasaki, K. An Attempt to Produce Electrical Discharges in Acoustic Cavitation Bubbles. *Plasma Fusion Res.* **2016**, *11*, 1406113. [CrossRef]

38. Yu, X.; Qiu, M.; Fu, J.; Kong, L.; Han, Y. Multi-channel aerosol dielectric electrical discharge machining ablation based on discrete electrode. *Int. J. Adv. Manuf. Technol.* **2018**, *99*, 1037–1045. [CrossRef]
39. Pinahin, I.A.; Chernigovskij, V.A.; Bracihin, A.A.; Yagmurov, M.A. Improvement of wear resistance of VK6, VK8, T5K10, and T15K6 hard alloys by volume pulsed laser hardening. *J. Frict. Wear* **2015**, *36*, 330–333. [CrossRef]
40. Blinkov, I.V.; Belov, D.S.; Volkhonskii, A.O.; Blinkov, V.I.; Shatalov, R.L. Structure of Nanocrystalline Arc-PVD (Ti, Al)N Coatings Modified with Nickel. *Russ. Metall.* **2015**, *5*, 421–427. [CrossRef]
41. Sheleg, V.K.; Shaturov, D.G.; Shaturov, G.F. Analysis and Selection of Rational Cutting Modes while Using Hard-Alloy Tool for Shaft Turning. *Sci. Tech.* **2018**, *17*, 14–20. [CrossRef]
42. Sokolov, A.G.; Bobylyov, E.E. The element-phase composition and properties of the surface layers of carbide-tipped tools made of TK and WC-Co alloys. *Lett. Mater.* **2017**, *7*, 222–228. [CrossRef]
43. Lin, Y.-C.; Hung, J.-C.; Lee, H.-M.; Wang, A.C.; Chen, J.T. Machining characteristics of a hybrid process of EDM in gas combined with ultrasonic vibration. *Int. J. Adv. Manuf. Technol.* **2017**, *92*, 2801–2808. [CrossRef]
44. Kozochkin, M.P.; Porvatov, A.N.; Grigor'ev, S.N. Vibroacoustic Monitoring of the Major Parameters of Electrical Discharge Machining. *Meas. Tech.* **2017**, *59*, 1228–1233. [CrossRef]
45. Zhang, Z.; Liu, M.; Liao, Y.; Su, Z.; Xiao, Y. Contact acoustic nonlinearity (CAN)-based continuous monitoring of bolt loosening: Hybrid use of high-order harmonics and spectral sidebands. *Mech. Syst. Signal. Proc.* **2018**, *103*, 280–294. [CrossRef]
46. Liang, J.F.; Liao, Y.S.; Kao, J.Y.; Huang, C.H.; Hsu, C.Y. Study of the EDM performance to produce a stable process and surface modification. *Int. J. Adv. Manuf. Technol.* **2018**, *95*, 1743–1750. [CrossRef]
47. Jain, P.S.; Ravindra, H.V.; Ugrasen, G.; Prakash, G.N.; Rammohan, Y.S. Study of Surface Roughness and AE Signals while Machining Titanium Grade-2 Material using ANN in WEDM. *Mater. Today Proc.* **2017**, *4*, 9557–9560. [CrossRef]
48. Grigoriev, S.N.; Kozochkin, M.P.; Porvatov, A.N.; Pinargote, S.; Washington, N. Features of Vibro-Acoustic Monitoring EDM of Conductive Ceramics. *Mater. Sci. Forum* **2016**, *876*, 8–13. [CrossRef]
49. Corný, I.; Pitel, J.; Hašová, S. Statistical Approach to Optimize the Process Parameters of HAZ of Tool Steel EN X32CrMoV12-28 after Die-Sinking EDM with SF-Cu Electrode. *Metals* **2017**, *7*, 35. [CrossRef]
50. Salinas, C.F.P.; Moya, E.; Coello, D. Use of an orthogonal arrangement for the analysis of the process of die sink electrical discharge machining with shape electrodes of graphite and copper on aluminum micro-casting. *Enfoque UTE* **2018**, *9*, 67–79.
51. Liu, Y.; Chang, H.; Zhang, W.; Ma, F.; Sha, Z.; Zhang, S. A Simulation Study of Debris Removal Process in Ultrasonic Vibration Assisted Electrical Discharge Machining (EDM) of Deep Holes. *Micromachines* **2018**, *9*, 378. [CrossRef]
52. Xiao, X.; Zheng, X.; Wang, Y.; Xu, S.; Zheng, Z. A Method for Utility Harmonic Impedance Estimation Based on Constrained Complex Independent Component Analysis. *Energies* **2018**, *11*, 2247. [CrossRef]
53. Herrojo, C.; Moras, M.; Paredes, F.; Núñez, A.; Ramon, E.; Mata-Contreras, J.; Martín, F. Very Low-Cost 80-Bit Chipless-RFID Tags Inkjet Printed on Ordinary Paper. *Technologies* **2018**, *6*, 52. [CrossRef]
54. Tamura, T.; Kobayashi, Y. Measurement of impulsive forces and crater formation in impulse discharge. *J. Mater. Process. Technol.* **2004**, *149*, 212–216. [CrossRef]
55. Mohri, N.; Yamada, H.; Furutani, K.; Narikiyo, T.; Magara, T. System identification of wire electrical discharge machining. *CIRP Ann.* **1998**, *47*, 173–176. [CrossRef]
56. Pang, L.; Hosseini, A.; Hussein, H.M.; Deiab, I.; Kishawy, H.A. Application of a new thick zone model to the cutting mechanics during end-milling. *Int. J. Mech. Sci.* **2015**, *96*, 91–100. [CrossRef]
57. Kozochkin, M.P.; Porvatov, A.N.; Sabirov, F.S. Vibration Testing of Technological Processes in Automated Machining Equipment. *Meas. Tech.* **2014**, *56*, 1414–1420. [CrossRef]
58. Artamonov, B.A.; Volkov, Y.S. *Analiz Modelei Elektrokhimicheskoi I Elektroerozionnoi Obrabotki. Chast' 2, Modeli Protsessov Elektroerozionnoi Obrabotki. Provolochnaya Vyrezka (Analysis of Models of Electrochemical and Electrodischarge Treatment, Part 2: Models of Electrodischarge Treatment. Wire Cutting)*, 1st ed.; Vseross. Nauchno-issled. Inst. Patent. Inform.: Moscow, Russia, 1991; pp. 98–101. (In Russian)
59. Gutkin, B.G. *Avtomatizatsiya Elektroerozionnykh Stankov (Automation of Electrical Discharge Machines)*; Mashinostroenie: Leningrad, USSR, 1971; pp. 28–29. (In Russian)
60. Maity, K.; Mishra, H. ANN modelling and Elitist teaching learning approach for multi-objective optimization of mu-EDM. *J. Intell. Manuf.* **2018**, *29*, 1599–1616. [CrossRef]

61. Grigor'ev, S.N.; Kozochkin, M.P. Improvement of Machining by the Vibroacoustic Diagnostics of Electrophysical Processes. *Rus. Eng. Res.* **2015**, *35*, 801–806. [CrossRef]
62. Fu, X.; Zhang, Q.; Gao, L.; Liu, Q.; Wang, K.; Zhang, Y.W. A novel micro-EDM-piezoelectric self-adaptive micro-EDM. *Int. J. Adv. Manuf. Technol.* **2016**, *85*, 817–824. [CrossRef]
63. Yue, X.; Yang, X.; Tian, J.; He, Z.; Fan, Y. Thermal, mechanical and chemical material removal mechanism of carbon fiber reinforced polymers in electrical discharge machining. *Int. J. Mach. Tools Manuf.* **2018**, *133*, 4–17. [CrossRef]
64. Abdelmalek, A.; Bedrane, Z.; Amara, E.-H.; Sotillo, B.; Bharadwaj, V.; Ramponi, R.; Eaton, S. Ablation of Copper Metal Films by Femtosecond Laser Multipulse Irradiation. *Appl. Sci.* **2018**, *8*, 1826. [CrossRef]
65. Isaev, A.; Grechishnikov, V.; Pivkin, P.; Mihail, K.; Yuriy, I.; Andrey, V. Machining of thin-walled parts produced by additive manufacturing technologies. *Proc. CIRP* **2016**, *41*, 1023–1026. [CrossRef]
66. Kozochkin, M.P.; Porvatov, A.N. Effect of adhesion bonds in friction contact on vibroacoustic signal and autooscillations. *J. Frict. Wear* **2014**, *35*, 389–395. [CrossRef]
67. Kozochkin, M.P.; Sabirov, F.S. Attractors in cutting and their future use in diagnostics. *Meas. Tech.* **2009**, *52*, 166–171. [CrossRef]
68. Schmitz, T.L.; Smith, K.S. *Machining Dynamics. Frequency Response to Improved Productivity*; Springer: New York, NY, USA, 2009; pp. 108–109. ISBN 978-0-387-09645-2.
69. Altintas, Y. *Manufacturing Automation: Metal Cutting Mechanics, Machine Tool Vibrations, and CNC Design*, 2nd ed.; Cambridge University Press: Cambridge, UK, 2012; pp. 68–70. ISBN 978-0-511-84372-3.
70. Webster, J.; Dong, W.P.; Lindsay, R. Raw Acoustic Emission Signal Analysis of Grinding Process. *CIRP Ann.* **1996**, *45*, 335–340. [CrossRef]
71. Grabec, L.; Leskovar, P. Acoustic emission of a cutting process. *Ultrasonic* **1977**, *15*, 17–20. [CrossRef]

© 2018 by the authors. Licensee MDPI, Basel, Switzerland. This article is an open access article distributed under the terms and conditions of the Creative Commons Attribution (CC BY) license (http://creativecommons.org/licenses/by/4.0/).

technologies

Article

Influence of WC-Based Pin Tool Profile on Microstructure and Mechanical Properties of AA1100 FSW Welds

Abbas Tamadon [1,*], Arvand Baghestani [2] and Mohammad Ebrahim Bajgholi [3,†]

1. Department of Mechanical Engineering, University of Canterbury, Christchurch 8041, New Zealand
2. Department of Mechanical Engineering, Kish International Campus, University of Tehran, Kish lsand 79416-55665, Iran; a.baghestani@alumni.ut.ac.ir
3. Department of Mechanical Engineering, École de Technologie Supérieure (ETS), Montreal, QC H3C 1K3, Canada; mohammadebrahim.bajgholi.1@ens.etsmtl.ca or mebajgholi@gmail.com

* Correspondence: abbas.tamadon@pg.canterbury.ac.nz; Tel.: +64-021-028-12680

† Current Address: Department of Materials Engineering, Science and Research Branch, Islamic Azad University, Tehran 1477893855, Iran.

Received: 4 April 2020; Accepted: 5 June 2020; Published: 15 June 2020

Abstract: The effect of various tungsten carbide (WC) pin tools and operating parameters on the material structure and properties of an AA1100 friction stir welding (FSW) weld were evaluated. Three different pin shapes were employed (conical, square and threaded). For each tool shape, welds were generated for a set of tool (revolutions per minute, RPM) (710, 1120 and 1400) and advancing speeds (150, 250 and 400 mm/min). Weld samples were tested for mechanical strength by tensile testing. Morphology was examined using optical microscopy, and weld composition with scanning electron microscopy (SEM), energy-dispersive X-ray spectroscopy (EDS) and X-ray diffraction (XRD). No weld contamination from the tools was observed. However, a number of structural defects, inherent to the FSW process, were observed (including tunnel voids, kissing bonds and swirling lines). These defects, associated with the stirring action, could not be eliminated. The results show how the operating parameters may be optimized to produce stronger welds.

Keywords: friction stir welding; WC; AA1100 aluminium plate; weld contamination; tunnel void; kissing bond

1. Introduction

Friction stir welding (FSW) is a solid-phase process [1] carried out at lower temperatures than conventional fusion welding [2]. This joining technique was introduced by The Welding Institute (TWI) [3], originally for aluminium alloys. The welding action is conducted by a non-consumable rotating tool, consisting of a penetrating pin at the centre supported by the tool shoulder [4]. The advancement of the tool through the weld-line results in severe plastic deformation [3,5,6], inducing frictional heating in the base material [7,8]. The stirring action mixes the softened material from both sides of the weld-line (advancing side (AS) and retreating side (RS)). The mixing is the result of both forging and extrusion forces underneath and around the tool [9–13]. As the rotating tool leaves the stirring position, material is deposited behind the tool, forming a continuous weld-line [13,14].

FSW can be used for a variety of ferrous and non-ferrous metallic materials [15–17], where solidification-related problems common to fusion welding can be eliminated from the weld structure. Mechanical properties may also be improved. In general, the severe deformation inherent to FSW causes grain fragmentation in the stir zone (SZ) and produces a fine-grained microstructure as the major region of the FSW weld [9,18]. In immediate proximity to the stir zone is the transition zone where the microstructure differs from the base material, due to dynamic recrystallization (DRX) caused by thermal flux and mechanical strain. These transitional regions are called the thermo-mechanically affected zone (TMAZ) and heat affected zone (HAZ) [19–21]. During cooling, microstructural evolution eventually leads to grain refinement and precipitation, which increases the mechanical strength of the weld [9].

The thermomechanical nature of the process induces DRX, which causes some grain-scale alterations within the weld texture. This microstructural evolution has considerable effects on the final properties of the FSW weld [20].

Depending on the welding parameters (welding speeds and the tool geometry), the frictional heat input and the subsequent flow-based stirring action can establish a relatively complex strain distribution which is stored in the fragmented grain structure [22,23]. By the beginning of the cooling stage, DRX leads to a variety of microscopic transformations throughout the polycrystalline weld structure [24,25].

These thermomechanical alterations need to be accurately studied at a grain-structure level to characterize the origin of microscopic transformations (e.g., grain refinement, morphological alterations and precipitation). However, the FSW process can introduce structural weld defects, such as micro-voids or micro-cracks due to improper welding parameters [26,27]. It is expected that the strength of the weld should reach the same as the parent metal [28] given the homogenous DRX microstructure in the SZ. However the mechanical strength of the weld can be easily affected by the welding speeds—advancing speed (V) and rotating speed (ω). Additionally, the geometry of the pin tool has an impact on the weld quality [13,29], it influencing the heating of the substrate and transportation of the plasticised mass flow during the stirring action. The pin tool is subject to large forces and at elevated temperatures may undergo abrasion, possibly contaminating the weld [30,31].

In metal forming processes, the surface quality can be changed during the plastic deformation. In this regards, by using some proper surface finishing procedures (such as burnishing), the high spots of the surface can be flattened by a smooth hard tool. The ball-burnishing [32] is a secondary process that is applied after the FSW to smooth the surface of the welding nugget by a radial feed and exerting a sufficient compressive force. Consequently, the blanked edge of the weld surface is smoothed out by a certain magnitude of compressible residual stress distribution. As a result, it improves the tensile strength or bending strength of the weld in the springback phenomenon.

The inherent formability of aluminium guarantees the applicability of FSW to Al alloys. Deformable non-ferrous materials, such as aluminium, are generally processed using tool steels for the pin-tools. Tool steels provide enough friction and thermal strength for the FSW process, however weld contamination is common. Contaminants displaced from the FSW tool can be trapped into the weld region causing impurity or void formation [33], both of which deteriorate the expected strength properties of the weld. These issues can be worse for FSW of reactive materials such as aluminium. Therefore, enhanced wear resistance and stability at elevated temperatures is required to promote the tool lifetime. Tungsten carbide (WC), which has high hardness (1650 HV), is a promising candidate. However, the performance of WC in FSW conditions [34–36], and its feasibility for FSW, is unknown.

The research objective of this study is to investigate the effect of a WC pin tool on the microstructure and mechanical properties of FSW AA1100 aluminium alloy under different welding conditions. Additionally, the impact of various FSW operating parameters, including tool advancing speed, rotational speed and tool geometry, were assessed. AA1100 aluminium is a commercial aluminium alloy metallurgically stabilized with a standard element composition to attain specific mechanical properties. Weld contamination may be easily detected due to the chemical purity. The alloy was chosen as the first material to test as it is easily deformable, and hence is a good candidate for FSW. If unsatisfactory results are obtained for this alloy, the method will not be suitable for other grades. Three tool geometries were used to generate weld samples: conical, threaded and square, manufactured by powder metallurgy [24,37]. Although WC-based tools are usually used in processing of high strength and high-temperature alloys (e.g., steels or super alloys), this research was designed to investigate the possibility of the tool material loss and contamination of the weld region arising from the tool material.

Different tool shapes can provide different mechanical stirring and engagement between the tool and substrate, allowing evaluation of the mechanical instability of the WC tool under the high-friction conditions of the stirring action. By applying different speeds (spindle rotational speed and advancing feed rate), the mechanical performance of the tool can be assessed better.

The thickness of the workpiece and the spindle load were kept constant. The samples were tested for mechanical strength, microscopic structure and weld contamination. Metallographic measurements and mechanical strength can be used as a reliable method to evaluate the optimum microscopic features.

2. Materials and Methods

FSW welding trials were conducted on AA1100 alloy plates with a thickness of 5 mm. The composition of the AA1100 aluminium plate (utilised as the parent metal in this work) is listed in Table 1, measured using an atomic absorption analysis quantometer (AA-6300, Shimadzu Corporation, Kyoto, Japan).

Table 1. Composition of the AA1100 Al alloy.

Element	Composition (wt %)
Al	99.3
Zn	0.006
Mn	0.016
Si	0.138
Cu	0.009
Cr	0.03
Fe	0.182
Mg	0.007

Three single-piece tools with different pin geometry designs but similar configuration and dimensions were utilised for the weld trials. With a constant shoulder diameter of 20 mm, pins were manufactured to three different geometries; conical, square and threaded (Figure 1). The pin tools were made of WC, by powder metallurgy sintering [38]. The composition of the powder mixture utilized was WC—0.3 wt % C—6 wt % Co. The free-carbon powder was used to prevent porosity, and to avoid formation of oxide phases (e.g., WO and WO_3) during the sintering process. Cobalt (Co) was employed as the binder phase to improve the wettability on micro-size WC particles, therefore promoting interfacial bonding strength and wear resistance of the segment. Sintering was done at 1800 °C for 90 min under a pressure of 70 MPa. The shoulders of the FSW tools were fabricated from H13 hot work tool steel (hardness 560 HV), a standardised grade suitable for hot working. For each tool, the pin and shoulder were joined by the transient liquid phase diffusion bonding (TLPDB) technique.

The weldment plates were cut into two plates with dimensions of 235 mm in length and 45 mm in width and set side-by-side for a butt weld. For this soft aluminium grade, initial trials showed that lower speeds could not provide a stable welding condition, while higher speeds led to poor weld quality with structural defects (e.g., flash, voids or discontinuities) occurring throughout the weld. Many welds were conducted, but the speed parameters (ω and V) were limited to the ones with better results in preliminary tests. Five feed rate speeds in the range of 100–400 mm/min and eight rotational speeds in the range of 600–1500 RPM (revolutions per minute) were applied to produce the FSW weldments. While some complete welds were produced, there was also a high proportion of only partially successful welds, hence indicating a quality problem. The issue was invariably a partially successful joint with a localised defect (discontinuity line or tunnel void defect). We aimed to better understand the causes of the quality issues for different pin tool profiles. Therefore, after some trial and error, the feed rates and rotational speeds were optimized to those given in Table 2.

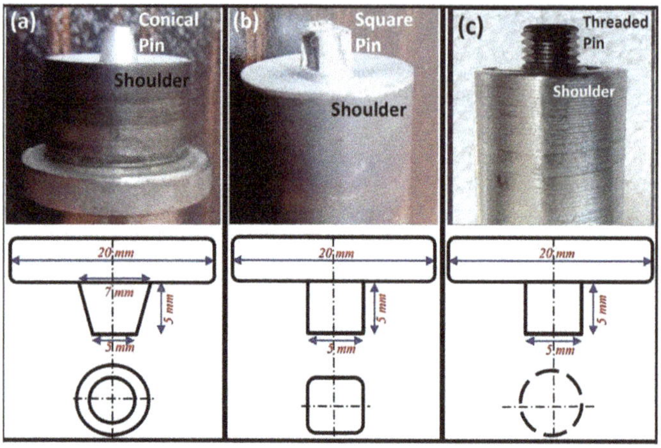

Figure 1. Configuration of the FSW tools utilized for the AA1100 weld trials ((**a**) conical, (**b**) square and (**c**) threaded pin tools) with geometry and dimension.

Table 2. The operational welding parameters of the friction stir welding (FSW) process for the AA1100 plates.

Welding Parameters			
ω (rpm)	710	1120	1400
V (mm/min)	150	250	315
Tilt Angle		3 degrees	
Plunge Depth (shoulder)		0.5 mm	

The aluminium welds were conducted using a 3-axis Computer Numerical Control (CNC) machine with a control unit and motor capacity of 10 kW. The direction of tool rotation was clockwise, and the plates were fixed rigidly by strap clamps at the corners. Clockwise rotation of the tool enables feeding of material into the centre of the weld during rotation (Figure 2). The welding was done without any initial pre-heating, and in an open air atmosphere at a temperature of 18 °C. Table 2 presents more details of the process parameters employed for the AA1100 aluminium weld trials. A schematic of the FSW set-up and process is shown in Figure 2. After the test, tensile test samples and metallographic specimens were cut from the middle of the plate, perpendicular to the weld-seam.

To evaluate weld quality, samples were primarily examined by the visual inspection. Weld surfaces were photographed using a high-resolution bridge digital camera (FinePix S9500 Model, Fujifilm brand, Tokyo, Japan), under the light of a fluorescent lamp. For metallographic analysis, mounted samples of the weld cross-sections were polished by standard preparation methods down to 0.5 μm to provide a smooth mirror finish. For microstructural observations, the cross-section samples were etched to identify grain boundaries and morphology. The specimens were etched in an ultrasonic bath using the Keller's reagents (95 mL H_2O, 2.5 mL HNO_3, 1.5 mL HCl and 1.0 mL HF), at 70 °C for 20 s. After etching, the specimens were washed with deionised water, rinsed with ethanol and dried with hot air. The macro- and microscopic observations of the etched cross-sections were done with a typical optical microscope, at magnifications of 50-1000×.

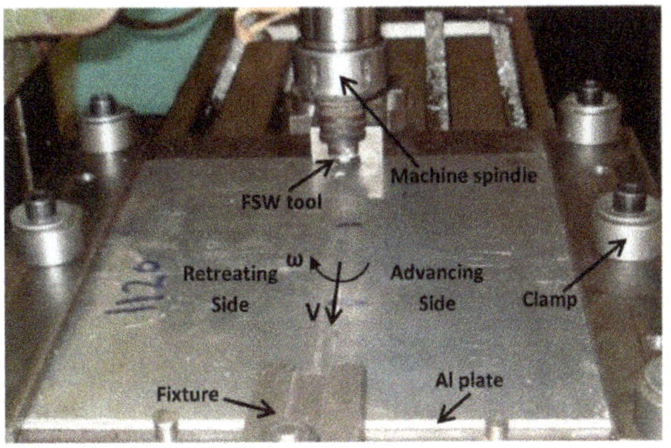

Figure 2. Schematic of the FSW process, the FSW tool in interaction with the AA1100 plates as the substrate.

For further microstructural characterisation, after etching samples were examined in a scanning electron microscope (SEM, MIRA model, TESCAN instrument, Brno, Czech Republic) with an energy-dispersive spectroscopy (EDS) detector. The accelerating voltage of the SEM was 20 kV. For phase characterization via X-ray diffraction (XRD), a Philips diffractometer instrument was used with Cu-Kα radiation at angles of 20–110°, and an accelerating voltage of 40 kV.

The tensile test samples, with a dimension of 120 mm gauge length and 30 mm gauge width (Figure 3), were fabricated according to the ASTM E8/E8M standard method. For FSW joints, the sampling position was selected perpendicular to the welding direction. In this configuration, the weld-seam is located in the centre of the dog-bone tensile specimen. Therefore, during the tensile testing, the strength of both the base metal and cross-weld are measured simultaneously [39,40]. Tensile tests were performed with an Instron-5967 tensile tester at a strain rate of 10^{-5} s^{-1}, at room temperature. For each sample, we repeated the tensile test five times and then reported the average value between the three most consistent samples as the selected tensile curve.

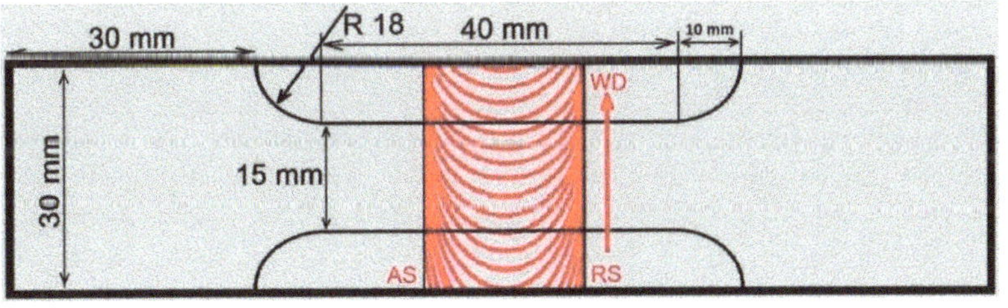

Figure 3. Schematic dimensions of the tensile test sample, made of FSW plates of AA1100.

3. Results

3.1. General Appearances of the FSW Weld

The general appearance of the FSW weld, along with typical structural defects formed during the process, is shown in Figures 4–6. In these figures, the weld appearance varied depending on welding speeds (ω and V) and tool features. It is believed that the heat input plays a key role in determining weld appearance, and insufficient heat input can produce defects in the stir zone. Figure 4 shows the crown view at the top surface of the FSW joints welded by the conical-shape tool, at a constant rotating speed (1120 rpm) and two different feed rates (150 and 315 mm/min). The first set of speeds (1120 rpm and 150 mm/min) in Figure 4a shows a uniform pitched surface without any visible defects at the surface. By increasing the feed rate at the same rotating speed (1120 rpm and 315 mm/min), the weld surface became more smooth, while the ripple features disappear and a cavity/groove defect emerges on the advancing side (AS) of the weld surface. Additionally, for both set of speeds, flash defects are visible on the retreating side (RS) edge of the weld-line.

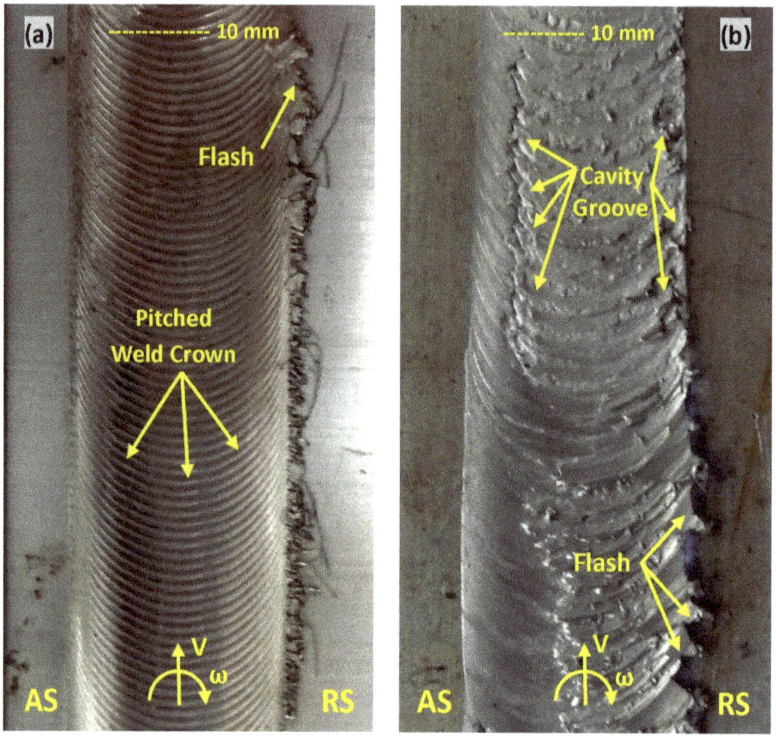

Figure 4. Surface appearance of the weld for conical FSW tool in constant feeding rate; (**a**) at the welding speeds of (1120 rpm and 150 mm/min) and (**b**) at the welding speeds of (1120 rpm and 315 mm/min).

For a better understanding of the surface appearance of the weld, different sets of speeds were studied with the same conical-shape tool, where the feed rate was kept constant and the RPM varied. Figure 5 shows the surface view of the FSW joint for two different speed sets with the same RPM. For the welding speeds set of (250 rpm and 710 mm/min), Figure 5a shows a relatively smooth surface, compared to Figure 5b at a higher feeding rate (250 rpm and 1400 mm/min). Similar to Figure 4, again flash defects were obvious in Figure 5a,b, but the size of the flash defects and the apparent surface roughness increased. In Figure 5, the keyhole position where the tool exits the workpiece was also visible. Clearly, by increasing the feed rate (Figure 5b) the size of the keyhole and the flash defect increased. Our interpretation is that inconsistency in welding speeds (ω and V) can intensify the slipping during the stirring action. Therefore, excessive circumferential inertia affects the flow distribution during the stirring, leading to interruption of the flow deposition.

The role of the tool geometry on the quality of the weld was studied in Figure 6, where the general appearance of the weld-seam is shown for the conical, square and threaded designed tool. Welding speeds (ω and V) remained constant. As is clear in Figure 6, for the same processing speeds (1120 rpm and 315 mm/min), the weld-line provided by the conical tool shows the best surface quality: a uniform pitched ripple pattern for the weld crest, a filled keyhole and minimal flash defects.

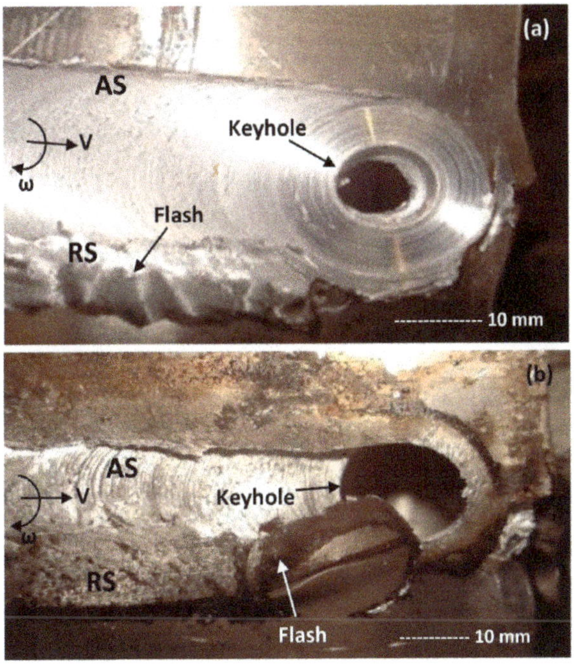

Figure 5. Surface appearance of the weld for conical FSW tool at constant RPM; (**a**) at welding speeds of (250 rpm and 710 mm/min) and (**b**) at welding speeds of (250 rpm and 1400 mm/min).

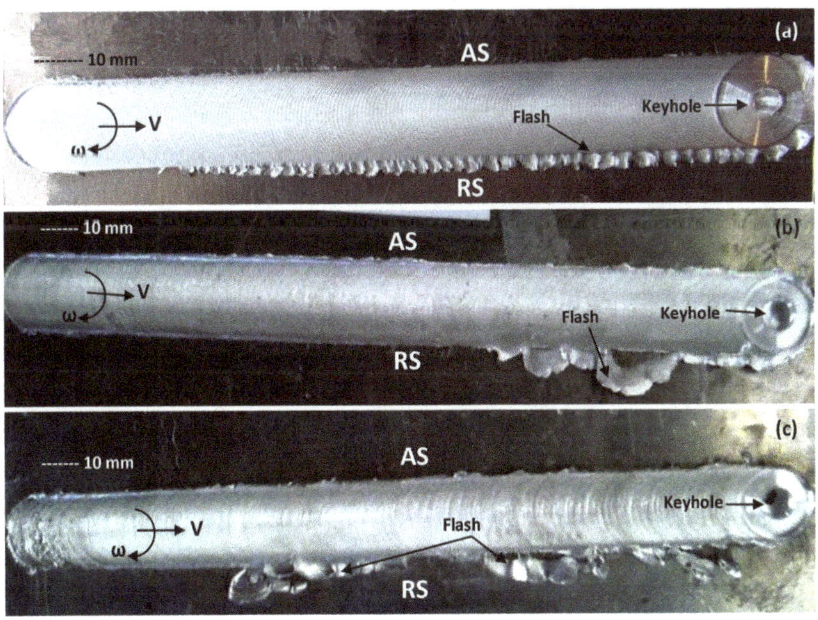

Figure 6. Surface appearance of the weld for different FSW tool in constant speed sets of (250 mm/min and 1400 rpm); (**a**) conical pin profile, (**b**) square pin profile and (**c**) threaded pin profile.

The performance of the conical tool can be attributed to the inherent lateral motion of the mass around the pin, which pumps material inwards and simultaneously in a spiralling manner to the top surface. This can avoid mass deficit or ejection as the tool moves forward along the welding locus. Therefore, material loss and discontinuity defects are minimized.

It should be noted that in the FSW process, the material flow plays a key role in the control of defects. In general, insufficient heat input can lead to failure of flow regimes in which the stirred mass is not strong enough to fill the discontinuity, and therefore, a defect occurs.

On the other hand, redundant heat input also can result in the emergence of defects. By increasing welding speeds (ω and V), excessive frictional heat is generated between the tool and the workpiece. The increased heat input can yield more plasticized mass under the shoulder. At higher speeds, this plasticized material flow can split out in the form of the excessive flash defects at the edge of the weld-line. Furthermore, the ejection of the mass results in mass deficit at the weld-seam, where groove defects appear underneath the shoulder, or other subsurface discontinuity defects form in weld.

The relationships between processing parameters and the apparent quality of aluminium welds are illustrated in Figure 5. The photographs, both taken from the top surface of the weld samples at the spindle side position, show a distinctive ejected tail as the flash defect, protruding from the retreating side (RS) of the weld-line. The exit point of the tool shows a keyhole feature where the tool leaves the workpiece. In both samples, the tool leaves the weld-seam by a disruption in the body of the workpiece.

The curved features at the weld surface reveal layered flow patterns that trace the plastic flow between the AS and RS. However, due to disruption at the sides of the weld, the circulation of material and therefore the primary stirring flow-lines are discontinuous.

As the front of the stirred mass around the pin loses stiffness because of the shortage of forwarding material, flow circulation becomes irregular. Thus due to the upward motion of the tool, the plasticized material bulges out and a hole is left at the position of ejection.

In a similar way, where the leading edge of the shoulder approaches the free surface of the workpiece, the stirred material at the retreating side of the tool creates a curve in the surface pattern of the weld-line. During rotation of the tool, the inconsistency in speed ratio (ω/V) can cause excessive stretching at the weld edge, more so than at the middle of the weld-line, whereby the compressed mass can be shredded outwards instead of being deposited at the trailing edge of the tool. This can form a flash sprayed tail as a defect at the RS position of the weld-line.

In Figure 6, constant welding speeds (250 mm/min and 1400 rpm) for different FSW tool geometry (conical, square and threaded shaped pins) show a similarity in the size of the flash defect and the keyhole. The weld samples indicate a similar amount of material loss as the flash defect, and similar shape of the keyhole at the surface. This suggests that the tool geometry plays the main role in the stirring conditions in proximity to the rotating tool.

The surface quality of the weld-seam shows a more uniform pitch-pattern for the conical pin tool (see Figure 6a) compared to the square pin and threaded pin (Figure 6b,c, respectively). This indicates that in similar welding speeds, the presence of the flats (in square pin tool) and the threads (in threaded pin tool) aggregates the plastic flow deposition within the stirring zone, in which needs a more accurate flow control during the FSW position. It should be noted that because of applying a rotating pin, the keyhole profiles could not show the exact flow geometry around the pin. Therefore, the rotating pin leaves a circular pattern at the keyhole by ejection from the weld-line. More specifically regarding the square pin, because of the sharp orthogonal edges of the flats, during the tool ejection, it might cause some material loss from the stirred Al-mass, stuck around the pin (see Figure 1b). Nevertheless, the WC-based square pin tool shows an acceptable mechanical instability during the stirring action without occurring any contamination or tool failure.

The surface features of the weld-line for different welding parameters were illustrated in Figures 4–6. However, the flow aspects of the FSW joints concerning various welding conditions need to be elucidated in more detail through microstructural study of the weld cross-sections. Microstructural observations can also reveal more flow features within the weld structure, where internal defects arise from flow failure within the stir zone.

3.2. Macrostructure of the Weld Region

The macrostructure of the cross-section of an FSW joint, processed using three different tools, is shown in Figure 7. The macrostructure readily revealed the SZ basin-shaped pattern in the cross-section of the weld, and onion-ring flow patterns were also visible. In Figure 7c, compared to Figure 7a,b, the weld border was more visible in the middle of SZ rather than at the sides of the weld. Even so, in Figure 7b there is a blurring of the weld border. This is attributed to the tendency for the tool to aggressively remove the substrate material from the base metal (BM) and subsequently backfill the region with heavily worked material, hence making the SZ borders more defined in comparison to the sides.

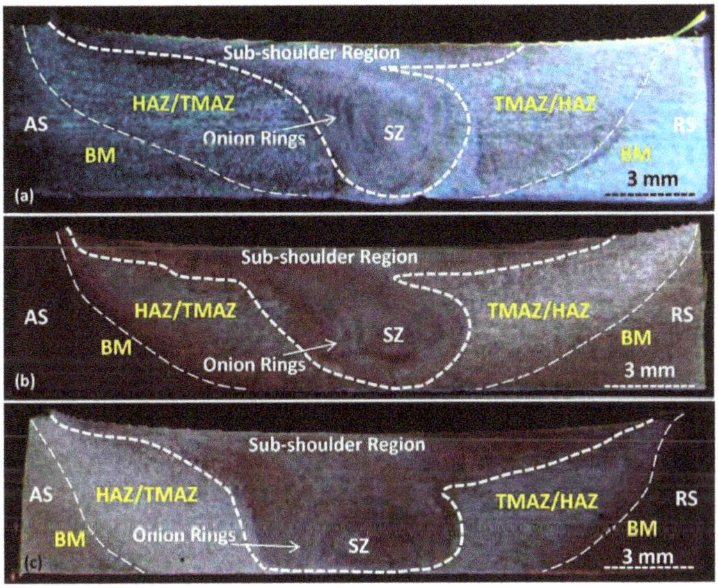

Figure 7. The macrostructure of the cross-section of the FSW weld for three different tools: (**a**) conical, (**b**) square and (**c**) threaded; all in speeds of 250 mm/min and 1400 rpm.

The flow eddy features known as onion rings at the bottom of the SZ were attributed to the mass transport mechanism inside the weld. The tool scouring action during stirring caused the plastically deformed material from the substrate to be aggressively conducted in flow path lines between the advancing side and retreating side. Consequently, because the combined linear and rotation motions of the tool were greater at the bottom surface, this is also the hottest part of the weld region. However, the material was also scoured from the leading edge towards the retreating side, and was deposited at the trailing edge of the tool. The rotation of the tool (clockwise from above) stirred the flowing mass in the same rotational direction, corresponding to a horizontal flow from right to left in the wake of the tool. Consequently, during mass deposition, the stirred flow-lines were packed into the downstream region, and the boundaries between these layered masses were believed to correspond to the onion rings in the cross-section samples shown in Figure 7a–c. The spacing between the boundary lines was also observed to be smaller closer to the centre of the pin, attributed to the flow mechanism during stirring. Additional complexity was introduced by the threads on the tool (Figure 7c). This transports the material downward, hence explaining why the onion rings were preferentially located at the bottom of the SZ.

Inconsistency in flow-lines transport can cause the emergence of internal void defects. As the macroscopic views of the FSW samples were not able to show the grain structure in detail, micro-etching and microscopic observation at higher magnifications is required. The delineation of fine grain size in proximity to probable defects can also explain the origin of defects based on the microstructural evolution of the weld region.

3.3. Microscopic Measurements of the Weld

The micro-etching of the weld cross-section in Figure 8a shows that the weld region comprised of the SZ and the transition region, which were distinct from the base metal (BM) region. The transition region, including the HAZ and TMAZ, was situated between BM and the SZ. The dynamically recrystallized nature of the FSW weld structure caused each layer to have a different grain size and morphology in comparison with the adjacent layers.

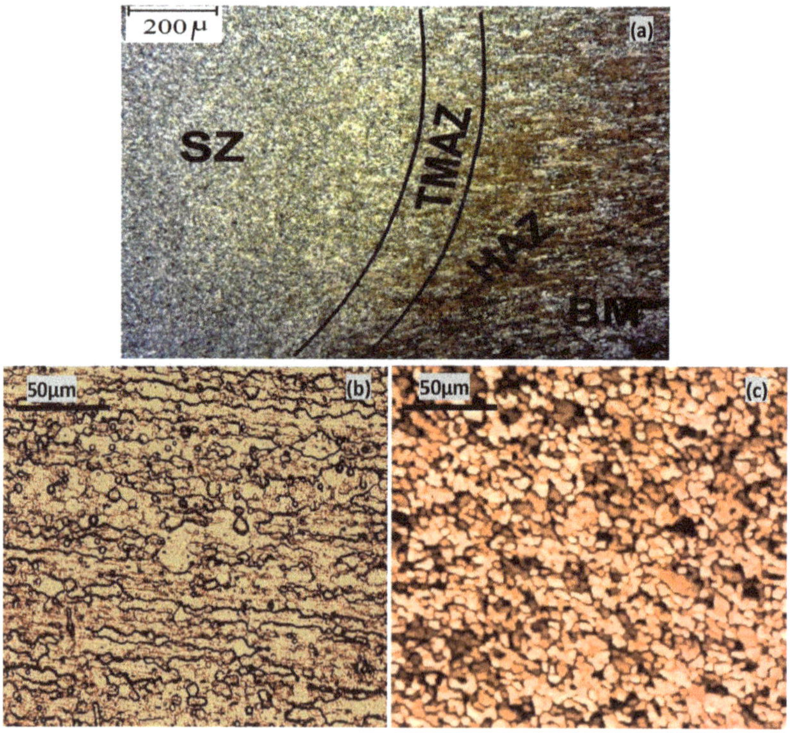

Figure 8. Microstructure of the grain distribution in the cross-section of the AA1100 FSW weld (processed at 1120 rpm and 250 mm/min); (**a**) general view of the cross-section, (**b**) stir zone (SZ) and (**c**) the base metal at higher magnification.

While the BM shows a directional grain morphology with the average grain size of 50 μm (Figure 8b), this changed to an equiaxed, ultrafine structure with an average grain size of 10–15 μm in the SZ (Figure 8c). This can be interpreted as the direct outcome of grain fragmentation by mechanical stirring, and subsequent post-welding dynamic recrystallization arising from frictional heat generation.

Entering into the SZ, the morphological flow features through the grain structure revealed some inhomogeneous transitional micro-patterns in the form of weld defects. Figures 9–11 demonstrated some of the typical weld defects in the weld region that emerge at the bottom of the SZ adjacent to the base metal (BM).

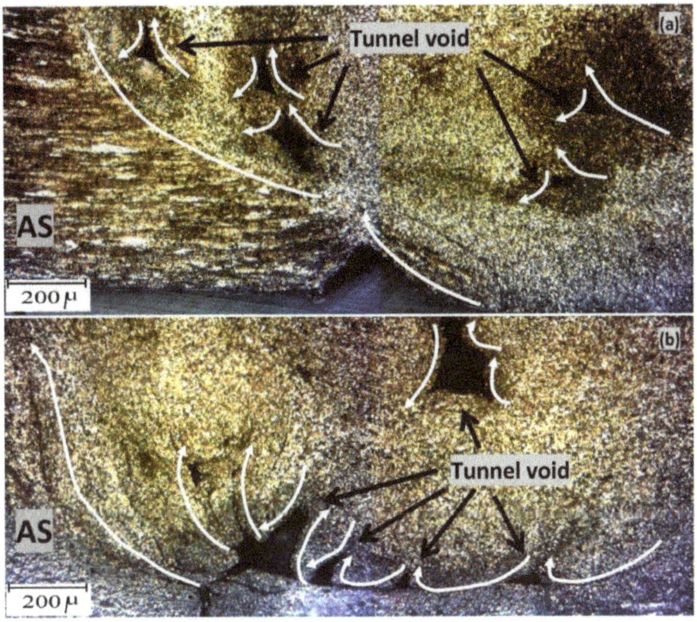

Figure 9. Micro-void features visible in stirring zone of the FSW weld, known as a tunnel void, observed in: (**a**) conical tool and (**b**) square tool welds.

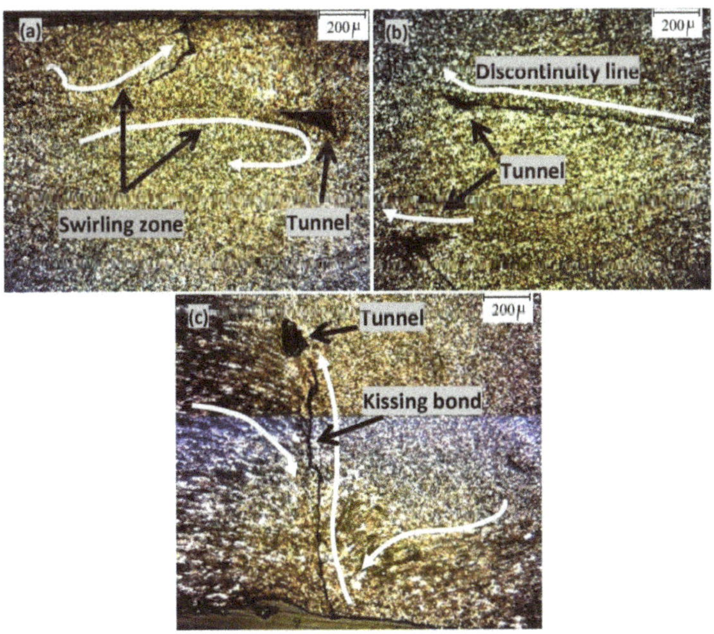

Figure 10. Formation of flow-based defects around tunnel voids formed by different FSW tools at constant speed sets of (250 mm/min and 1400 rpm); (**a**) conical pin profile, (**b**) square pin profile and (**c**) threaded pin profile.

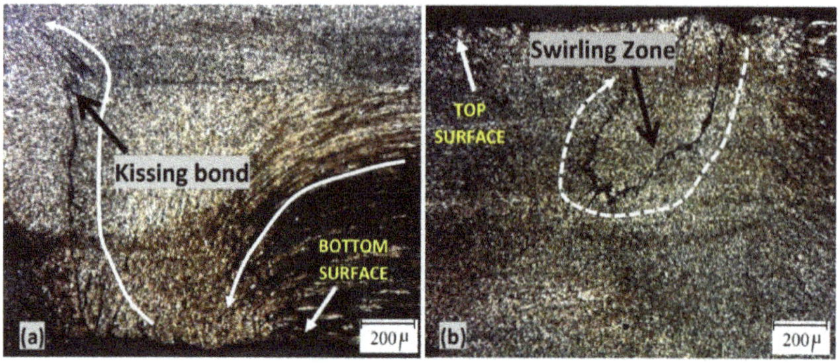

Figure 11. Formation of the flow-based defects at the structure of the FSW weld; (**a**) kissing bond by conical pin profile and (**b**) swirling zone by threaded pin profile, in speed sets of (250 mm/min and 1400 rpm).

Figure 9 shows the flow patterns around tunnel voids at the cross-section of the FSW weld for the conical and the square tool geometries.

The drawn flow-lines around the tunnel voids show a layered structure representing different gradients in the deposited flow at the BM regions.

Comparison between the samples shows that while the number of voids in the weld sample from the conical-shaped tool (Figure 9a) was greater than for the square-shaped tool (Figure 9b), the size of the voids were greater when processed using the square tool. This can be related to the coarse performance of the square-shaped tool and the sharpness of the square corners compared to the conical geometry.

Further study of the micro-features around tunnel voids is shown in Figure 10. While all three different tool geometries produce some tunnel voids within the stirring zone, the conical (Figure 10a) and threaded (Figure 10c) pin profiles exhibited more complex flow patterns around the tunnel voids compared to the square-shaped pin profile (Figure 10b). The swirling zone in Figure 10a for the conical-shaped pin profile and the kissing bond defect in Figure 10c for the threaded-pin were the direct outcome of flow failure around the formed tunnel void. These flow-based defects could be attributed to the complexities of the stirring flow mechanisms driven by the pin geometry.

Figure 10 also revealed that the swirling zone was situated at the top surface, and the kissing bond had propagated from the bottom surface. Microscopic observation of the weld cross-section at a higher magnification revealed these flow-based features as the kissing bond by conical pin profile (Figure 11a), and the swirling zone by threaded pin profile (Figure 11b). In both cases, the microstructural patterns show a change in the grain direction from the weld surface towards the body of the weld in the ultrafine structure in the SZ.

The fine grains of the SZ reveal the changes in the flow direction, with narrow flow-lines delineating the kissing bond (Figure 11a) and the swirling zone (Figure 11b) as a direct outcome of the tool action within the stirring zone of the weld. These features were indicative of the mechanical stirring action and the subsequent flow inconsistencies and dynamic recrystallization experienced during the FSW welding process.

These dark flow-lines and the existing tunnel voids are representative of flow failure as the microstructure stabilised during cooling to room temperature. The flow inhomogeneity can result in the deterioration in mechanical properties of the weld. Therefore, it is necessary to measure the mechanical properties of the weld to see how the internal defects could affect the strength of the weld.

To investigate the possibility of contamination of the welding region by the tool, the composition of the stir zone texture should be measured. Micromorphology of the grain structure was observed via

SEM, and composition was analysed using EDS, as shown in Figure 12. Tungsten atoms have a high atomic number (74) and atomic weight (183.84 g/mol), which can be easily characterized by EDS.

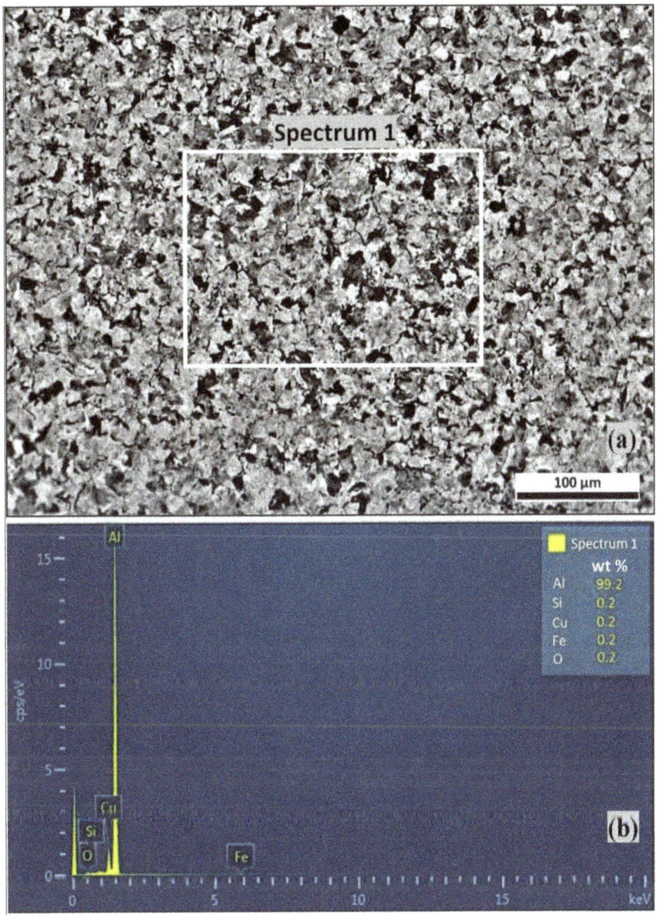

Figure 12. Schematic plot of the SEM results; (**a**) micromorphology of the stirring zone and (**b**) EDS composition for the selected region of the stirring zone processed by the square-shape pin tool.

The microstructure of the stirring zone (processed by the square-shape tool) in Figure 12a confirms that the DRX equiaxed grains are uniformly distributed through the weld. Moreover, it can be seen that there were no irregular morphologies indicating agglomerate phases, or any specific inclusions formed due to possible dispersed contamination within the microstructure. In Figure 12b, the EDS spectrum of the selected region from Figure 12a (Spectrum 1) could be seen. This confirms that the composition did not contain any W or C impurities. The presence of Fe (0.2 wt %) was consistent with the initial composition of the parent metal AA1100 (containing 0.182 wt % of Fe in quantometer analysis), so was unlikely to represent contamination from the H13 steel shoulder of the tool.

To provide a better correlation between the FSW processing conditions and the phases formed within the weld region, further phase characterization was conducted by X-ray diffraction (XRD) analysis (Figure 13), where the parent metal AA1100 was compared with the SZ material.

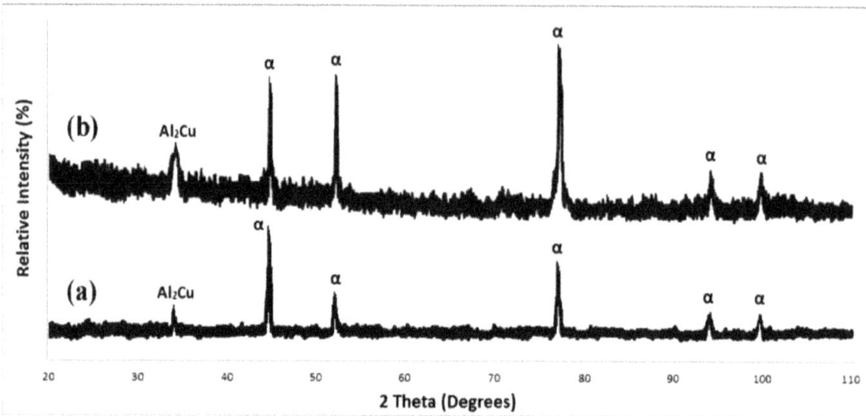

Figure 13. XRD analysis of AA1100 aluminium plate, before and after FSW processing by the square-shape pin tool; (**a**) parent metal and (**b**) stirring zone.

Figure 13 illustrates the XRD patterns for the AA1100 alloy parent metal (Figure 13a) and the SZ (Figure 13b) processed by the FSW tool (square-shaped tool). As shown in Figure 13a, the α-Al phase (matrix) and Al_2Cu precipitate were identified from the peaks of the diffraction pattern for the parent metal.

According to Figure 13b, the XRD analysis of the SZ region after the FSW processing shows the same phases (α-Al matrix and Al_2Cu precipitate), compared to the parent. However, a significant increase in the intensity of the diffraction peaks was evident for both phases of the matrix (α-Al) and precipitate (Al_2Cu), upon stirring. This can be attributed to phase stabilization after the stirring process and also an increase in precipitate distribution density due to DRX.

Consistent with the EDS analysis, the XRD analysis also confirmed that the SZ of the weld region was free of any W and C impurities. This shows that the FSW processing provides proper phase homogeneity in the SZ, without any contamination from the tool. This is based on examination of the weld sample processed using the square-shaped pin geometry, which has the highest degree of stress concentration at the sharp corners of the tool.

3.4. Mechnical Strength of the Weld

Figure 14 shows the tensile strength graphs (stress–strain) for the three sets of weld samples processed by varying different conditions; alteration of RPM (Figure 14a), alteration of feed rate (Figure 14b), and changing of the tool geometry (Figure 14c). As observed in Figure 14a,b, by increasing RPM or decreasing feed rate, the ultimate strength of the weld slightly decreased compared to the base metal. However, at the same time the elongation rate of the sample increased. Regarding the dependence of tensile strength on the geometry of the pin profile, Figure 14c confirms that threaded > square > conical.

Lower rotation speed during FSW results in less frictional heat generation, and consequently poor plastic flow of material. Therefore the tensile strength can be considered a function of the process speeds.

It was observed in Figure 14a that the ultimate tensile stress (UTS) for the base metal (340 MPa) was higher than the weld samples. It can be attributed to the formation of the HAZ region with a negative impact on the strength of the weldment, because of the coarsened grains associated with this region [41].

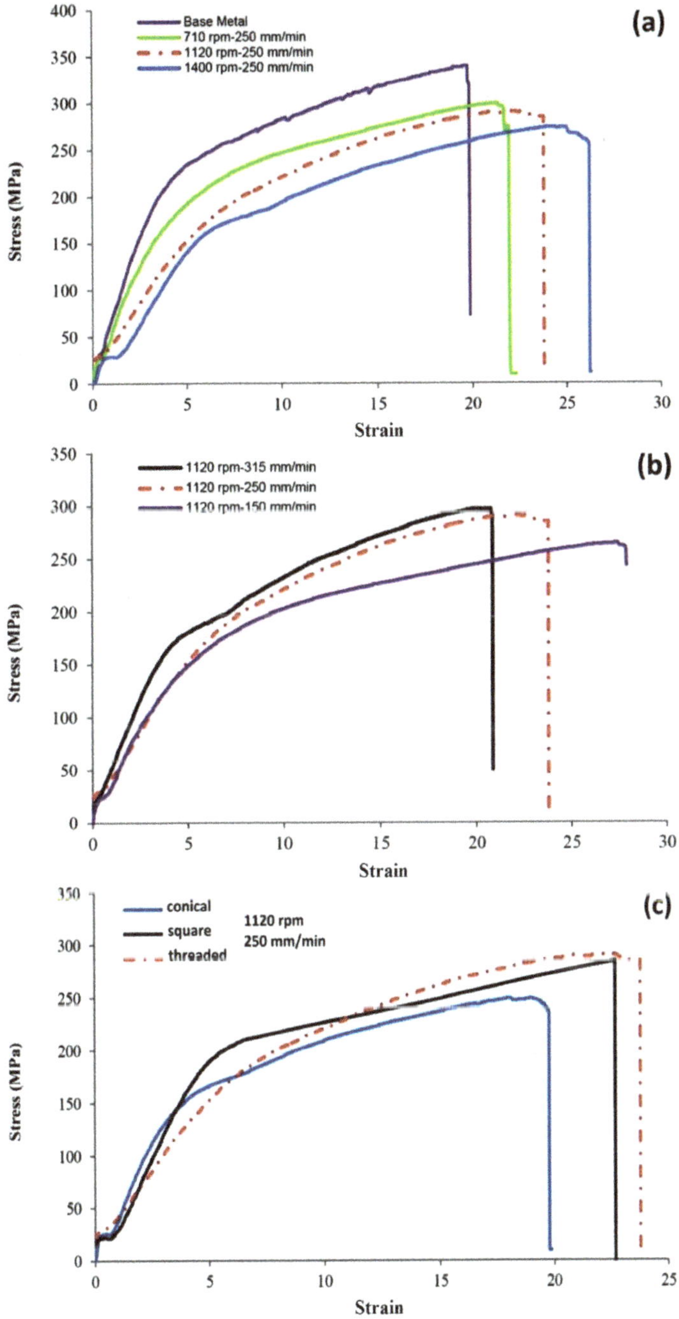

Figure 14. Tensile test results for the AA1100 aluminium welds; (**a**) RPM = 710, 1120 and 1400 at feed rate of 250 mm/min, (**b**) feed rate = 150, 250 and 315 mm/min at 1120 RPM and (**c**) comparison between conical, square and threaded pin tools, processed in constant welding speeds (1120 RPM and 250 mm/min).

However, the percentage elongation of the weld sample was more than the base metal, representative of the ductile behaviour of the weld. Our interpretation is that the processing parameters (rotational speed and feed rate) increase the homogeneity of the weld region during the stirring process, in which lead to the ductility of the stirred material, compared to the parent metal [42]. More specifically, the comparison between the tensile test–strain curve for the samples processed by different tools in constant speeds (1120 RPM and 250 mm/min), shows a higher strength and the percentage elongation achieved by the threaded pin tool, see Figure 14c. This also implied the flow uniformity induced by the tool, leading to an increase of the ductility of the weld [39,43].

It should be noted that all the tensile curves show a kink [44] at the stress level of 25 MPa, approximately. This behaviour is explained as the toe region at the beginning of the tensile test [45,46], where the mechanical loading is less than 2% nonlinear strain for the stress–strain curve [47]. There is not a clear explanation for this behaviour, however, it mainly occurs for the soft materials [48,49]. When the stiffness is reduced, or in the presence of micro-defects, a micro-failure accumulates within the material, which was shortly released as the loading continues. Regarding the AA1100 FSW samples, existing of some flow-based discontinuity or micro-porosities like tunnel voids cause a self-interaction inside the material texture during the first steps of the tensile test loading, therefore the toe region occurs. However, the main aim was to compare the strength of the weld in different processing parameters. Additionally, because the tensile testing method was conducting in the same way, and the toe region was observed at stress level approximately 25 MPa for all samples, it did not affect the results of the tensile test.

4. Discussion

This work makes the following original contributions to improve the understanding of the FSW welding processed by the WC-based tool for the aluminium alloy, regarding the microstructural evolution and mechanical properties of the weld.

According to the microscopic observations made, at higher speeds the frictional heat generated is higher, which leads to an increase in the plastic flow of material. This can intensify the emergence of flow-based defects (e.g., tunnel voids, kissing bonds and swirling zone) and subsequently results in poor stirring conditions [33,50,51]. This is consistent with observations in the literature for the Al alloys processed by the FSW, which show the control of the plastic flow deformation as the main factor for improvement of the weld quality [42,52,53]. Alternatively, the interruption of the flow integrity can easily deteriorate the quality of the weld structure.

Similar to the speed condition, the pin geometry also can affect the tensile strength. The pin profile in different geometrical configurations (conical, square and threaded pin) can affect the frictional heat. The tensile strength of the threaded tool weld was higher than that of both other types (conical and square). This is because of greater material softening obtained at higher temperatures, as the threaded pin enhances the plastic flow of material due to higher engagement between the pin and material, and extra material contact. According to the literature, the pin profile for the Al FSW welds show a distinct shearing effect, mainly an ultrafine grain refinement at the mid-SZ, altered to the transition region borders (TMAZ and HAZ) towards the base metal. It is also supported by the previous works that the shearing field induced by the rotating tool into the proximity mass flows is the main factor in the microstructure alteration of the weld texture [39,54–56].

As a key finding of the phase characterization, from the XRD and SEM/EDS analysis, it was confirmed that there was no evidence of weld contamination from the WC pin tool. However, optical microscopy revealed some defects, (e.g., kissing bonds, swirling patterns and tunnel voids), which were identified as being inherent to the FSW process, rather than being due to tool contamination. Furthermore, the absence of carbon-containing compounds in XRD and EDS analysis eliminates the possibility of formation of carbon monoxide (CO) gas due to oxidation of the WC at the elevated temperatures occurring during the FSW [24]. Therefore, the stability of the WC pin tool was confirmed, as no oxidative wear degradation occurred during the severe frictional deformation.

This has been affirmed in the literature that the formation of oxide inclusions or other external contaminations can strongly disrupt the flow homogeneity within the plasticized stirred mass [57–59]. Subsequently, the recrystallization and grain refinement also can be affected. The microscopic observations in this study indicate the absence of the formation of the oxide layers of contamination or any other W-based compounds within the weld structure, which confirms the main idea of this research in the successful performance of the WC-based pin tools for FSW processing of the Al alloy as an active material, without occurrence of any metallurgical issue.

5. Conclusions

The effect of a WC pin in the microstructural quality of the AA1100 FSW weld was evaluated using different sets of welding speeds (RPM and feeding rate) and different pin geometries (conical, square and threaded shapes). Metallurgical analysis confirmed that the weld structure was free of possible W and/or C contamination from the pin. Nevertheless, microstructural observations revealed the formation of flow-based defects (tunnel voids, kissing bonds and swirling patterns) within the weld structure. Since the quality of the weld was directly affected by these structural defects, the tensile strength of the welded sheets was reduced, as the plasticized material of the weld could not sufficiently achieve the strength of the base metal. Moreover, the welded samples experienced a clear decrease in elongation rate by the increase in the feeding rate. The metallographic observations confirmed that the emerging internal defects could not be completely avoided. Therefore, the specific application should adjust the welding parameters according to the required strength.

Author Contributions: Conceptualization, A.T. and A.B.; Formal analysis, A.T. and A.B.; Investigation, A.T. and A.B.; Methodology, A.T. and A.B.; Project administration, A.T. and A.B.; Resources, A.T. and A.B.; Supervision, A.T.; Validation, A.T. and A.B.; Visualization, A.T. and A.B.; Writing—original draft, A.T. and A.B.; Writing—review and editing, A.T. and M.E.B. All authors have read and agreed to the published version of the manuscript.

Funding: This research received no external funding.

Acknowledgments: The authors would like to thank Alice Young (University of Canterbury, NZ) for her assistance in proofreading the manuscript.

Conflicts of Interest: The authors declare no conflict of interest.

References

1. Thomas, W.; Nicholas, E.; Needham, J.; Murch, M.; Temple-Smith, P.; Dawes, C. Friction Stir Butt Welding. International Patent Application No. PCT/GB92 Patent Application 9125978.8, 6 December 1991.
2. Murr, L.; Flores, R.; Flores, O.; McClure, J.; Liu, G.; Brown, D. Friction-stir welding: Microstructural characterization. *Mater. Res. Innov.* **1998**, *1*, 211–223. [CrossRef]
3. Thomas, W.; Nicholas, E. Friction stir welding for the transportation industries. *Mater. Des.* **1997**, *18*, 269–273. [CrossRef]
4. Threadgill, P.; Leonard, A.; Shercliff, H.; Withers, P. Friction stir welding of aluminium alloys. *Int. Mater. Rev.* **2009**, *54*, 49–93. [CrossRef]
5. Tamadon, A. Characterization of Flow-Based Bobbin Friction Stir Welding Process. Ph.D. Thesis, University of Canterbury, Christchurch, New Zealand, 2019.
6. Tamadon, A.; Pons, D.; Sued, M.; Clucas, D.; Wong, E. Analogue Modelling of Bobbin Tool Friction Stir Welding. In Proceedings of the International Conference on Innovative Design and Manufacturing (ICIDM2016), Auckland, New Zealand, 24–26 January 2016.
7. Barcellona, A.; Buffa, G.; Fratini, L.; Palmeri, D. On microstructural phenomena occurring in friction stir welding of aluminium alloys. *J. Mater. Process. Technol.* **2006**, *177*, 340–343. [CrossRef]
8. Tamadon, A.; Pons, D.J.; Clucas, D. Afm characterization of stir-induced micro-flow features within the aa6082-t6 bfsw welds. *Technologies* **2019**, *7*, 80. [CrossRef]
9. Tamadon, A.; Pons, D.; Sued, K.; Clucas, D. Thermomechanical grain refinement in aa6082-t6 thin plates under bobbin friction stir welding. *Metals* **2018**, *8*, 375. [CrossRef]

10. Tamadon, A.; Pons, D.; Sued, K.; Clucas, D. Formation mechanisms for entry and exit defects in bobbin friction stir welding. *Metals* **2018**, *8*, 33. [CrossRef]
11. Tamadon, A.; Pons, D.; Sued, K.; Clucas, D. Development of metallographic etchants for the microstructure evolution of a6082-t6 bfsw welds. *Metals* **2017**, *7*, 423. [CrossRef]
12. Sued, M.; Tamadon, A.; Pons, D. Material flow visualization in bobbin friction stir welding by analogue model. *Proc. Mech. Eng. Res. Day* **2017**, *2017*, 368–369.
13. Colligan, K. Material flow behavior during friction welding of aluminum. *Weld. J.* **1999**, *75*, 229s–237s.
14. Tamadon, A.; Pons, D.; Sued, M.; Clucas, D.; Wong, E. Preparation of plasticine material for analogue modelling. In Proceedings of the International Conference on Innovative Design and Manufacturing (ICIDM2016), Auckland, New Zealand, 24–26 January 2016.
15. Elrefaey, A.; Gouda, M.; Takahashi, M.; Ikeuchi, K. Characterization of aluminum/steel lap joint by friction stir welding. *J. Mater. Eng. Perform.* **2005**, *14*, 10–17. [CrossRef]
16. Watanabe, T.; Takayama, H.; Yanagisawa, A. Joining of aluminum alloy to steel by friction stir welding. *J. Mater. Process. Technol.* **2006**, *178*, 342–349. [CrossRef]
17. Kimapong, K.; Watanabe, T. Friction stir welding of aluminum alloy to steel. *Weld. J.* **2004**, *83*, 277.
18. Tamadon, A.; Pons, D.J.; Clucas, D. Microstructural study on thermomechanical behaviour of 6082-t6 aluminium bfsw weld plates. In *Materials@UC 2018*; University of Canterbury: Christchurch, New Zealand, 2018.
19. Murr, L.; Liu, G.; McClure, J. Dynamic recrystallization in friction-stir welding of aluminium alloy 1100. *J. Mater. Sci. Lett.* **1997**, *16*, 1801–1803. [CrossRef]
20. Tamadon, A.; Pons, D.J.; Clucas, D. Thermomechanical performance of bobbin tool design as an innovative variant for friction stir welding. In *Manufacturing and Design Conference (MaD 2019)*; NZ's Manufacturing, Design and Entrepreneurship (MaDE) Network: Auckland, New Zealand, 2019.
21. Tamadon, A.; Pons, D.J.; Clucas, D.; Sued, K. Internal material flow layers in aa6082-t6 butt-joints during bobbin friction stir welding. *Metals* **2019**, *9*, 1059. [CrossRef]
22. Hattingh, D.; Blignault, C.; Van Niekerk, T.; James, M. Characterization of the influences of fsw tool geometry on welding forces and weld tensile strength using an instrumented tool. *J. Mater. Process. Technol.* **2008**, *203*, 46–57. [CrossRef]
23. Tamadon, A.; Pons, D.; Clucas, D. Analogue modelling of flow patterns in bobbin friction stir welding by the dark-field/bright-field illumination method. *Adv. Mater. Sci.* **2020**, *20*, 56–70. [CrossRef]
24. Siddiquee, A.N.; Pandey, S. Experimental investigation on deformation and wear of wc tool during friction stir welding (fsw) of stainless steel. *Int. J. Adv. Manuf. Technol.* **2014**, *73*, 479–486. [CrossRef]
25. Dialami, N.; Cervera, M.; Chiumenti, M.; de Saracibar, C.A. A fast and accurate two-stage strategy to evaluate the effect of the pin tool profile on metal flow, torque and forces in friction stir welding. *Int. J. Mech. Sci.* **2017**, *122*, 215–227. [CrossRef]
26. Ramulu, P.J.; Narayanan, R.G.; Kailas, S.V.; Reddy, J. Internal defect and process parameter analysis during friction stir welding of al 6061 sheets. *Int. J. Adv. Manuf. Technol.* **2013**, *65*, 1515–1528. [CrossRef]
27. Kumar, S.S.; Ravisankar, B.; Raviram, R.D. Evaluation of mechanical properties of friction stir welded commercially pure aluminium. In *MATEC Web of Conferences*; EDP Sciences: Les Ulis, France, 2018; p. 04003.
28. Uthayakumar, M.; Balasubramanian, V.; Rani, A.M.A.; Hadzima, B. Effects of Welding on the Fatigue Behaviour of Commercial Aluminum aa-1100 Joints. In *IOP Conference Series: Materials Science and Engineering*; IOP Publishing: Bristol, UK, 2018; p. 012065.
29. Tamadon, A.; Pons, D.J.; Clucas, D.; Sued, K. Texture evolution in aa6082-t6 bfsw welds: Optical microscopy and ebsd characterisation. *Materials* **2019**, *12*, 3215. [CrossRef]
30. Tamadon, A.; Pons, D.J.; Clucas, D. Flow-based anatomy of bobbin friction-stirred weld; aa6082-t6 aluminium plate and analogue plasticine model. *Appl. Mech.* **2020**, *1*, 3–19. [CrossRef]
31. Tamadon, A.; Pons, D.J.; Clucas, D. Structural anatomy of tunnel void defect in bobbin friction stir welding, elucidated by the analogue modelling. *Appl. Syst. Innov.* **2020**, *3*, 2. [CrossRef]
32. Egea, A.S.; Rodriguez, A.; Celentano, D.; Calleja, A.; De Lacalle, L.L. Joining metrics enhancement when combining fsw and ball-burnishing in a 2050 aluminium alloy. *Surf. Coat. Technol.* **2019**, *367*, 327–335. [CrossRef]
33. Kolbeck, C.; Pitonak, R.; Weissenbacher, R. Friction Stir Welding Tool. U.S. Patent Application No. 12/742,760, 14 October 2010.

34. Ye, F.; Fujii, H.; Tsumura, T.; Nakata, K. Friction stir welding of inconel alloy 600. *J. Mater. Sci.* **2006**, *41*, 5376–5379. [CrossRef]
35. Seighalani, K.R.; Givi, M.B.; Nasiri, A.; Bahemmat, P. Investigations on the effects of the tool material, geometry, and tilt angle on friction stir welding of pure titanium. *J. Mater. Eng. Perform.* **2010**, *19*, 955–962. [CrossRef]
36. Tolle, C.R.; Clark, D.E.; Barnes, T.A. Friction Stir Welding Tool. U.S. Patent No. 7,357,292, 15 April 2008.
37. Park, H.K.; Oh, I.H.; Yoon, H.J.; Son, H.T.; Lee, K.J.; Bang, H.S.; Bang, H.S. Preparation Method of Tungsten Carbide Sintered Body for Friction Stir Welding Tool. U.S. Patent No. 9,580,361, 28 February 2017.
38. Bahaaddini, M.; Baharvandi, H.R.; Ehsani, N.; Khajehzadeh, M.; Tamadon, A. Pressureless sintering of lps-sic (sic-al2o3-y2o3) composite in presence of the b4c additive. *Ceram. Int.* **2019**, *45*, 13536–13545. [CrossRef]
39. Machniewicz, T.; Nosal, P.; Korbel, A.; Hebda, M. Effect of fsw traverse speed on mechanical properties of copper plate joints. *Materials* **2020**, *13*, 1937. [CrossRef]
40. Torzewski, J.; Grzelak, K.; Wachowski, M.; Kosturek, R. Microstructure and low cycle fatigue properties of aa5083 h111 friction stir welded joint. *Materials* **2020**, *13*, 2381. [CrossRef]
41. Msomi, V.; Mbana, N. Mechanical properties of friction stir welded aa1050-h14 and aa5083-h111 joint: Sampling aspect. *Metals* **2020**, *10*, 214. [CrossRef]
42. Li, Y.; Sun, D.; Gong, W. Effect of tool rotational speed on the microstructure and mechanical properties of bobbin tool friction stir welded 6082-t6 aluminum alloy. *Metals* **2019**, *9*, 894. [CrossRef]
43. Wang, F.; Li, W.; Shen, J.; Hu, S.; Dos Santos, J. Effect of tool rotational speed on the microstructure and mechanical properties of bobbin tool friction stir welding of al–li alloy. *Mater. Des.* **2015**, *86*, 933–940. [CrossRef]
44. Wang, L.; Sabisch, J.; Lilleodden, E. Kink formation and concomitant twin nucleation in mg–y. *Scr. Mater.* **2016**, *111*, 68–71. [CrossRef]
45. Korhonen, R.K.; Saarakkala, S. Biomechanics and modeling of skeletal soft tissues. In *Theoretical Biomechanics*; InTech: Rijeka, Croatia, 2011; pp. 113–132.
46. Roeder, B.A.; Kokini, K.; Sturgis, J.E.; Robinson, J.P.; Voytik-Harbin, S.L. Tensile mechanical properties of three-dimensional type i collagen extracellular matrices with varied microstructure. *J. Biomech. Eng.* **2002**, *124*, 214–222. [CrossRef]
47. Zitnay, J.L.; Weiss, J.A. Load transfer, damage, and failure in ligaments and tendons. *J. Orthop. Res.* **2018**, *36*, 3093–3104. [CrossRef]
48. Mitsuhashi, K.; Ghosh, S.; Koibuchi, H. Mathematical modeling and simulations for large-strain j-shaped diagrams of soft biological materials. *Polymers* **2018**, *10*, 715. [CrossRef]
49. Lee, B.; Zhou, X.; Riching, K.; Eliceiri, K.W.; Keely, P.J.; Guelcher, S.A.; Weaver, A.M.; Jiang, Y. A three-dimensional computational model of collagen network mechanics. *PLoS ONE* **2014**, *9*, e111896. [CrossRef]
50. Khan, N.Z.; Siddiquee, A.N.; Khan, Z.A.; Shihab, S.K. Investigations on tunneling and kissing bond defects in fsw joints for dissimilar aluminum alloys. *J. Alloy. Compd.* **2015**, *648*, 360–367. [CrossRef]
51. Kang, M.; Yoon, J.; Kim, C. Hook formation and joint strength in friction stir spot welding of al alloy and al–si-coated hot-press forming steel. *Int. J. Adv. Manuf. Technol.* **2020**, *106*, 1671–1681. [CrossRef]
52. Krishnan, M.M.; Maniraj, J.; Deepak, R.; Anganan, K. Prediction of optimum welding parameters for fsw of aluminium alloys aa6063 and a319 using rsm and ann. *Mater. Today* **2018**, *5*, 716–723. [CrossRef]
53. Liu, X.; Wu, C.; Padhy, G.K. Characterization of plastic deformation and material flow in ultrasonic vibration enhanced friction stir welding. *Scr. Mater.* **2015**, *102*, 95–98. [CrossRef]
54. Mahto, R.P.; Anishetty, S.; Sarkar, A.; Mypati, O.; Pal, S.K.; Majumdar, J.D. Interfacial microstructural and corrosion characterizations of friction stir welded aa6061-t6 and aisi304 materials. *Met. Mater. Int.* **2019**, *25*, 752–767. [CrossRef]
55. Nasir, S.N.N.M.; Sued, M.K.; Abidin, M.Z.Z. The effects of rotational tool speed on mechanical properties of bobbin friction stir welded aa1100. In *Advances in Material Sciences and Engineering*; Springer: Berlin, Germany, 2020; pp. 337–342.
56. Shen, Z.; Li, W.; Ding, Y.; Hou, W.; Liu, X.; Guo, W.; Chen, H.; Liu, X.; Yang, J.; Gerlich, A. Material flow during refill friction stir spot welded dissimilar al alloys using a grooved tool. *J. Manuf. Process.* **2020**, *49*, 260–270. [CrossRef]

57. Teimournezhad, J.; Masoumi, A. Experimental investigation of onion ring structure formation in friction stir butt welds of copper plates produced by non-threaded tool pin. *Sci. Technol. Weld. Join.* **2010**, *15*, 166–170. [CrossRef]
58. Wen, Q.; Li, W.; Gao, Y.; Yang, J.; Wang, F. Numerical simulation and experimental investigation of band patterns in bobbin tool friction stir welding of aluminum alloy. *Int. J. Adv. Manuf. Technol.* **2019**, *100*, 2679–2687. [CrossRef]
59. Wiedenhoft, A.G.; Amorim, H.J.D.; Rosendo, T.D.S.; Tier, M.A.D.; Reguly, A. Effect of heat input on the mechanical behaviour of al-cu fsw lap joints. *Mater. Res.* **2018**, *21*. [CrossRef]

© 2020 by the authors. Licensee MDPI, Basel, Switzerland. This article is an open access article distributed under the terms and conditions of the Creative Commons Attribution (CC BY) license (http://creativecommons.org/licenses/by/4.0/).

technologies

MDPI

Communication

AFM Characterization of Stir-Induced Micro-Flow Features within the AA6082-T6 BFSW Welds

Abbas Tamadon *, Dirk J. Pons * and Don Clucas

University of Canterbury, Department of Mechanical Engineering, Christchurch 8140, New Zealand; don.clucas@canterbury.ac.nz

* Correspondence: abbas.tamadon@pg.canterbury.ac.nz (A.T.); dirk.pons@canterbury.ac.nz (D.J.P.); Tel.: +64-021-028-12680 (A.T.)

Received: 4 October 2019; Accepted: 6 November 2019; Published: 7 November 2019

Abstract: Bobbin Friction Stir Welding (BFSW) is a thermomechanical process containing severe plastic deformation by mechanical stirring and Dynamic Recrystallization (DRX) during recooling. Here we report the three-dimensional characteristics of the micro-flow patterns within the aluminium weld structure. The Surface topography observations by Atomic Force Microscopy (AFM) show the stirred-induced microstructural evolution where the rearrangement of dislocations at the sub-grain scale, and the subsequent High- and Low-Angle Grain Boundaries (HAGBs, LAGBs) exhibit specific alterations in grain size and morphology of the weld texture. The dislocations interaction in different regions of the weld structure also was observed in correlation to the thermomechanical behaviour of the BFSW process. These micro-flow observations within the weld breadth give a new insight into the thermomechanical characteristics of the FSW process during the stirring action where the plastic flow has a key role in the formation of the weld region distinct from the base metal.

Keywords: thermomechanical processing; bobbin friction stir welding; atomic force microscopy; AA6082-T6 aluminium alloy; dynamic recrystallization; precipitation

---

1. Introduction

Bobbin Friction Stir Welding (BFSW) is a modified variant of Friction Stir Welding (FSW) [1] where the conventional tool is replaced by a bobbin-shaped double-sided configuration [2,3]. The rotating double-shoulder bobbin tool penetrates from the edge through the interface of the side-by-side plates, and mixes the materials into a butt-shaped joint [4–6]. The heat input [4] generated by the friction between the rotating tool and the workpiece plasticises the material from both sides of the interface, Advancing Side (AS) and Retreating Side (RS), and stirs them together to form a bonded structure [6]. The stirring action causes severe plastic deformation [7] at temperatures well below the usual melting point [4]. Hence, the process is suitable for the joining of low temperature deformable alloys [8]. Aluminium is an ideal material for successful processing under BFSW [2,9]. AA6082-T6 aluminium is an industrial marine grade alloy with good machinability which has recently become attractive for FSW processing [10–14].

To achieve a defect-free weld in FSW processes, the material flow regimes have a higher priority than the metallurgical details [15]. Therefore, it is necessary for the continued improvement of the BFSW process to determine the plastic flow patterns in the weld region, which has received minimal attention in the literature compared with metallurgical aspects.

The severe plastic deformation during friction stir welding is the main cause of alteration in grain size and morphology [16–18]. The sub-grain scale analysis of microstructure can elucidate the relationship between the microstructure and the thermomechanical nature of the FSW process, especially shearing and heat generation [19]. The deformation-induced texture varies across the weld, as the shear is the function of the distance of the stirred flow mass from the position of the rotating tool [16,20]. Hence, a better understanding of the grain structure has the potential to contribute to knowledge of the evolution of the thermomechanical mechanism.

This paper presents an innovative study of the three-dimensional topology of the material flow features of BFSW weld texture. By utilising the Atomic Force Microscopy (AFM), the surface topography in the microscopic scale reveals the flow-based characteristics of the weld arising from the stirring action as a severe plastic deformation. This has the potential to give a better understanding of the effect of the microscopic flow regimes on the thermomechanical properties of the BFSW weld texture.

In this work, a high-magnification microscopic measurement was used to observe how the micro-scale plastic deformation affects the microstructure. In this regard, microstructural changes with a focus on dislocation and flow-induced imperfections were analysed in diverse regions of the AA6082-T6 BFSW weld structure.

AFM analysis was used to identify the microscopic details of the weld texture. This provides a greater resolution—to the atomic level—compared to other microstructure measurements. More specifically, the dislocation network and the precipitation evolution within the weld texture can be observed with AFM.

2. Materials and Methods

The BFSW welding test was conducted with the AA6082-T6 aluminium alloy (Standard; EN AW-6082, ISO: Al Si1MgMn) rolled plates (Aalco Metals Ltd, UK) as the workpiece. The analysed chemical composition of the AA6082-T6 plate with the elemental details is listed in Table 1.

Table 1. Chemical composition of the AA6082-T6 aluminium alloy, in elemental detail (wt.%).

AA6082-T6 Aluminium Alloy	
Chemical Element	% Present
Silicon (Si)	(0.70–1.30)
Magnesium (Mg)	(0.60–1.20)
Manganese (Mn)	(0.40–1.00)
Iron (Fe)	(0.0–0.50)
Chromium (Cr)	(0.0–0.25)
Zinc (Zn)	(0.0–0.20)
Titanium (Ti)	(0.0–0.10)
Copper (Cu)	(0.0–0.10)
Other (Each)	(0.0–0.05)
Other (total)	(0.0–0.15)
Aluminium (Al)	Balance

The weld samples were in a butt joint configuration comprising two similar plates (250 mm × 75 mm × 6 mm). The geometrically full-featured bobbin tool (included threads, flats and scrolls) was made of H13 tool steel with a hardness of 560 HV [5,6,21]. The schematic of the BFSW process is shown in Figure 1. The BFSW experiments were performed on a 3-axis CNC machining centre (2000 Richmond VMC Model, 600 Group brand, Sydney, Australia) with a Fanuc control unit and 14-horsepower spindle motor capacity. The simultaneous operation speeds consisted of spindle rotational speed (ω = 650 rpm), and weld travel rate (V = 400 mm/min). The direction of tool rotation was clockwise viewed from above.

The welded plates were cross-sectioned perpendicular to the weld-seam and were micro-polished to a mirror level, according to the standard metallographic preparation [22]. The samples then were etched by using of two modified reagents; (Reagent A: 2 mL HF + 2 mL HBF_4 + 10 mL HNO_3 + 20 mL CH_3COOH + 33 mL H_2O + 33 mL ethanol), and (Reagent B: 0.5 g $(NH_4)_2MoO_4$ + 3.0 g NH_4Cl + 1 mL HF + 18 mL HNO_3 + 80 mL H_2O) per [6,21]. Previous research has shown these etchants successfully delineate grain-boundary microstructure (Reagent A) and micro-flow patterns (Reagent B) [6,21]. Both reagents were used for similar etching time and temperature (90 s, 70 °C). For a better cleaning of the samples, the specimens were cleaned in the acetone bath (60 s, 18 °C), under an ultrasonic field of 40 KHz frequency. There is no known corrosive effect of acetone on this material.

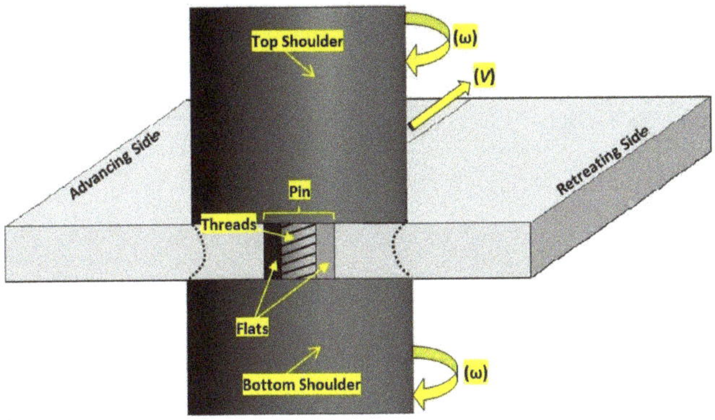

Figure 1. Schematic of the Bobbin Friction Stir Welding process; Bobbin-Tool in interaction with the workpiece as the substrate.

To observe the macro-flow patterns within the BFSW weld structure, the etched samples were studied with a stereoscopic microscope (Olympus Metallurgical Microscope, Tokyo, Japan). Microscopic observations of the morphological features were conducted with an atomic force microscope (Veeco Digital Instruments Dimension 3100, from Bruker). The in-situ observations were done in contact mode, equipped with TAP300-G (PDMS imprint and replica) tips (BudgetSensors, USA), in dry conditions at ambient conditions (in air, at room temperature). Visualisation of the recorded mapping data and the three-dimensional topography analysis were processed by Gwyddion software (VERSION 2.45).

3. Results

The macro-etched cross-section of the AA6082-T6 BFSW sample (etched by Reagent A) and micrographs of the relevant regions of the weld are shown in Figure 2. The microscopic images distinguish different morphological microstructure within the weld region from the Base Metal (BM), through to the transition region; Heat-Affected Zone (HAZ) and Thermomechanically Affected Zone (TMAZ), onwards to the Stirring Zone (SZ); flow arm patterns and middle of SZ. Compared to the base metal, the grain size in other regions of the weld shows an extensive grain refinement (reduction of grain size), and grain morphology.

The thermomechanical nature of the FSW process and the subsequent mechanical and thermal behaviours of the weld texture are believed to be the major activators for this microstructural evolution [21,23].

The AFM images of the weld region for the AA6082-T6 BFSW sample (etched by Reagent B) are demonstrated in Figure 3. After FSW treatment, there are microscopic changes that have taken place in different regions of the weld. The topographic features denote that the BM region (Zone 1) is smoother than the SZ (Zone 5), as well as the transition region at the AS/RS borders of the weld (Zones 3 and 4) where bulging lamellar microbands are evident.

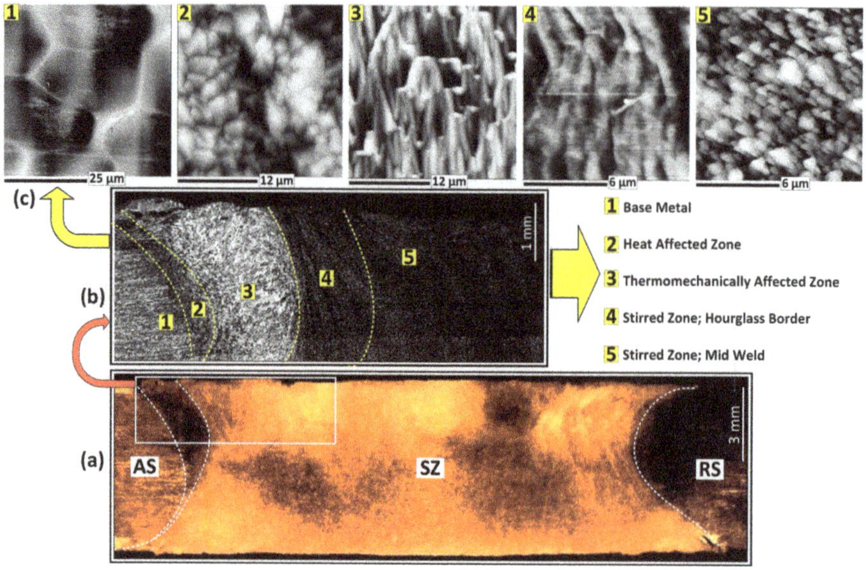

Figure 2. Macroscopic and microscopic features of the BFSW weld for the etched cross-section of the AA6082-T6 plate (reagent A); (**a**) Macro-etched cross-section of the AA6082-T6 sample, representative of the hourglass shaped weld structure, (**b**) the selected region from AS region at Figure 2a, in higher magnification, distinguishing five different regions for the weld breadth, (**c**) SEM images of 5 different regions of the weld texture, demonstrated in Figure 2b. (AS; Advancing Side, RS; Retreating Side, BM; Base Metal, SZ; Stirring Zone).

The topographic features are interpreted as micro-flow patterns caused by plastic deformation during stirring. The peak-and-valley-like micropattern is apparent throughout the weld section. The figure also shows roughness and texture measures derived from the AFM process. Also given in the figure are descriptions of the grain morphology, and the evolution between these states.

The figure is left for inspection, but there are several broad comments to be made. Inspection of Zones 3 and 4 (SZ) (Figure 3) shows the dominance of wrinkled and linear features, unlike the other regions. Shear is only active in Zones 3 and 4: it does not exist in Zone 1 due to the T6 heat treatment, nor in Zone 2 since this is only affected by heat flux. The shear is less apparent in mid-SZ (Zone 5), which is consistent with a stress relieving process. The Dynamic Recrystallization (DRX) mechanism uses stored strain and heat to nucleate and recrystallize the microstructure into ultrafine grains, thereby erasing the shear bands that might otherwise be expected at this location [23].

We attribute the wrinkled features in Zones 3 and 4 to activation of dislocations due to shear. Zone 3 shows activation at main grain boundaries (High-Angle Grain Boundaries, HAGBs). There is possible evidence for accumulation of dislocations at grain boundaries. In contrast in Zone 4 the dislocations are arrays within the grains (Low-Angle Grain Boundaries, LAGBs) forming sub-grain boundaries. This rearrangement of dislocations is a typical thermomechanical characteristic of DRX.

The 3D topography maps and roughness profiles are not all the same scale, so care must be taken in interpretation of microscopic features. Overall it is apparent that there is a large change in microstructure from the bases metal to the other zones. The base metal is characterised, as expected, by large grains, absence of sub-grain boundaries, an evidence of precipitation inside the grains (because of the artificial ageing T6 cycle [21]). The AFM method is not ideally suited to the larger scale of 50µm where there may be voids and other surface defects that may introduce noise, hence caution is required in the interpretation of some the features at the larger scale. In the transition regions (Zones 2–4) there is evidence of similar roughness and size of features as quantified by the surface profile, but the 3D images show that the morphology is very different across these zones. In the stirred region (Zone 5) the roughness increases again. We believe this represents the density of grain boundaries, which is confirmed by the 3D image which shows more homogeneity and fineness in morphology.

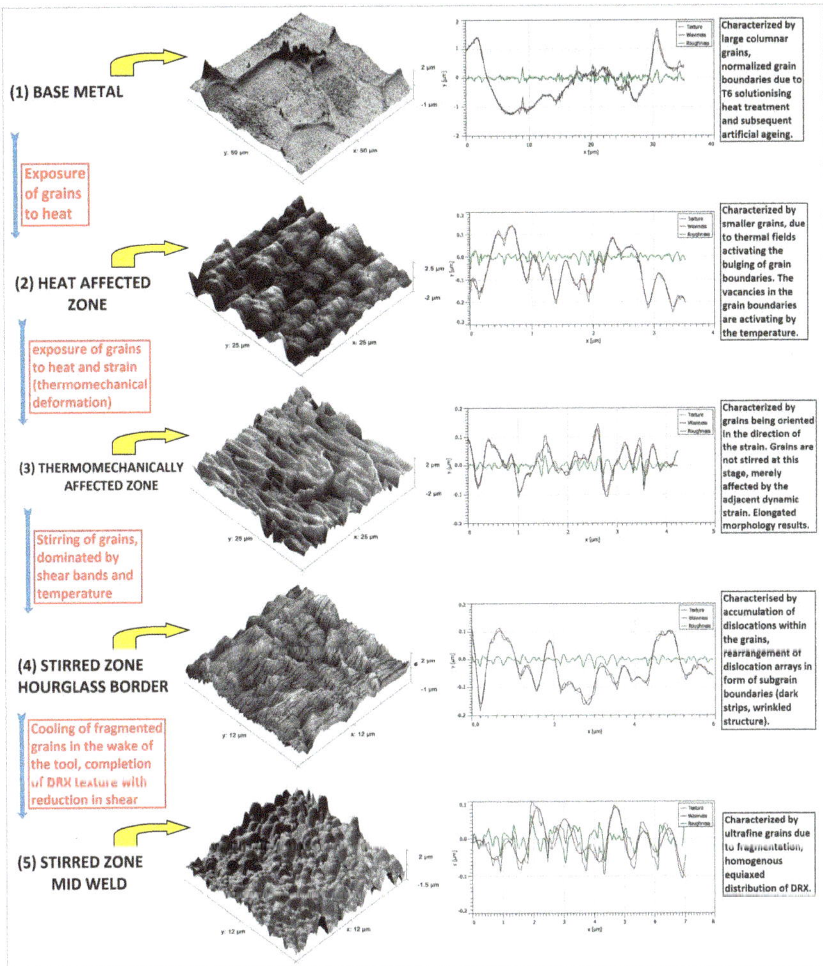

Figure 3. 3D topographic AFM images of different regions of the BFSW weld texture measured by the relevant height profiles of the surface roughness and the in detail explanation of the morphological evolution of the microstructure.

The question might be asked whether the observations might be explained by corrosion (from the etching process) rather than shear. We suggest that the effects are not due to corrosion, for the following reasons. First, the same etchants have previously been shown not to result in corrosion artefacts [6,21]. Second, there is no evidence of pitting or a local depletion of the matrix phase around the precipitate particles, as might be expected from an aggressive etching. Neither do the local line scan profiles show any evidence of pitting or protrusions.

We propose the following interpretation of the process. The mechanical stirring induces a shearing distortion to the lattice of the plasticised mass. To mitigate this during DRX, the microshrinkage positions evolve to be at the location of the accumulated shear bands. The etching procedure leads to more clarity of these shearing microbands in 3D surface tomography scanned by the AFM (Zones 3 and 4 in Figure 3, apparent at nanoscale). It is evident that the etchant has a very different effect on the material in Zone 3, compared to the other zones. The grain boundaries are pronounced, creating a three-dimensional interconnected set of ridges. The inside of the grain is locally dissolved to a greater extent. We proposed that this arises from the selective etching features of the reagent solution, and we attribute this to the shear bands being more sensitive to this etchant. Specific sample preparation was used to avoid over-etching and production of corrosion artefacts.

The grain boundaries and morphologies were revealed because of different reactions of each region of the weld texture to the applied reagent. In this regards, the surface roughness measurements indicate the morphology alteration in different regions after etching [6,24]. All samples were treated with the same polishing and etching procedure.

The surface morphology in the flow arm region (Zone 4) is in agreement with the theory of the stirring action in the FSW process. We suggest that the periodicity of the flow arms (visible in Figure 2b) is caused by the rotary motion of the pin and its flats create. This causes the deposition, in the wake of the tool at the advancing side, of the parallel arm shapes [24–27]. The microstructure of Zone 4 (in Figure 3) was selected from one of these arms, and shows that the sub-grain boundaries (dark lines in Figure 3) are arranged parallel to the flow arm.

The actual roughness in the mid-SZ (Zone 5) was reduced, as the grain size has decreased compared to other regions of the weld. The micrograph shows larger variations, which is because of the higher magnification (smaller scale). The graph of surface roughness confirms that the distance of peaks and valleys as a measure of the surface roughness, is minimized for Zone 5.

3.1. *Observation of Precipitate in Mid-SZ*

The most plausible area for location of the precipitate particles is mid-SZ, which experiences a full DRX compared to other regions of the weld. However, from an imaging perspective there is a risk that the etching might eject the fine precipitates from their position within the microstructure. To minimise this risk, an unetched micropolished sample was used for the precipitation analysis. Furthermore, a variety of channelling modes were used for the AFM analysis.

In addition to the usual height channel mapping, frequency channel was used to provide a better resolution for the free edges of the precipitate in contrast with the matrix. Also to indicate the localized torsional stress field at the interface the precipitate-matrix phase, dissipation channel was used. These different channeling modes provide a more comprehensive observation.

The micrographs (Figure 4a,d) and corresponding line profiles (Figure 4e) show an ultrafine particle of size less than 100 nm, a platelet morphology, and an embedment in the matrix. The height channel (Figure 4a) identifies a platelet shape particle without surface etching. The frequency channel (Figure 4b) provides better sharpness of contrast, and shows a localized cleaved pattern at the boundary of the precipitate-matrix. The dissipation channel (Figure 4c) also highlights the dissipated energy from the tapping tip of the AFM probe onto the sample surface. The red contrast, constructed at the interface of the precipitate-matrix phase, shows higher density of dissipated surface energy, attributed to the mechanical torsion at the border of the particle [28]. Frequency channel and dissipation channel both show that the border of the particle and the matrix possesses a distortion which is due to the embedment of the precipitate into the matrix [29,30]. Therefore, it is not an external particle or sediment at the surface.

In general, precipitation requires diffusion of alloying elements, and is function of time and temperature. DRX occurs within seconds during FSW, not enough time for diffusion to occur to produce precipitates and therefore the precipitating phase particles are expected to be ultrafine size [21,29], which indeed is demonstrated here (<100nm). The mechanism of precipitation here in FSW process is attributed to the severe shearing in the SZ and the heat. In the T6 artificially ageing process the cycle time is longer, the temperature is higher, and the process is closely controlled, resulting in control over the precipitates [21,29–31].

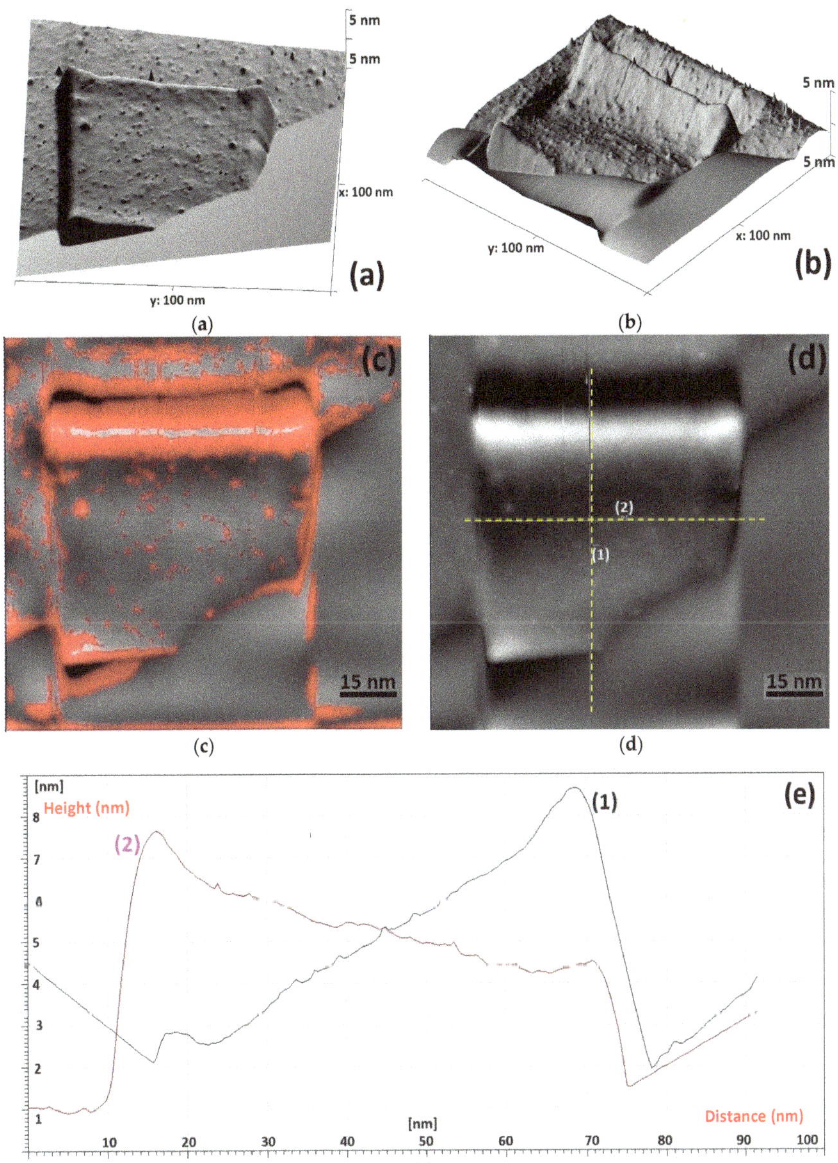

(**a**) (**b**) (**c**) (**d**) (**e**)

Figure 4. Selected surface area demonstrating the platelet shape precipitate, scanned by AFM using different channel modes; (**a**) height channel exposure, (**b**) frequency channel, (**c**) dissipation channel, (**d**) phase contrast micrograph, (**e**) line profiles corresponding to the surface roughness of the particle.

The observed platelet morphology has a maximum surface-to-volume, compared to other precipitate morphologies such as spherical or needle shaped. The platelet precipitation can occur in a relatively fast cooling rate at lower reaction temperatures [32,33]. Hence the observation of such a precipitate is consistent with the FSW condition. The density of precipitates was relatively low in the observed samples, less than might be expected from say fusion welding. This and the shape observed, suggest that the low temperature conditions in FSW result in a reduced precipitation.

3.2. Dislocations

Dislocations are the out-of-position of atoms in the crystal structure of grains. The dislocation patterns within different region of the weld are shown in Figure 5. At this high magnification the misarrangement of the crystal layers becomes apparent. The edge dislocations are evident as the crystalline defect in the structure of the aluminium. In general, in polycrystalline structures under shearing (with an intrinsic misorientational angle >0.99) it is expected to observe dislocations as a structural defect through the lattice [19,23].

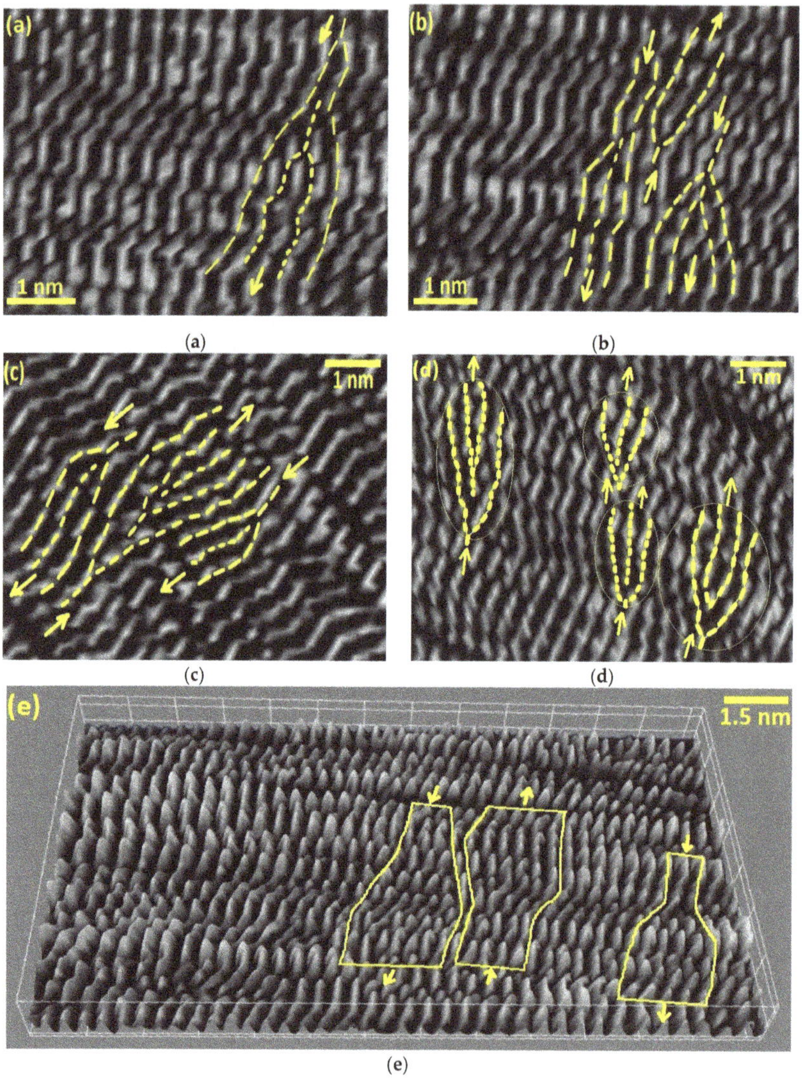

Figure 5. Stripped patterns of the crystalline lattice of AA6082-T6 BFSW weld; (**a**–**e**) The landmarks indicates some of edge dislocations between the lattice planes (crystal layers), in different regions of the weld; (**a**) BM, (**b**) HAZ, (**c**) TMAZ, (**d**) Hourglass-border of SZ. (**e**) AFM map with very high resolution crystal structure with corresponding edge Dislocations (mid-SZ).

During plastic deformation and the subsequent DRX, dislocations are formed in specific preferential orientations within the crystal lattice. Furthermore, during the recooling, interaction and annihilation of dislocation results in rearrangement of dislocation arrays with different density in different regions. This can be related to the amount of the stored strain releasing during the recovery process, also the absorbed heat which is different for each region based on the distance from the frictional stirring action. The dislocations originate from the applied stress during stirring causing shear between the crystal layers. After stirring the DRX process involves movement of the dislocations. They may aggregate to form Low Angle Grain Boundaries (LAGBs) within the grains, or transfer to the grain boundaries and contribute to formation of new High Angle Grain Boundaries (HAGBs) and motion of grain boundaries [21,23].

4. Discussion

This paper describes joining of an aluminium alloy by bobbin FSW, and investigation of weld microstructural features using the Scanning Electron Microscopy (SEM) and AFM.

A key finding is the identification of three-dimensional micro-flow features with specific changes in grain size and morphology attributed to the stirring action. This has not previously been shown for friction stir welding, and the literature is silent on this aspect.

The AFM technique provides a visualization of features within grains, and by inference provides a record of the flow patterns occurring in a solid-state mechanical stirring by the bobbin-tool FSW [21,23,24]. The surface topographic features for different regions of the weld can also be measured quantitatively to compare the surface roughness corresponding to the shearing regime. A tentative flow-induced thermomechanical mechanism has been suggested for the SZ and the transition region, where the stirring-induced shearing stress affects the weld structure through the thermomechanical behaviour of the BFSW process.

A metallurgical transformation during the DRX process is identified by its effects on changing of grain size to ultrafine. This is evident as grain refinement by increasing density of grain boundaries (see Zone 5, Figure 3). Precipitation is also expected from DRX, but was not readily observed in etched surfaces. Dislocation interaction in the sub-grain scale was visible in the transition regions (Zones 3 and 4, Figure 3), as were HAGBs (Zone 3, Figure 3) and LAGBs (Zone 4, Figure 3). Hence the transformations preceding and caused by DRX have been observed.

4.1. Welding Parameters

The formation of the weld texture is because of the mechanical stirring action at the proximity of the rotating tool [16]. The welding process causes fragmentation, severe plasticizing, shearing deformation and frictional heating. Therefore, the welding process parameters (tool geometry and welding speeds) can have a major effect on the final microscopic characteristics of the weld texture [34,35]. The complexity of the tool geometry increases the frictional heating generated at the position of the tool-material, inducing more plastic flow through the softened mass [36].

Similarly, the welding speeds (ω, V) also can induce more fragmentation and subsequent plasticising, resulting in more strain and hence DRX during the stirring action [37].

All these can intensify the shearing flow during the mass transportation, and potentially elevate the generated heat useable for the DRX mechanism. Therefore, this might be worthwhile to investigate the role of optimised welding parameters in the microscopic evolution of the weld texture characteristics.

4.2. Limitations of this Work and Implications for Future Research

Our 3D visualization analysis of the microscopic features was limited to the ultrafine microstructural details at the scale of the grain structure of the weld. However, there are some macro-size defects such as tunnel void or cracks which may also have shearing-flow effects. In this regard, because of the limitation of the AFM analysis to ultrafine magnification, the macroscopic defects are better analysed by other microscopic measurements, such as optical microscopy or electron microscopy. Furthermore, fractography analysis could evaluate the crack propagation and failure mechanisms. The formation of these macro-size tunnel void and the micro-cracks adversely affects the strength of the final weld, therefore is unacceptable to industry users.

Another possible future research opportunity could be to use AFM to quantify the grain characteristics for the different weld regions. It may be possible to characterise the surface features, and quantify sub-grain boundaries, and mathematical link these metrics to the weld process. Complementary methods such as electron microscopy (e.g., Electron Backscatter Diffraction (EBSD) and Transmission Electron Microscopy (TEM)) might be considered.

5. Conclusions

This research determined a physical measurement for describing micro-flow features within the BFSW weld breadth using a three-dimensional surface topography by AFM. It was revealed that the mechanical stirring was associated with complex flow regimes through the stirring zone, also induced shear features at the microscale. These add stored strain to the texture which appears to lead to physical alteration in recrystallization of the weld texture during the post-welding cooling. Therefore, different regions of the BSFW (SZ, TMAZ, HAZ) are identified in different microscopic patterns corresponding to thermomechanical behaviour of the weld.

A key outcome is the use of AFM to better understand the grain structure of the AA6082-T6 material under solid-state friction-stir welding. This is an important industrial material but its thermomechanical behaviour has been poor in this type of welding. The results of this paper elucidate the grain boundaries and precipitates, and thereby show the results of the thermomechanical processes. AFM has been shown to be a useful tool to better understand the gain boundary engineering, dislocation behaviour, and precipitation of this material.

Author Contributions: Conceptualization, A.T. and D.J.P.; Methodology, A.T.; Validation, A.T.; Formal analysis, A.T. and D.J.P.; Writing—original draft preparation, A.T.; Writing—review and editing, A.T. and D.J.P.; Supervision, D.J.P. and D.C.

Funding: This research received no external funding.

Conflicts of Interest: The authors declare no conflict of interest.

References

1. Thomas, W.; Nicholas, E.; Needham, J.; Murch, M.; Temple-Smith, P.; Dawes, C. Friction Stir Butt Welding. International Patent No. PCT/GB92/02203 27 November 1991.
2. Threadgill, P.L.; Ahmed, M.; Martin, J.P.; Perrett, J.G.; Wynne, B.P. The use of bobbin tools for friction stir welding of aluminium alloys. In *Materials Science Forum*; Trans Tech Publications: Zurich, Switzerland, 2010; pp. 1179–1184.
3. Sued, M.; Pons, D.; Lavroff, J.; Wong, E.-H. Design features for bobbin friction stir welding tools: Development of a conceptual model linking the underlying physics to the production process. *Mater. Des.* **2014**, *54*, 632–643. [CrossRef]
4. Hilgert, J.; Schmidt, H.; Dos Santos, J.; Huber, N. Thermal models for bobbin tool friction stir welding. *J. Mater. Process. Technol.* **2011**, *211*, 197–204. [CrossRef]
5. Tamadon, A.; Pons, D.; Sued, K.; Clucas, D. Formation mechanisms for entry and exit defects in bobbin friction stir welding. *Metals* **2018**, *8*, 33. [CrossRef]
6. Tamadon, A.; Pons, D.; Sued, K.; Clucas, D. Development of metallographic etchants for the microstructure evolution of a6082-t6 bfsw welds. *Metals* **2017**, *7*, 423. [CrossRef]
7. Hilgert, J.; Hütsch, L.L.; dos Santos, J.; Huber, N. Material flow around a bobbin tool for friction stir welding. In Proceedings of the COMSOL Conference, Paris, France, 29–30 October 2010.
8. Wang, F.; Li, W.; Shen, J.; Wen, Q.; dos Santos, J. Improving weld formability by a novel dual-rotation bobbin tool friction stir welding. *J. Mater. Sci. Technol.* **2018**, *34*, 135–139. [CrossRef]
9. Shen, J.; Wang, F.; Suhuddin, U.F.; Hu, S.; Li, W.; Dos Santos, J.F. Crystallographic texture in bobbin tool friction-stir-welded aluminum. *Metall. Mater. Trans. A* **2015**, *46*, 2809–2813. [CrossRef]

10. Huang, Y.; Wan, L.; Lv, S.; Feng, J. Novel design of tool for joining hollow extrusion by friction stir welding. *Sci. Technol. Weld. Join.* **2013**, *18*, 239–246. [CrossRef]
11. Węglowska, A. The use of a bobbin tool in the friction stir welding of plates made of aluminium alloy en aw 6082–t6. *Biuletyn Instytutu Spawalnictwa Gliwicach* **2018**, *62*. [CrossRef]
12. Wan, L.; Huang, Y.; Guo, W.; Lv, S.; Feng, J. Mechanical properties and microstructure of 6082-t6 aluminum alloy joints by self-support friction stir welding. *J. Mater. Sci. Technol.* **2014**, *30*, 1243–1250. [CrossRef]
13. Chen, S.; Lu, A.; Yang, D.; Lu, S.; Dong, J.; Dong, C. Analysis on flow pattern of bobbin tool friction stir welding for 6082 aluminum. In Proceedings of the 1st International Joint Symposium on Joining and Welding, Osaka, Japan, 6–8 November 2013; Elsevier: Amsterdam, The Netherlands, 2013; pp. 353–358.
14. Wan, L.; Huang, Y.; Lv, Z.; Lv, S.; Feng, J. Effect of self-support friction stir welding on microstructure and microhardness of 6082-t6 aluminum alloy joint. *Mater. Des.* **2014**, *55*, 197–203. [CrossRef]
15. Hilgert, J.; Dos Santos, J.F.; Huber, N. Shear layer modelling for bobbin tool friction stir welding. *Sci. Technol. Weld. Join.* **2012**, *17*, 454–459. [CrossRef]
16. Dialami, N.; Cervera, M.; Chiumenti, M. Effect of the tool tilt angle on the heat generation and the material flow in friction stir welding. *Metals* **2019**, *9*, 28. [CrossRef]
17. Egea, A.S.; Rodriguez, A.; Celentano, D.; Calleja, A.; de Lacalle, L.L. Joining metrics enhancement when combining fsw and ball-burnishing in a 2050 aluminium alloy. *Surf. Coat. Technol.* **2019**, *367*, 327–335. [CrossRef]
18. He, X.; Gu, F.; Ball, A. A review of numerical analysis of friction stir welding. *Prog. Mater. Sci.* **2014**, *65*, 1–66. [CrossRef]
19. Tayon, W.A.; Domack, M.S.; Hoffman, E.K.; Hales, S.J. Texture evolution within the thermomechanically affected zone of an al-li alloy 2195 friction stir weld. *Metall. Mater. Trans. A* **2013**, *44*, 4906–4913. [CrossRef]
20. Fonda, R.; Bingert, J. Texture variations in an aluminum friction stir weld. *Scr. Mater.* **2007**, *57*, 1052–1055. [CrossRef]
21. Tamadon, A.; Pons, D.; Sued, K.; Clucas, D. Thermomechanical grain refinement in aa6082-t6 thin plates under bobbin friction stir welding. *Metals* **2018**, *8*, 375. [CrossRef]
22. Vander Voort, G.F.; Lampman, S.R.; Sanders, B.R.; Anton, G.J.; Polakowski, C.; Kinson, J.; Muldoon, K.; Henry, S.D.; Scott Jr, W.W. Asm handbook. *Metallogr. Microstruct.* **2004**, *9*. Available online: https://www.asminternational.org/documents/10192/1849770/06044G_Frontmatter.pdf (accessed on 11 November 2019).
23. Tamadon, A.; Pons, D.J.; Clucas, D.; Sued, K. Texture evolution in aa6082-t6 bfsw welds: Optical microscopy and ebsd characterisation. *Materials* **2019**, *12*, 3215. [CrossRef] [PubMed]
24. Tamadon, A.; Pons, D.J.; Clucas, D.; Sued, K. Internal material flow layers in aa6082-t6 butt-joints during bobbin friction stir welding. *Metals* **2019**, *9*, 1059. [CrossRef]
25. Tamadon, A.; Pons, D.; Sued, M.; Clucas, D.; Wong, E. Preparation of plasticine material for analogue modelling. In Proceedings of the International Conference on Innovative Design and Manufacturing (ICIDM2016), Auckland, New Zealand, 24–26 January 2016.
26. Tamadon, A.; Pons, D.; Sued, M.; Clucas, D.; Wong, E. Analogue modelling of bobbin tool friction stir welding. In Proceedings of the International Conference on Innovative Design and Manufacturing (ICIDM2016), Auckland, New Zealand, 24–26 January 2016.
27. Sued, M.; Tamadon, A.; Pons, D. Material flow visualization in bobbin friction stir welding by analogue model. *Proc. Mech. Eng. Res. Day* **2017**, *2017*, 368–369.
28. Barényi, I.; Eckert, M.; Majerík, J.; Bezecný, J. Afm and nanoindentation study of selected aluminium alloys. *Zeszyty Naukowe Politechniki Rzeszowskiej. Mechanika* **2018**, 143–152. [CrossRef]
29. Chrominski, W.; Lewandowska, M. Precipitation phenomena in ultrafine grained al–mg–si alloy with heterogeneous microstructure. *Acta Mater.* **2016**, *103*, 547–557. [CrossRef]
30. Marioara, C.; Andersen, S.; Jansen, J.; Zandbergen, H. Atomic model for gp-zones in a 6082 al–mg–si system. *Acta Mater.* **2001**, *49*, 321–328. [CrossRef]
31. Dadbakhsh, S.; Karimi Taheri, A. Study on static strain aging of 6082 aluminium alloy. *Mater. Sci. Technol.* **2010**, *26*, 169–175. [CrossRef]
32. Gubicza, J.; Krállics, G.; Schiller, I.; Malgyn, D. Evolution of the microstructure of al 6082 alloy during equal-channel angular pressing. In *Materials Science Forum*; Trans Tech Publications: Zurich, Switzerland, 2005; pp. 453–458.

33. Marioara, C.; Andersen, S.; Jansen, J.; Zandbergen, H. The influence of temperature and storage time at rt on nucleation of the β phase in a 6082 al–mg–si alloy. *Acta Mater.* **2003**, *51*, 789–796. [CrossRef]
34. Fonda, R.; Knipling, K.; Bingert, J. Microstructural evolution ahead of the tool in aluminum friction stir welds. *Scr. Mater.* **2008**, *58*, 343–348. [CrossRef]
35. Fonda, R.; Knipling, K. Texture development in friction stir welds. *Sci. Technol. Weld. Join.* **2011**, *16*, 288–294. [CrossRef]
36. Fuse, K.; Badheka, V. Bobbin tool friction stir welding: A review. *Sci. Technol. Weld. Join.* **2019**, *24*, 277–304. [CrossRef]
37. Fonda, R.; Reynolds, A.; Feng, C.; Knipling, K.; Rowenhorst, D. Material flow in friction stir welds. *Metall. Mater. Trans. A* **2013**, *44*, 337–344. [CrossRef]

© 2019 by the authors. Licensee MDPI, Basel, Switzerland. This article is an open access article distributed under the terms and conditions of the Creative Commons Attribution (CC BY) license (http://creativecommons.org/licenses/by/4.0/).

technologies

MDPI

Article

Comparison of Different Hydrotalcite Solid Adsorbents on Adsorptive Desulfurization of Liquid Fuel Oil

Mozammel Mazumder [1], Rajib Das [1], Md Symon Jahan Sajib [2], Andrew Jewel Gomes [1,†], Mohammad Islam [3], Thinesh Selvaratnam [4] and Ashiqur Rahman [1,*]

1 Department of Chemical Engineering, Lamar University, Beaumont, TX 77710, USA; mmazumder@lamar.edu (M.M.); mrdasceps@gmail.com (R.D.); andrew.gomes@gmail.com (A.J.G.)
2 Department of Chemical Engineering, Howard University, Washington, DC 20059, USA; symon.sajib@howard.edu
3 ConocoPhillips Company, Houston, TX 77252, USA; rafiq.r.Islam@conocophillips.com
4 Department of Civil and Environmental Engineering, Lamar University, Beaumont, TX 77710, USA; tselvaratnam@lamar.edu
* Correspondence: arahman2@lamar.edu
† Current Address: Silver Spring, MD 20906, USA, Former Research Faculty of Lamar University.

Received: 28 February 2020; Accepted: 25 April 2020; Published: 27 April 2020

Abstract: With increasingly stringent environmental regulations, desulfurization for gasoline oil production has become an important issue. Nowadays, desulfurization technologies have become an integral part of environmental catalysis studies. It is also important for processing of fuel for fuel-cells, which has a strict requirement for sulfur content for internal combustion engines. In this study, we focused on the preparation and characterization of magnesium hydroxide/aluminum supported NiO, ZnO, ZrO_2, NiO-ZnO, NiO-ZrO_2, adsorbents for the adsorptive desulfurization of liquid fuels. These hydrotalcite adsorbents were prepared by co-precipitation method and used for adsorption of thiophene (in n-pentane, as model fuel) and dibenzothiophene at ambient temperature and pressure. The physicochemical behaviors of the fresh adsorbents such as structure, composition, and bonding modes were determined using X-ray diffraction (XRD), Raman spectroscopy, Fourier-transform infrared spectroscopy (FTIR), energy dispersive X-Ray analysis (EDAX), scanning electron microscopy (SEM), X-ray photoelectron spectroscopy (XPS) and thermogravimetric analysis (TGA). The sulfur concentration in the mixture (thiophene and n-pentane) was measured by UV-Vis spectrophotometry. The percentages of thiophene removal and the adsorption capacity (mg of sulfur per g of adsorbent) of the five adsorbents were compared. The adsorption performance confirmed that NiO-ZrO_2 and NiO-ZnO adsorbents are more efficient in removing thiophene/dibenzothiophene than that of three other adsorbents. The qualitative studies using XPS confirmed the efficient adsorption nature of modified hydrotalcite adsorbents on dibenzothiophene.

Keywords: adsorption; hydrotalcite; thiophene/dibenzothiophene; n-pentane; desulfurization

1. Introduction

The combustion of fuels containing sulfur produces sulfur dioxide, which is liable to a series of air pollution events. Human exposure to sulfur dioxide in the ambient air has been related to respiratory system diseases and even lung cancer [1]. Therefore, in 2006 the U.S. Environmental Protection Agency (EPA) reduced the allowable sulfur levels in liquid fuels. The gasoline sulfur limit was reduced to 30 ppm, while the diesel fuel sulfur limit was reduced to 15 ppm [2]. With time, EPA regulations became more stringent. However, the existing desulfurization technologies failed to reduce the sulfur level of gasoline or diesel fuel to less than 10 ppm [1,3].

Various methods such as adsorption desulfurization (ADS) [3], biodesulfurization (BDS) [4], extraction desulfurization (EDS) and oxidative desulfurization (ODS) have been reported for the removal of sulfur compounds from fuels [1,5,6]. Hydrodesulfurization (HDS) is the most popular and effective sulfur removing technology in refineries. However, the production of ultra-low sulfur fuels requires a large volume of catalyst [7]. Adsorptive desulfurization can provide low sulfur fuel for fuel cells and catalyst beds. Many technologies have diverged from HDS to produce low sulfur products; however, sorption, catalytic oxidation, and evaporation show the most potential among them [8]. Oxidative desulfurization (ODS) is an alternative technique for adsorptive sulfur removal. Normally an oxidative reagent is used in combination with the catalyst to oxidize the sulfur. In recent years, several kinds of ODS systems were developed successfully such as H_2O_2 organic matrixes, H_2O_2/Ti-modified zeolites, H_2O_2/polyoxometalates (POMs), H_2O_2/ionic liquid, H_2O_2/polyoxometalate-based ionic liquid, WO3/TiO_2 and CeO_2/TiO_2 [9,10]. The oxidation products are generally removed by an extraction process using solvents. Recently Ullah et al. reported on the adsorptive removal of benzothiophene (BT) from liquid fuel using a highly porous metal-organic framework based on a bicomponent zirconium (IV) benzene-tricarboxylate Zr (BTC), and its post-synthetically modified hybrid form with dodeca-tungstophosphoric acid (HPW/Zr(BTC) [11]. Additionally, different microorganisms have been used to remove sulfur. Both aerobic and anaerobic microorganisms prove to be effective desulfurization agents while maintaining aliphatic and aromatic content in the fuel [12]. Raj et al. investigated the effect of temperature, time and mass ratio for extractive desulfurization [13].

The objective of the present study is to identify an adsorbent that selectively removes sulfur from transportation fuels. The candidate adsorbents contain nickel, zinc, and zirconia. Five different types of adsorbents have been examined. Identical support of magnesium and aluminium hydroxides (hydrotalcites) was used for each of these adsorbents in order to determine the influence of the main component (Ni, Zn and ZrO_2). Nickel is relatively inexpensive and has previously shown to be more promising for sulfur removal [14]. Sulfur molecules from liquid fuels are adsorbed by direct interaction. Zinc oxide (ZnO), a highly active component used to remove sulfur species from refinery liquids [15]. Zirconium is used as an efficient adsorbent for the desulfurization because of its moderate surface area, bifunctional properties of acid and base [14]. The addition of nickel on ZnO and ZrO_2 is an innovative approach that takes advantage of the selectivity of Ni towards S-species and the high adsorptive capacity of ZnO and ZrO_2 support. It changes the nature of the active metal sites and increases the sulfur removing capacity of the adsorbent. Therefore, the current study compares different types of adsorbents in order to pave the way to remove sulfur to ultra-low levels by the selective adsorption of sulfur from liquid fuels.

2. Materials and Methods

Five modified hydrotalcite adsorbents were prepared by the combination of Mg, Al, Ni, Zn, and ZrO_2. The first adsorbent is the combination of Mg: Al: Ni in the approximate molar ratio of 4:1:3. It is a highly active nickel adsorbent supported on magnesium and aluminium. Increasing the capacity of a nickel adsorbent is highly beneficial, as nickel adsorbents are very effective at removing sulfur compounds. It is believed that the addition of some substance to the fuel can help increase the capacity of the nickel adsorbent. The second adsorbent is the combination of Mg: Al: Zn in the approximate molar ratio of 4:1:3. The third adsorbent is prepared with the combination of Mg: Al: Ni: Zn in the approximate molar ratio of 4:1:3:3. The fourth one is Mg: Al: ZrO_2 in the molar ratio of 4:1:7. The fifth adsorbent is Mg: Al: ZrO_2: Ni in the mole ratio 4:1:7:3. All the adsorbents are prepared by the co-precipitation method. They are used for adsorption of thiophene (in n-pentane) and dibenzothiophene at ambient temperature and pressure. A model fuel consisting of 50 ml of n-pentane (C_5H_{12}) and thiophene (100 ppm) was prepared, and adsorption experiments were performed under ambient conditions. The physicochemical characterizations of the fresh adsorbents such as structure, component, bond, elements, and composition were conducted using X-ray diffraction (XRD), Raman spectroscopy, Fourier-transform infrared spectroscopy (FTIR), energy dispersive X-Ray analysis (EDAX), scanning electron microscopy (SEM), X-ray photoelectron spectroscopy (XPS) and thermogravimetric analysis (TGA). The sulfur concentration in the mixture was monitored by UV-Vis spectrophotometry. Finally, the thiophene removal efficiency of the five adsorbents were compared. In addition, qualitative studies using XPS were performed to investigate the sulfur removal efficiency and adsorptive nature of adsorbents on dibenzothiophene.

2.1. Adsorbent Preparation

Five hydrotalcites with Mg: Al ratio 3:1 were prepared by the co-precipitation method. In this method, two solutions, A and B, were added at the same rate 50 ml h^{-1} to a beaker containing 100 ml of deionized water while stirring.

Solution A was prepared by mixing an equimolar solution of Mg and Al metal nitrates (200 ml) in the 3:1 molar ratio [16]. For making the adsorbent, Mg and Al metal nitrates were used as supporting metals and Zn, Ni, Zr were used as active metal. To make Ni adsorbent, Mg: Al: Ni molar ratio were used as 4:1:3. For making Zn adsorbent, Mg: Al: Zn molar ratio were used as 4:1:3. For making Zn and Ni combined adsorbent, Mg: Al: Ni: Zn molar ratio were used as 4:1:3:3. Zr adsorbent was prepared by maintaining Mg: Al: Zr molar ratio 4:1:7. Lastly, Ni and Zr adsorbent were prepared by maintaining Mg: Al: Ni: Zr molar ratio 4:1:3:7.

Solution B was prepared by dissolving 14 g sodium hydroxide (0.35 mol) and 15.9 g sodium carbonate (0.15 mol) in 200 ml deionized water. The pH of the suspensions was around 10. The precipitates were aged at 75 °C for 18 h in a dryer. The resulting product was filtered, washed thoroughly with deionized water until the filtrate showed no presence of NaOH, and subsequently dried at 95 °C for 24 hours. Part of the samples was heated at 450 °C for 12 h in a furnace for further calcinations and catalytic activity study.

2.2. Model Fuels using Thiophene (in Pentane) and Dibenzothiophene

Model fuels were used in some of the adsorbent testing to determine the selectivity of the adsorbent towards certain compounds. The model fuel consisted of pentane (C_5H_{12}) and thiophene, an organosulfur compound with the chemical formula C_4H_4S. The composition was 50 ml of pentane and 5 microliters of thiophene. It contained 100 ppm of thiophene in 50 ml of pentane. The sulfur compounds found in the model fuel contained highly substituted sulfur compounds, which tended to be more difficult to remove due to the steric hindrance around the sulfur atom. For making the calibration curve, five different concentrations of thiophene and pentane solution (20 ppm, 15 ppm, 10 ppm, 5 ppm, 0 ppm) were prepared. In addition, for the adsorbent performance testing, five different concentrations (100 ppm, 50 ppm, 25 ppm, 12 ppm, 6 ppm of thiophene in pentane solution) were prepared. A five-point calibration curve was made in order to retain the quality of analysis. For more accuracy and to understand the error, the response at each concentration was repeated to obtain the error bar from the responses. Additionally, the adsorbents were mixed with the dibenzothiophene for a qualitative study using the XPS. Percentages of sulfur removal capacity by the five adsorbents were measured by these five different concentrations (100 ppm, 50 ppm, 25 ppm, 12 ppm, 6 ppm of thiophene (in pentane solution) and dibenzothiophene.

2.3. Characterization Techniques

The powder X-ray diffraction (XRD) experiment of the sample was performed by the Bruker AXS D8 Discover diffractometer with GADDS (General Area Detector Diffraction System, Bruker Corporation, Billerica, MA, USA); operated by a Cu-Kα radiation source and filtered with a graphite monochromator (l = 1.5406 A°). A HI STAR two-dimensional multi-wire area detector was used. All the samples were first ground to make very fine particle powder. The X-ray beam was 40 kV and 40 mA power. The incident ω angle was 5°. A laser system was used to ensure the alignment of the sample position on the instrument. XRD scans were recorded from 5° to 77° for 2θ with a 0.050° step-width and 180 step times. Further, the crystal sizes of different adsorbents were determined by Scherrer equation (L = kλ/(FWHM)Cosθ) with dimensionless shape factor, k = 0.94 and x-ray wavelength, λ = 1.5406 A°. The full width at half maximum (FWHM) was determined using OriginPro software (version 9.0, Northampton, MA, USA). The multiple peak fit function was used in the case of nonlinear curve fitting.

Fourier-transform infrared spectroscopy (FTIR) spectra were recorded on a BRUKER TENSOR 27 FTIR Spectrometer (Bruker Corporation, Billerica, MA, USA). This machine works in the range of 400–4000 cm^{-1} wave number. In order to minimize the amount of bound water, the samples were kept in an air-tight container at room temperature until measurement, however, the possibility of water absorption from the atmosphere is not entirely excluded.

Hitachi S-3400N scanning electron microscope (SEM), (Hitachi, Ltd., Tokyo, Japan) was used to investigate the morphology and the composition of the adsorbents. Automatic beam axis alignment functions like auto beam setting and auto axial alignment were used. For this experiment, high SE resolution of 10 nm at 3 KV and 5-axis motorized stage with high tilt (−20 to +90 degree) and allowance for tall samples up to 80 mm high were used.

The sample's thermal stability was studied by the thermogravimetric analysis (TGA) by the TGA instrument (Netzsch, STA 449C Jupiter, and TA Instruments SDT Q-600, Erich NETZSCH GmbH & Co. Holding KG, Selb, Germany). All the samples were heated from 25 to 1200 °C, and an airflow rate of 60 mL min^{-1} was maintained. The heating rate was 20 °C min^{-1}. Differential scanning calorimetry (DSC) (Netzsch, STA 449C, and TA Instruments SDT Q-600, Erich NETZSCH GmbH & Co. Holding KG, Selb, Germany) measurements were carried out by maintaining the nitrogen flow rate of about 60 mL min^{-1}.

The Raman spectra of the sample were measured using Perkin Elmer, RAMAN FLEX 400 Raman spectrometer (Perkin Elmer, Waltham, MA, USA). This machine probes in the spectral range from 230 cm^{-1} to 3,500 cm^{-1} Raman shift.

The UV visible measurements were performed on a Varian Cary 50 Version 3 UV Visible Spectrometer (Agilent Technologies, Santa Clara, CA, USA) coupled with the potentiostat for applying electrochemical potentials. To study the evolution, the difference between the maximum peak absorbance at a particular wavelength (λ_{max}), and the absorbance at the initial scanning wavelength (λ_0) was accounted and plotted vs. time [17].

3. Result and Discussions

From the UV spectroscopy, pure pentane absorbance value of 0.564 at the 230 nm wavelength was determined. For every 20 ppm, 15 ppm, 10 ppm, 5 ppm concentration of thiophene in 50 ml pentane was measured at the 230 nm wavelength and then pure pentane value was subtracted from the measured value to determine the thiophene concentration. The absolute values for different thiophene concentrations are shown in Figure 1. The calibration curve and equation are used to calculate the thiophene concentration in pentane after using the adsorbent.

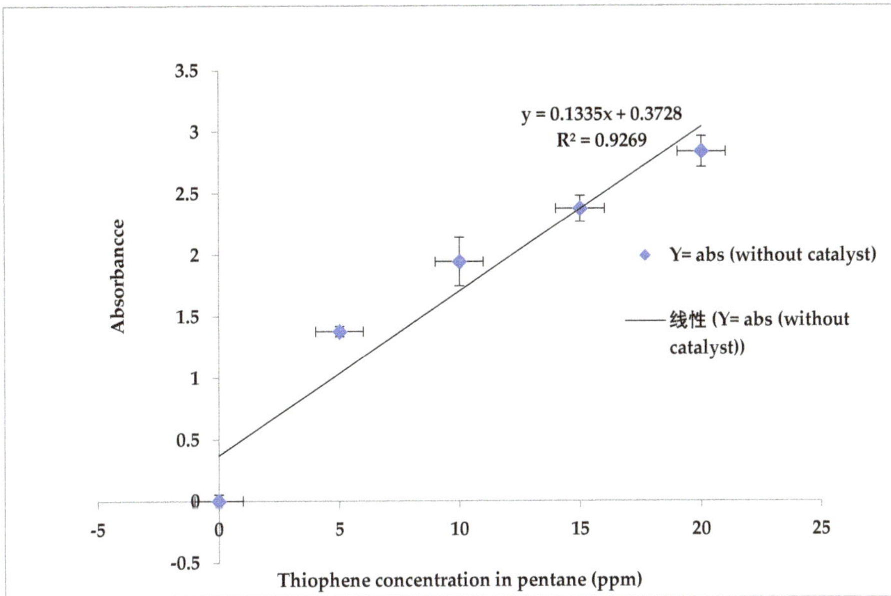

Figure 1. Calibration curve for thiophene concentration in pentane.

3.1. XRD Analysis of the Adsorbents

Figure 2a–e shows the X-ray diffractogram of the five adsorbents. From Figure 2a, it is clear that there has been the presence of hydrotalcite in the Ni adsorbent. Hydrotalcite peaks were found at 11.35°, 22.4°, 34.34°, 38.49°, 61.72°, 65.41° 2-theta angles. Figure 2a shows the presence of aluminium oxide, magnesium oxide, nickel oxide, magnesium aluminates, nickel aluminates, and magnesium nickel. Moreover, the formation of the hydrotalcite in the samples indicates the presence of carbonate in the interlayer of hydrotalcite. Hydrotalcite peaks are also observed at the same 2-theta angles for the Zn adsorbent (Figure 2b) indicating the presence of aluminium oxide, magnesium oxide, zinc oxide, magnesium aluminates, zinc aluminates, and magnesium zinc etc. Hydrotalcite peaks were found at 22.4°, 34.34°, 61.72°, 71.6° 2 theta angle (JCPDS file no 14-0191) for the Ni and Zn adsorbent (Figure 2c). The XRD patterns of Zr, and Zr and Ni (Figure 2d,e) indicate the presence of zirconia majorly in the form of the metastable tetragonal phase and minorly in the form of the monoclinic phase [18]. Generally, the tetragonal phase of zirconia can be stabilized by incorporating the promoters into the zirconia lattice. It is well known that the tetragonal phase of Zr is more active in catalysis [14]. The average crystalline sizes were calculated as 9.1 ± 2.8 nm for Ni adsorbent, 14.0 ± 3.9 nm for Zn adsorbent, 14.9 ± 7.6 nm for Ni and Zn adsorbent, 12.5 ± 1.0 nm for Zr adsorbent, 16.0 ± 7.5 nm for Zr and Ni adsorbent. The peak centers and corresponding FWHM are presented in the Supplementary Materials (Table S1).

Overall, the XRD patterns show both sharp peaks and broad humped peaks, as are presented by the Figure 2a–e, which indicate that the adsorbent materials are partially crystalline and partially amorphous.

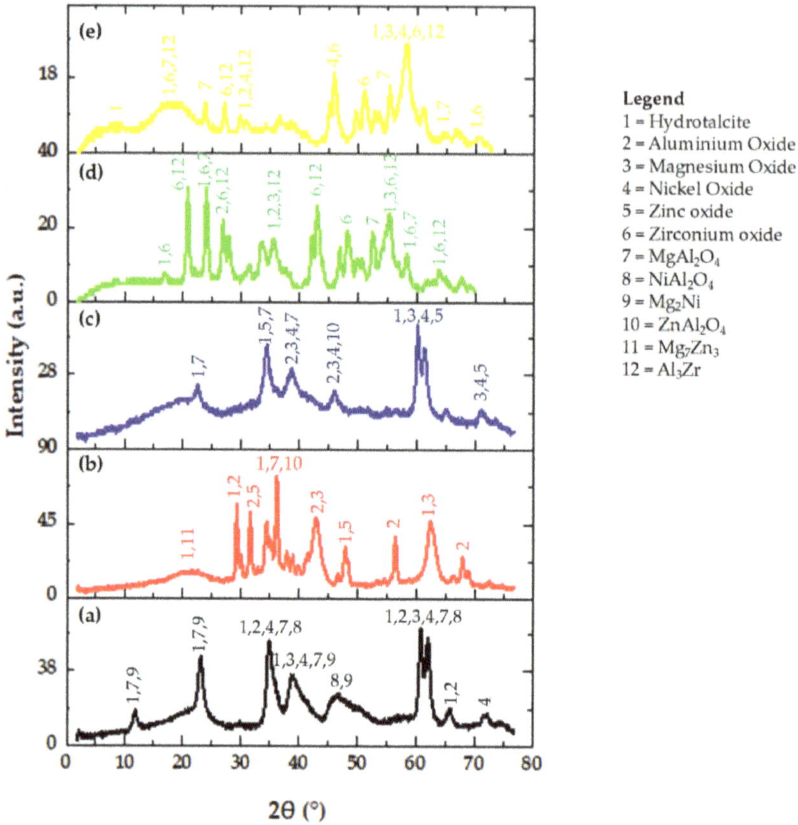

Figure 2. XRD analysis of calcined (**a**) Ni adsorbent; (**b**) Zn adsorbent (**c**) Ni and Zn adsorbent, (**d**) Zr adsorbent (**e**) Zr and Ni adsorbent (legends indicated on top of the peaks correspond to the various crystalline and amorphous phases identified from JCPDS file no. 14-0191).

3.2. SEM Analysis of the Adsorbents

The SEM images of the adsorbents are shown in Figure 3a–e. These figures show the surface texture of the adsorbents. The micrographs reveal that the materials have definite crystalline structures and suggest that no deformation of the metal oxides occurred during the preparation. Hydrotalcite is frequently reported as the substituted form of brucite [$Mg(OH)_2$] with the related hexagonal crystal shape. For instance, Brady et al. reported a hexagonal/rhombohedral crystal shape for the hydrotalcite [19]. In addition, our support materials; magnesium and aluminium interlocked with the active materials such as nickel and zinc that lead to a definite crystal shape and an increase in the adsorbent volume. The elemental composition of the five calcined adsorbents from EDAX analyses are presented in wt% in Figure 4. This figure clearly shows the relative abundance of elements in the adsorbent. Additionally, the elements and their corresponding atomic percentage (at. %) are presented in the Supplementary Materials (Table S2).

Figure 3. SEM analysis of calcined (**a**) Ni adsorbent; (**b**) Zn adsorbent (**c**) Ni and Zn adsorbent, (**d**) Zr adsorbent (**e**) Zr and Ni adsorbent.

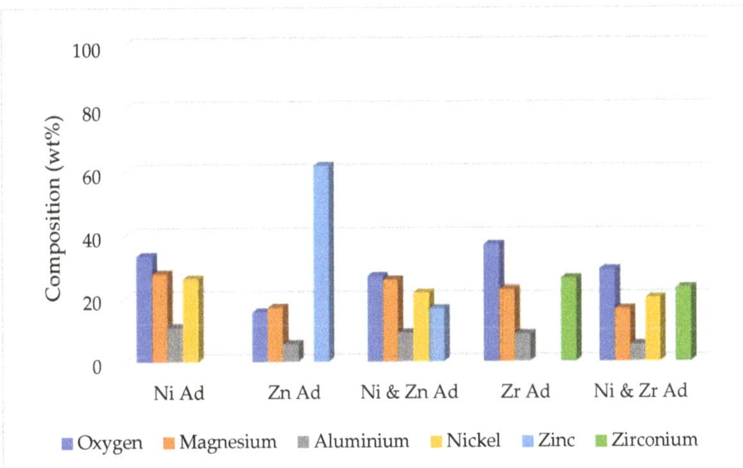

Figure 4. Elemental composition of the five calcined adsorbents from energy dispersive X-Ray analysis (EDAX).

3.3. FTIR Analysis of the Adsorbents

The recorded peaks corresponding to various functional groups of the carbonate intercalated hydrotalcite are presented in Table 1. The recorded spectra are presented by the Figure S1 in the Supplementary Materials.

When compared to the infrared spectrum of the sample and hydrotalcite reference materials, a number of similarities, as well as some differences, are observed. For the Ni adsorbent, the metal-hydroxide peaks present in the hydroxyl stretching region (3000 cm^{-1}–4000 cm^{-1}) are similar to the bands found in the hydrotalcite materials. The formation of a number of C-O stretches that either correlates with carbonate or oxalate appears at 1700 cm^{-1} and 1400 cm^{-1}. The expected peaks for carbonate in hydrotalcite are 1640 cm^{-1}, 1365 cm^{-1}, and 1313 cm^{-1}. However, it is difficult to accurately determine the bands of carbonate or oxalate in this range because of the similarities of carbonate and oxalate stretches due to the carbon-oxygen stretching vibrations.

The Zn adsorbent shows the metal-oxide and metal-hydroxide vibration between 3000 cm^{-1} to 3500 cm^{-1} wavenumbers [20]. There is a close correlation between the sample and reported peaks in the

1350 cm^{-1}–1380 cm^{-1} range which ascribed to the carbonate antisymmetric increase, which indicates the hydrotalcite has carbonate between the layers.

At the Ni-Zn adsorbent the presence of water stretching bands at 1700 cm^{-1} is clear. This presence is mostly due to the existence of a water bridging mode at 3401 cm^{-1} in the hydroxyl stretching region. The presence of a number of C-O stretches that also associate with carbonate or oxalate comes out at 1700 cm^{-1} and 1400 cm^{-1}. The probable peaks for carbonate in hydrotalcite are 1640 cm^{-1}, 1365 cm^{-1}, and 1313 cm^{-1}.

Zirconium adsorbent corresponds to the stretching vibration νOH at the broad bands ranging between 3000 and 3400 cm^{-1}. This indicates the presence of both free and hydrogen-bonded OH groups on the sample. The peak in the region at 1620 cm^{-1} is given to the δHOH of quasi-free H_3O^+ group. The presence of the sharp band at 1620 cm^{-1} is due to the residual presence at the surface and is typically zirconia and water connected. Bands around 700–500 cm^{-1} correspond to Zr–O_2–Zr asymmetric and the formation of ZrO_2 phases is confirmed by the Zr–O stretching modes.

Table 1. Infrared peak of the five adsorbents.

Adsorbent Name	Sample Peak (cm^{-1})	Reported Peak (cm^{-1})	Assignment	References
Ni supported adsorbent	3500 1700 1400 550	3467 1640 1365 635	Mg_2Al-OH stretch C=O stretch CO_3^{2-} Ni-O stretch	[19,20]
Zn supported adsorbent	3400 1700 1400 500	3467 1642 1401 635	Mg_2Al-OH stretch H_2O-OH bending mode CO_3^{2-} Zn-O stretch	[19,20]
Ni and Zn supported adsorbent	3550 1700 1400 900 700	3467 1640 1365 870 733	Mg_2Al-OH stretch C=O stretch CO_3^{2-} M-OH deformation M-OH translation	[19,20]
Zr supported adsorbent	3500 1700 1400 600	3467 1642 1401 635	Mg_2Al-OH stretch H_2O -OH bending mode CO_3^{2-} Zr-O stretch	[19,20]

3.4. Raman Analysis of the Five Adsorbents

Table 2 presents the peak list of the Raman spectroscopy carried out on the adsorbent samples (the recorded spectra are presented by Figure S2 in the Supplementary Materials). The spectra of carbonate intercalate hydrotalcite with the hydrotalcite show the correlation between the modified hydrotalcite and the reference hydrotalcite. In addition, the spectra show clear proof of the presence of carbonate in the adsorbents. This corroborates with the absence of the peaks that are associated with the free sodium carbonate as presented by powder XRD pattern in Figure 2 (cf. vide supra). These results further confirm that the carbonate is successfully intercalated into the interlayer spaces of the hydrotalcite.

For the Ni adsorbent in the hydroxyl stretching region (3000–4000 cm^{-1}) there is some variation between the carbonate intercalated hydrotalcite and the reference hydrotalcite. The first of these variations are in the metal hydroxide bands, which shifted to lower wavenumbers, which is evocative to a variation in the chemical environment because of the presence of carbonate [22,23].

For the Zn adsorbent, there is some peak at higher wavenumbers for the metal hydroxide stretching vibrations which take place at 3389 cm^{-1} and 3200 cm^{-1} as reported previously [24]. In addition, there is a peak for the interlayer water/water-carbonate bridging mode at 3239 cm^{-1}. For this atom, this is the indication of some control of freedom due to hydrogen bonding (which is seen around 200 cm^{-1}). The aluminium hydroxide deformation at 999 cm^{-1} is at the same wavenumber as reported in the literature. This indicates that the metal-oxygen and metal hydroxide bonds form in a similar way as those in the other hydrotalcites reported before [24].

Table 2. Raman peak list of the five adsorbents.

Adsorbent Name	Sample Peak (cm^{-1})	Reported Peak (cm^{-1})	Assignment	References
Ni supported adsorbent	3200	3389	Mg_2Al-OH stretch	[19]
	1800	1660	Water bend	
	900	887	Sodium carbonate	
	100	122	Hydrogen bonding	
Zn supported adsorbent	3200	3389	$MgAl_2$-OH stretch	[21]
	1800	1660	Water bend	
	950	999	Al-OH deformation	
	200	200	Hydrogen bonding	
Zr supported adsorbent	3250	3361	Interlayer H_2O	[19]
	3200	3239	Interlayer H_2O	
	1750	1660	Water bend	
	400	377	Zr-O bonding	
	100	122	Hydrogen bonding	
Ni and Zn supported adsorbent	3000	3389	Mg_2Al-OH stretch	[19,21]
	1750	1660	Water bend	
	950	887	Sodium carbonate	
	200	144	M O bonding	

For the Ni-Zn adsorbent some differences are observed between the carbonate intercalated modified hydrotalcite and the hydrotalcite in the hydroxyl stretching region (3000–3500 cm^{-1}). The first difference is in the metal hydroxide bands, which shifted to lower wavenumbers. It indicates the difference in the chemical environment, due to the presence of carbonate. This twist is caused by the presence of the carbonate anions in the structure of the material. There is a band present at 950 cm^{-1}, which is a carbon-oxygen bond in the carbonate anions. The expected carbonate peak appeared at 1750 cm^{-1} instead of 1660 cm^{-1}. This difference in peak position is due to the presence of carbonate anions, which is opposing to the carbonate anions within the interlayer. Furthermore, there is a peak visible at 200 cm^{-1}, which disperses as a metal-oxygen band [24], and there is also a band at 100 cm^{-1}, which is likely a hydrogen bonding band [24]. The presence of these peaks suggests that there is a little dissimilarity between the metal cation layers of the carbonate and modified hydrotalcites.

For the Zr adsorbent, there is some peak at higher wavenumbers for the metal hydroxide stretching vibrations, which take place at 3289 cm^{-1} and 3239 cm^{-1} as reported in the literature [24]. Furthermore, there is also a peak for the interlayer water/water-carbonate bridging mode at 3239 cm^{-1}. For this atom, this indicates some control of freedom for these atoms due to hydrogen bonding (which is observed around 100 cm^{-1}). This is an indication of the metal-oxygen and metal hydroxide bonds forming in a similar way like the other reported hydrotalcites [24].

3.5. TGA and DSC Analyses of the Five Adsorbents

The thermogravimetric study of the Ni adsorbent shows a number of features (Figure 5a–e). The weight loss has been observed to take place in four steps; loosely bound water, tightly bound water, dehydroxylation, and decarbonization. The first weight loss is a mass loss step around 110 °C, which is due to the loss of loosely bound surface water. The large proportion of the mass loss indicates that this sample also has a large portion of tightly bound water molecules when analyzed, though it is calcined to 450 °C. The 6.73% mass loss around 400 °C is due to the loss of water molecules that are most tightly bound to the hydrotalcite and in the interlayer spaces. There are a number of simultaneous dehydroxylation steps occurring around 400 °C. At around 750 °C, weight loss is due to the loss of carbon monoxide and water from the sample, indicating simultaneous dehydroxylation and decarbonization. The mass loss at 1050 °C indicates the completion of the decarbonization step. The percentage mass loss and the calculated decomposition steps are shown in Table 3.

Table 3. Mass loss step for calcined five adsorbents.

Adsorbent Name	Temperature (°C)	Mass Loss (wt%)	Proposed Formula
Ni supported adsorbent	110	8.03	$Mg_6Al_2(OH)_{16}(CO_3)\cdot 2H_2O$ (Brady, et al., 2011)
	400	6.73	$Mg_6Al_2O(OH)_{14}(CO_3)$
	750	4.38	$Mg_6Al_2O_2(OH)_6(CO_3)$
	1050	0.64	$5MgO + MgAl_2O_4$
Zn supported adsorbent	200	8.03	$Mg_6Al_2(OH)_{16}(CO_3)\cdot H_2O$ (Brady, et al., 2011)
	400	4.02	$Mg_6Al_2O_2(OH)_{12}(CO_3)$
	750	7.35	$Mg_6Al_2O_3(OH)_{10}$
	950	1.74	$5MgO + MgAl_2O_4$
Zr supported adsorbent	400	5.82	$Mg_6Al_2(OH)_{16}(CO_3)H_2O$ (Brady, et al., 2011)
	600	1.51	$Mg_6Al_2O_2(OH)_{12}(CO_3)$
	700	4.6	$Mg_6Al_2O_3(OH)_{10}$
	900	1.93	$5MgO + MgAl_2O_4$
Ni and Zn supported adsorbent	110	7.91	$Mg_6Al_2(OH)_{16}(CO_3)\cdot 2H_2O$ (Brady, et al., 2011)
	400	8.55	$Mg_6Al_2O(OH)_{14}(CO_3)$
	750	3.79	$Mg_6Al_2O_2(OH)_6(CO_3)$
	950	0.57	$5MgO + MgAl_2O_4$

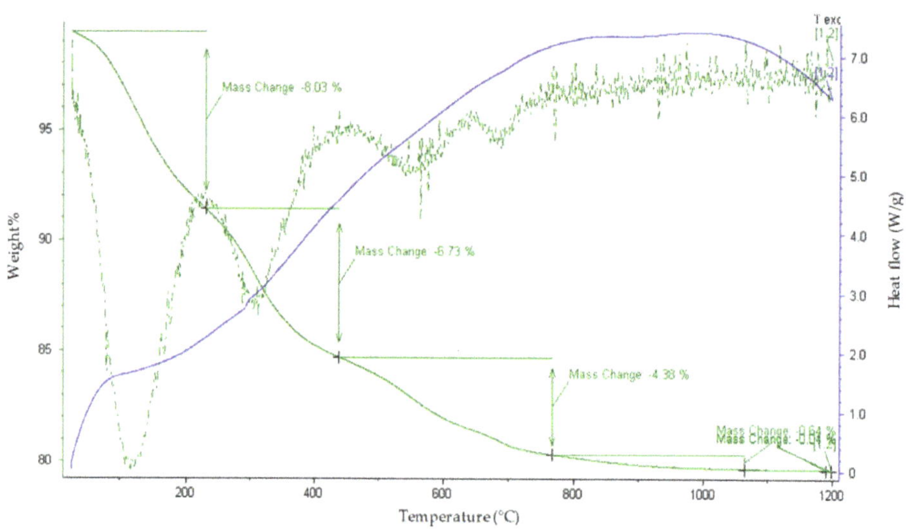

(a)

Figure 5. *Cont.*

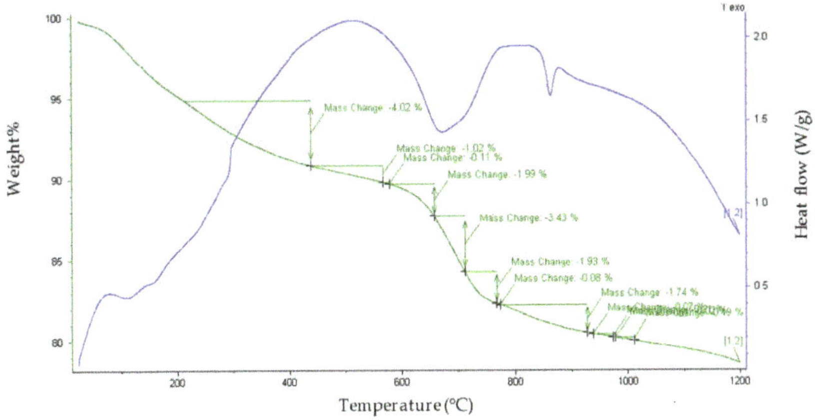

(b)

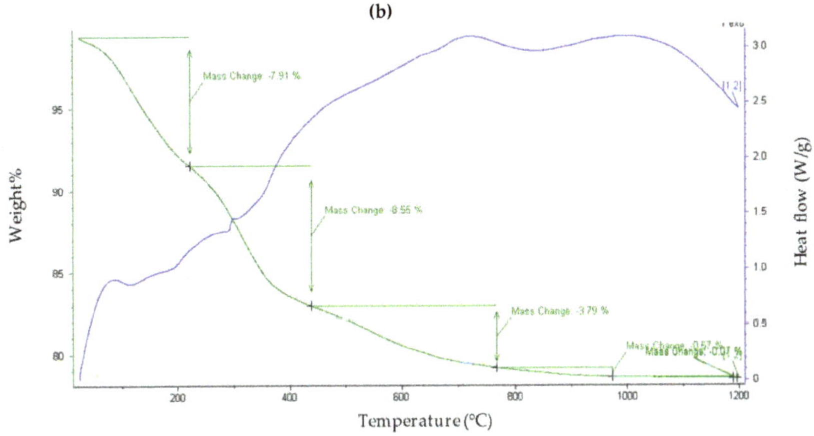

(c)

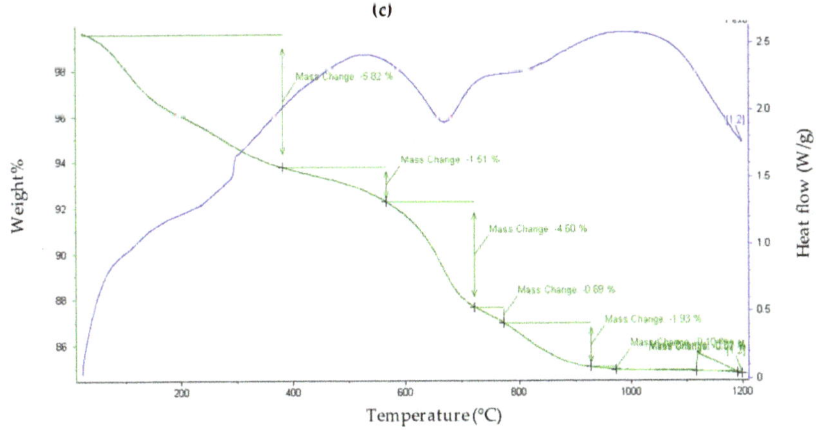

(d)

Figure 5. *Cont.*

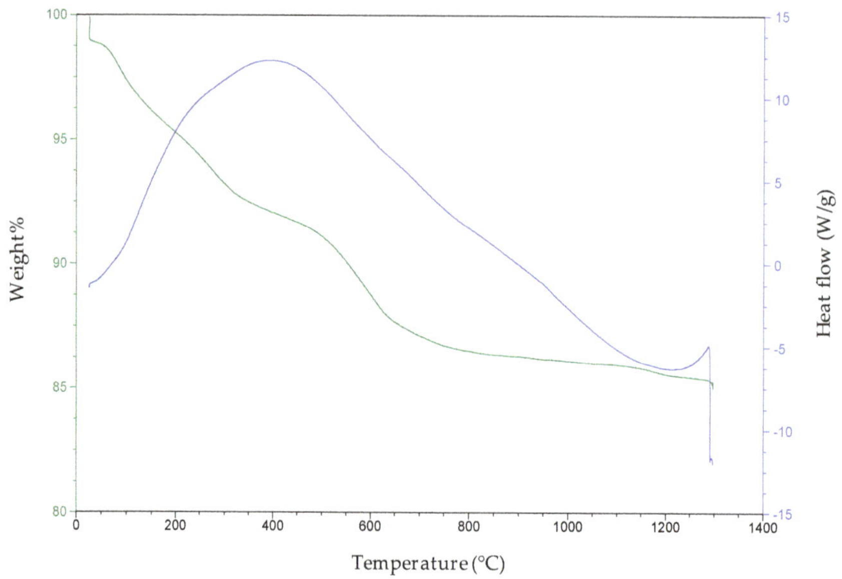

(e)

Figure 5. TGA of calcined (**a**) Ni adsorbent; (**b**) Zn adsorbent (**c**) Ni and Zn adsorbent, (**d**) Zr adsorbent (**e**) Zr and Ni adsorbent.

For the Zn adsorbent, from the TGA data in Figure 5b, the decomposition steps are the loss of some interlayer water up to 200 °C. The next decomposition step takes place at 280 °C. The first of these steps is the loss of the last interlayer water by the relocation of the brucite-like layer. This is followed by the change in the structure of the material by the partial dehydroxylation of the brucite layer and the partial merging of the carbonate group into the structure. The third decomposition step around 400 °C signals the end of the dehydroxylation, followed by the finishing point of decarbonization, which finishes around 750 °C [20]. The final dehydroxylation of the products to form a mixed metal solid solution occurs at almost 950 °C.

For the Ni and Zn adsorbent (Figure 5c), the 8.55 % mass loss around 400 °C corresponds to the loss of water molecules that are most tightly bound to the hydrotalcite and formed in the interlayer spaces. The weight loss around 750 °C corresponds to the loss of carbon monoxide and water from the sample and indicates a simultaneous dehydroxylation and decarbonization. The mass loss finishes at 950 °C, which completes the decarbonization step.

At the Zr adsorbent (Figure 5d) the next decomposition step takes place at 600 °C. The first of these is the loss of the last interlayer water through the relocation of the brucite-like layer. This is followed by the change in the structure of the material by the partial dehydroxylation of the brucite layer and the partial merging of the carbonate group into the structure. The third decomposition step occurs around 600 °C, which is the end of the dehydroxylation followed by the finishing point of decarbonization, which finishes around 700 °C [20]. The final dehydroxylation of the products to form a mixed metal solid solution occurs at almost 900 °C.

According to the mass losses recorded from the TGA for the Ni and Zr adsorbent, it is clear that there are major mass losses occurring in the sample. These are for the loosely bound water, tightly bound water, dehydroxylation, and decarbonization. The figure also shows an exothermic reaction, and the sample is gradually degraded. The sample is heated to 1200 °C, whereas the major decomposition occurs up to 600 °C.

3.6. XPS Analysis of the Three Adsorbents

The three adsorbents were mixed with the dibenzothiophene. Then the samples were tested by the XPS to check the bonding between the sulfur molecule and the adsorbents. Figure 6 below shows the clear evidence of bonding between sulfur molecule and the adsorbents.

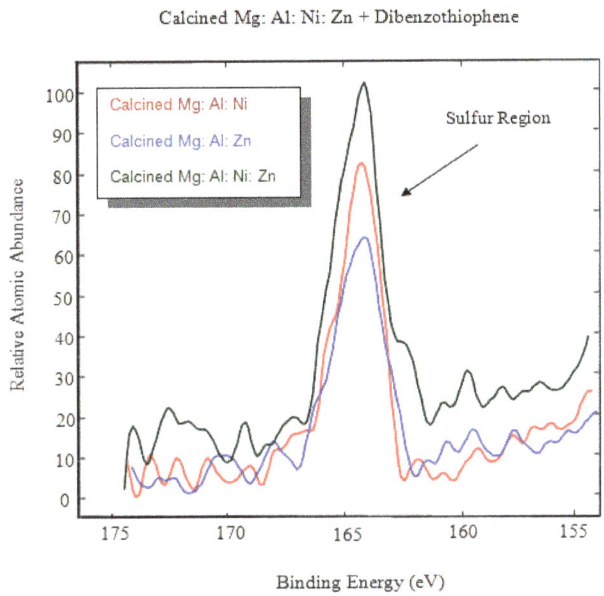

Figure 6. Adsorptive studies of dibenzothiophene.

Table 4 shows percentages of relative abundances of species from XPS. It shows clear evidence that sulfur is observed on substrates immersed in dibenzothiophene, irrespective of the formulation of the hydrotalcites preparations. Substrates with nickel showed increased sulfur content over that without nickel: Mg: Al: Ni: ≡ 1.57%, Mg: Al: Ni: Zn ≡ 1.48%, Mg: Al: Zn: ≡ 0.87%.

Table 4. Relative percent abundances of species from XPS.

	Zinc	Aluminium	Magnesium	Nickel	Sulfur
Ni Adsorbent	-	34.80	44.17	19.47	1.57
Zn Adsorbent	6.87	37.25	55.00	-	0.87
Ni and Zn Adsorbent	6.19	23.97	53.53	14.83	1.48

3.7. Performance Comparison of All Five Types of Adsorbents

Five different types of adsorbents were examined. Each of these adsorbents has the same support of magnesium and aluminium; therefore, an influence specific to the main component is determined.

Figure 7 shows that, after an hour, fresh thiophene reaches 71 ppm, and when hydrotalcite without any metal is used. The zirconia adsorbent reacts slowly in comparison to the other four adsorbents. On the other hand, nickel-based adsorbents react more quickly compared to the other three adsorbents. As shown in Figure 7, all five adsorbents work within an hour until the thiophene concentration in the pentane becomes almost constant. It turns out that the deposition of nickel on zirconia is an innovative approach that takes advantage of the selectivity of Ni towards S-species and the high adsorptive capacity of zirconia support. It is observed that adding zirconia into the adsorbent structure increase the sulfur removal capacity. Possibly, the added zirconia changes the nature of the active metal sites into the structure and increases its sulfur removal capacity. Figure 8 shows the percent removal of thiophene with respect to different time interval. It is evident that for all the adsorbents the maximum percentage of the removal occurred in the 0–10 min interval. It is also worth noting that more than 70% removal for both NiO-ZrO_2, and NiO-ZnO was obtained in the first 10 min. This finding can be used to investigate the reaction kinetics in future studies.

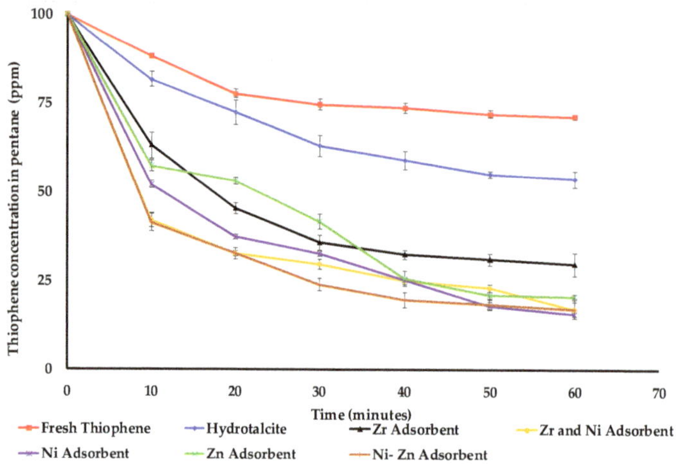

Figure 7. Thiophene concentrations in pentane with time.

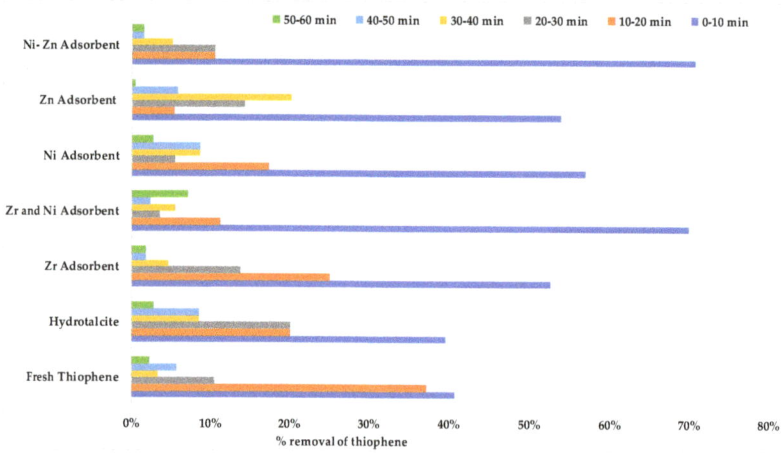

Figure 8. Percent removal of thiophene with respect to different time interval.

4. Conclusions

In the current study, we prepared hydrotalcite by the co-precipitation method and characterized using various analytical techniques. The analysis matches the typical behavior of hydrotalcites of the same composition reported in the literature. From the analysis of the modified hydrotalcites, it can be concluded that carbonate was successfully intercalated into the structure of hydrotalcite using co-precipitation methods. It was also evidenced that there is an intermediate structure for carbonate intercalated hydrotalcite. X-ray diffractograms showed the formation of hydrotalcites in the synthesized materials, while EDAX analysis confirmed the expected elemental ratio present in the modified hydrotalcite adsorbents. We further confirmed the presence of ZnO, NiO, and zirconia on the respective adsorbents. FTIR and Raman spectroscopy showed the presence of carbonate and water in the adsorbents. The removal efficiency of thiophene (100 ppm) from the model fuel containing n-pentane by the five adsorbents was in between 70%–85%. However, only NiO-ZrO_2 adsorbent could remove thiophene from 100 ppm to a 15-ppm level. The thiophene removing capacity of NiO-ZrO_2, and NiO-ZnO adsorbents is superior to the other three absorbents (ZnO, NiO, and ZrO_2). NiO-ZnO is a well-known adsorbent, however, the adsorption capacity of NiO-ZrO_2 presented in this work can an have potential advantages over other adsorbents including the former one. Additionally, XPS showed clear evidence of the retention of sulfur from dibenzothiophene when interacted with the NiO-ZnO, NiO, and ZnO containing modified hydrotalcite adsorbents. Our results also indicated that the deposition of nickel on zirconia changed the nature of the active metal sites into the structure that successfully increased the sulfur removal capacity of the adsorbents.

Supplementary Materials: The following are available online at http://www.mdpi.com/2227-7080/8/2/22/s1, Figure S1: Infrared peaks of (a) Ni adsorbent; (b) Zn adsorbent (c) Ni and Zn adsorbent, (d) Zr adsorbent, Figure S2: Raman peaks of calcined (a) Ni adsorbent; (b) Zn adsorbent (c) Ni and Zn adsorbent, (d) Zr adsorbent, Table S1: Peak center, FWHM and crystal size from X-ray diffractograms, Table S2: Composition of the five calcined adsorbents from EDAX.

Author Contributions: Conceptualization, M.M., R.D., M.I. and A.J.G.; methodology, M.M. and A.R.; software, R.D. and A.R.; validation, M.S.J.S., A.J.G. and A.R.; formal analysis, M.M., R.D. and M.I.; investigation, A.G.; resources, A.G.; data curation, M.M. and R.D.; writing—original draft preparation, M.M., R.D., M.S. and A.R.; writing—review and editing, M.M., R.D., M.S.J.S., T.S. and A.R.; visualization, M.M., R.D., M.S.J.S., T.S. and A.R.; supervision, A.J.G. and A.R.; project administration, A.J.G. All authors have read and agreed to the published version of the manuscript.

Acknowledgments: This work has been funded in part with funds from the State of Texas as part of the program of the Texas Hazardous Waste Research Center, Texas Air Research Center, Lamar University, and Prairie View A&M University. The contents do not necessarily reflect the views and policies of the sponsor nor does the mention of trade names or commercial products constitute endorsement or recommendation for use. The authors would like to extend special thanks to Hylton G. McWhinney, David Cocke, Paul Bernazzani, Hylton McWhinney, Tony Graddy and Tracy Benson for their analytical support.

Conflicts of Interest: The authors declare no conflict of interest.

References

1. Yu, X.; Shi, M.; Yan, S.Q.; Wang, H.; Wang, X.H.; Yang, W. Designation of choline functionalized polyoxometalates as highly active catalysts in aerobic desulfurization on a combined oxidation and extraction procedure. *Fuel* **2017**, *207*, 13–27. [CrossRef]
2. Yu, F.; Wang, R. Deep oxidative desulfurization of dibenzothiophene in simulated oil and real diesel using heteropolyanion-substituted hydrotalcite like compounds as catalysts. *Molecules* **2013**, *18*, 13691–13704. [CrossRef] [PubMed]
3. Miao, G.; Ye, F.; Wu, L.; Ren, X. Selective adsorption of thiophenic compounds from fuel over TiO_2/SiO_2 under UV-irradiation. *J. Hazard Mater.* **2015**, *300*, 426–431. [CrossRef] [PubMed]
4. Ismi, H.; Yustin, P.; Syoni, S.; Siti, K.C. Biodesulfurization of organic sulfur in Tondongkura coal from Indonesia by multi-stage bioprocess treatments. *Hydrometallurgy* **2017**, *168*, 84–93.
5. Hao, L.; Wang, M.; Deng, C.; Ren, W.; Shi, Z.; Lu, H. L-proline-based deep eutectic solvents (DESs) for deep catalytic oxidative desulfurization (ODS) of diesel. *J. Hazard Mater.* **2017**, *339*, 216–222. [CrossRef]
6. Liao, J.; Zhang, Y.; Fan, L.; Chang, L.; Bao, W. Insight into the acid sites over modified Na-Y zeolite and their adsorption mechanisms for thiophene and benzene. *Ind. Eng. Chem. Res.* **2019**, *58*, 4572–4580. [CrossRef]
7. Behnejad, B.; Abdouss, M.; Tavasoli, A. Comparison of performance of Ni-Mo/γ alumina catalyst in HDS and HDN reactions of main distillate fractions. *Petrol. Sci.* **2019**, *16*, 645–656. [CrossRef]
8. Dizaji, A.K.; Mokhtarani, B.; Mortaheb, H.R. Deep and fast oxidative desulfurization of fuels using graphene oxide-based phosphotungstic acid catalysts. *Fuel* **2019**, *236*, 717–729. [CrossRef]
9. Hossain, M.N.; Park, H.C.; Choi, H.S. A comprehensive review on catalytic oxidative desulfurization of liquid fuel oil. *Catalysts* **2019**, *9*, 229. [CrossRef]

10. Yue, D.; Lei, J.; Lina, Z.; Zhenran, G.; Du, X.; Li, J. Three-dimensional ordered macroporous HPW/titania–alumina catalysts for catalytic oxidative desulfurization of fuels. *J. Porous Mater.* **2019**, *26*, 133–144. [CrossRef]
11. Ullah, L.; Zhaoa, G.; Hedin, N. Highly efficient adsorption of benzothiophene from model fuel on a metalorganic framework modified with dodeca-tungstophosphoric acid. *Chem. Eng. J.* **2019**, *362*, 30–40. [CrossRef]
12. Nair, S.A. Desulfurization of Hydrocarbon Fuel at Ambient Conditions using Supported Silver Oxide-Titania Sorbents. Doctor Thesis, Auburn University, Auburn, AL, USA, 10 November 2010.
13. Raj, J.J.; Magaret, S.; Pranesh, M. Dual functionalized imidazolium ionic liquids as a green solvent for extractive desulfurization of fuel oil: Toxicology and mechanistic studies. *J. Clean Prod.* **2019**, *213*, 989–998. [CrossRef]
14. Kumar, S.; Srivastava, V.C. Oxidative desulferization by chromium promoted sulfated zirconia. *Fuel Process. Technol.* **2012**, *93*, 18–25. [CrossRef]
15. Zhang, S.; Zhang, Y. Mechanistic investigation on the adsorption of thiophene over Zn_3NiO_4 bimetallic oxide cluster. *Appl. Surf. Sci.* **2012**, *258*, 10148–10153. [CrossRef]
16. Parida, K.; Das, J. Mg/Al hydrotalcites: Preparation, characterization & ketonisation of acetic acid. *J. Mol. Catal. A-Chem.* **2000**, *151*, 185–192.
17. Rahman, A.; Kumar, S.; Bafana, A.; Dahoumane, S.A.; Jeffryes, C. Biosynthetic conversion of Ag^+ to highly stable Ag^0 nanoparticles by wild type and cell wall deficient strains of *Chlamydomonas reinhardtii*. *Molecules* **2019**, *24*, 98. [CrossRef]
18. Islam, M.W. A review for dolomite catalyst for biomass gasification tar removal. *Fuel* **2020**, *267*, 117095. [CrossRef]
19. Brady, J.P. An Examination of the Applicability of Hydrotalcite for Removing Oxalate Anions from Bayer Process Solution. Master Thesis, Queensland University of Technology, Brisbane, Australia, 26 September 2011.
20. Kloprogge, J.T.; Hickey, L.; Frost, R.L. Synthesis & spectroscopic characterization deuterited hydrotalcites. *J. Mater. Sci.* **2002**, *21*, 603–605.
21. Kloprogge, J.T.; Weier, M.L.; Crespo, I.; Ulibarri, M.A.; Barriga, C.; Rives, V. Intercalation of iron hexacano complexes in Zn, Al hydrotalcites and mid-infrared and Raman spectroscopy study. *J. Solid-State Chem.* **2004**, *177*, 1382–1387. [CrossRef]
22. Rahman, A.; Kumar, S.; Bafana, A.; Lin, J.; Dahoumane, S.A.; Jeffryes, C. A mechanistic view of the light-induced synthesis of silver nanoparticles using extracellular polymeric substances of *Chlamydomonas reinhardtii*. *Molecules* **2019**, *24*, 3506. [CrossRef]
23. Rahman, A.; Kumar, S.; Bafana, A.; Dahoumane, S.A.; Jeffryes, C. Individual and combined effects of extracellular polymeric substances and whole cell components of *Chlamydomonas reinhardtii* on silver nanoparticle synthesis and stability. *Molecules* **2019**, *24*, 956. [CrossRef] [PubMed]
24. Walfaran, G.E.; Pugh, E. Raman combination and stretching overtones from water, heavy water, and NaCl in water at shifts to 7000 cm^{-1}. *J. Solution Chem.* **2004**, *33*, 81–97.

© 2020 by the authors. Licensee MDPI, Basel, Switzerland. This article is an open access article distributed under the terms and conditions of the Creative Commons Attribution (CC BY) license (http://creativecommons.org/licenses/by/4.0/).

technologies

Article

Process Development of CO_2-Assisted Polymer Compression for High Productivity: Improving Equipment and the Challenge of Numbering-Up

Takafumi Aizawa

Research Institute for Chemical Process Technology, National Institute of Advanced Industrial Science and Technology, 4-2-1 Nigatake, Miyagino-ku, Sendai 983-8551, Japan; t.aizawa@aist.go.jp; Tel.: +81-22-237-5211

Received: 12 April 2019; Accepted: 7 May 2019; Published: 8 May 2019

Abstract: The CO_2-assisted polymer compression method is used herein to prepare porous polymer materials by bonding laminated polymer fiber sheets using a piston in the presence of CO_2. In this work, the CO_2 flow line connections were moved from the pressure vessel to the piston to increase productivity, which makes the pressure vessel free-moving and the processing time of sample introduction and removal seemingly zero. In addition, a numbering-up method suitable for CO_2-assisted polymer compression is proposed and verified based on the variability of the products. The variability of the product was evaluated using porosity, which is one of the most important properties of a porous material. It is found that the CO_2 exhaust process, specific to this method, that uses high-pressure CO_2, causes product variation, which can be successfully suppressed by optimizing the CO_2 exhaust process.

Keywords: CO_2-assisted polymer compression; numbering-up; high productivity; CO_2; polymer; porous material; process improvement

1. Introduction

Polymers are an essential material in everyday life due to their light and durable characteristics [1,2]. In particular, porous polymer materials are very lightweight and exhibit various functionalities [3] such as sound absorption, heat insulation, adsorption, filtering, moisture permeation, water absorption, drug loading, and sustained release [3]; their applications vary from mass production to advanced functional materials [4].

Considering the advantages of such materials, CO_2-assisted polymer compression (CAPC), a method that uses CO_2 to adhere fibrous sheets to create a porous polymer material, has been developed [5]. Because the method is a room temperature process and no heater is required, it is a low energy process. As the CO_2 used is released into the atmosphere and does not remain in the polymer, impurities do not remain. In addition, CO_2 is a very low toxic substance; it is also used in the food industry, e.g., carbonated beverages. CO_2 is known as an environmentally friendly solvent; in particular its supercritical state is recognized as a green solvent [6], and its application in various processes has been researched [7,8]. CO_2 has also been practically used as an extraction solvent for biomass in the food industry [9]. The demonstrated CO_2 process is a suitable method for manufacturing parts used for the food industry, medical treatment, and living environments that demand high purity. Among gases, CO_2 is known to dissolve well in polymers and has been reported as a specific interaction between the CO_2 and polymer [10], estimation of the solubility of CO_2 in polymers [11], measurement of solubility of CO_2 in crystalline polymers [12], and measurement of solubility of CO_2 in amorphous polymers [13]. CO_2 is known to be capable of plasticizing polymers and decreasing their viscosity [14], melting point [15], and glass transition temperature [16]. Many CO_2-based processes with polymers have been proposed such as dyeing and polymer particle synthesis [17], foam molding [18], fabrication of carbon nanotube composite foam [19], foam molding of polymer blends [20], molecularly imprinted polymer development in supercritical CO_2 [21], and fabrication of two-dimensional porous polymers [22]. As the CAPC process is performed at room temperature with no temperature controller and the pressure of CO_2 is also introduced at the vapor pressure, which is the pressure inside the liquid CO_2 cylinder, the CO_2 is conveniently introduced in the gas phase by opening a valve. As no pump is required, the process is simplified. This technology reduces equipment cost and processing time. Previous work has also shown that material properties, such as porosity, were easily tuned by changing the process conditions, even when using the same raw material [23]. The adhesion strength of the CAPC sample was also evaluated [24]. Others have demonstrated that a drug can be carried in the porous material very easily by loading the drug in the state of the fiber sheet of the raw material, and the sustained release property of the drug when the porous material is immersed in water can be controlled [25]. There are many applications of drug-loaded porous materials in the advanced materials and medical fields, but it is challenging to insert the drug into the center of a thick porous material. The CAPC process facilitates drug carrying by supporting drug insertion at the raw fiber stage.

In my previous studies, only one sample was fabricated in one batch, which is a disadvantage remedied in this work. The CAPC method is used to prepare a porous polymer material by adhering laminated fibrous sheets through a simple apparatus with a short pressing time. Even if it is an overwhelmingly fast process at the laboratory level, when considering mass production, the improvement in the production per unit time is indispensable. Historically, scaling chemical processes up is often challenging. One of the main research goals in chemical engineering is developing ways to smoothly scale up flask-scale reactions to be successful in large reactors.

For processes that cannot be scaled up, numbering-up was proposed. The concept of numbering-up has been structured with a chemical process using a microreactor as shown by numbering-up for multi-phase (gas–liquid) flow microreactor [26], numbering-up for liquid–liquid two-phase capillary microstructured reactor [27], and external numbering-up (parallelized microdevices) [28]. Using a small reactor (microreactor) and a small mixer (micromixer), the microscale chemical process maximizes the heat exchange rate and mixing speed and realizes a high-speed reaction that suppresses by-products. Because the reactors and the mixers must be small, scaling up cannot be performed. Numbering-up that increases the number of reaction lines is important. In manufacturing, it has been common practice to approximate numbering-up conventionally. Adding production lines to increase production is one possible numbering-up strategy. However, a new concept that has emerged in microchemistry is the introduction of the number of reaction lines in an apparatus, wherein numbering-up is considered inside the apparatus.

The micro chemical process is a flow process, but the concept of numbering-up is also adaptable to a batch-type production process in that there is a limit to the amount that can be produced in one reactor. In a batch production process, if the size of the batch is optimized, it is impossible to scale up, and the idea of numbering-up is necessary. When considering production by the CAPC method, scaling up can be applied when large porous polymer materials are fabricated, and the final products are cut into pieces. When it is desired to mold the polymer in the form of a final product, for example, when preparing a drug-loaded tablet, it is necessary to consider numbering-up to increase the production per unit time. The chemical reaction in the flow process and the compression of the polymer fiber sheets are completely unrelated phenomena. Therefore, only the concept of numbering-up for simultaneous production in the same equipment was considered herein.

This study is the first trial to enhance the productivity of the CAPC process. Herein, increasing the productivity of the CAPC process is investigated through two strategies. The flow line of CO_2 is improved to hide both sample preparation time and sample removal time. The application of numbering-up such that the cost of equipment is suppressed is also investigated. Product uniformity is evaluated using the porosity of the samples when numbering-up is incorporated. Porosity is an important property of porous materials, and previous research has shown that the sustained release rate of the drug trapped in the porous material is greatly influenced by the fabricated material's porosity [25]. The cause of sample variation is identified and a solution to suppress sample variation is presented.

2. Materials and Methods

2.1. Materials

A nonwoven fabric spun with a fiber diameter of 8 μm and a basis weight of approximately 30 g/m^2 using a melt-blown method (Nippon Nozzle Co., Ltd., Kobe, Japan) was used as a raw material; a polyethylene terephthalate (PET) pellet (Model number: TK3) of Bell Polyester Products Inc. (Yamaguchi, Japan) was used. The procedure of the melt-blown process is very simple. The PET pellets were melted, extruded from small nozzles, and stretched using high-speed blowing air to create a nonwoven fabric. Circular cut samples (18 mm diameter) of the nonwoven fabric were selected such that the same weight (0.257 g) was obtained for each set of 32 sheets.

2.2. Equipment

The experimental equipment used in this study is presented in Figure 1. The major difference from previously reported research [5,23–25] is found in the connections of the CO_2 introduction exhaustion tubes. The position of the tubes was moved from the high-pressure vessel (B1) to the piston (P). The piston area includes a CO_2 introduction port and an exhaust port at the top, a flow path for carrying CO_2 at the center, and diagonally positioned CO_2 introduction and exhaustion ports into the gap between the high-pressure vessel and piston at the bottom. Owing to the difficulty in creating such a shape with one piece, the system comprised mainly three pieces as shown in Figure 2. The inner diameter of component (P2) through which CO_2 flows is 4 mm, and the outer diameter of component (P3) is 19.5 mm. Black circles in Figure 2 indicate the O-rings. Component (P1) and component (P2) are sealed with an O-ring. An O-ring between components (P2) and (P3) seals the pressure vessel and component (P2). There is no gas seal between component (P2) and component (P3). The piston was attached to the press machine (Model number: JP-1504), manufactured by Janome Sewing Machine Co., Ltd (Hachioji, Japan). The inner diameter of the pressure vessel is 20.0 mm. Both the piston and the pressure vessel are made of stainless steel (SUS316).

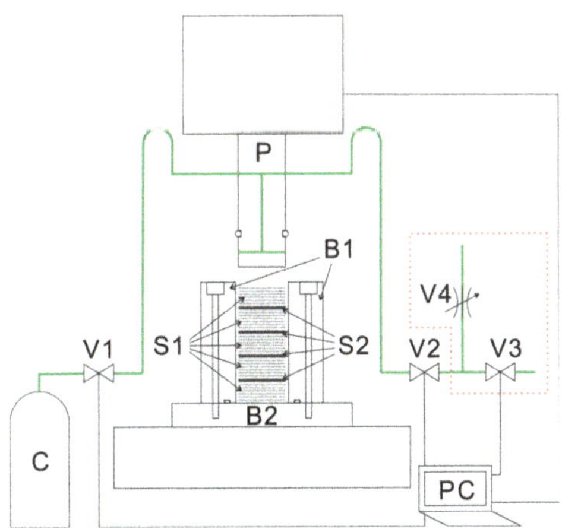

Figure 1. Schematic illustration of the cross-section of the high-pressure vessel used for CO_2-assisted polymer compression. B1: Body of the high-pressure vessel, B2: Base of the high-pressure vessel, C: CO_2 cylinder, P: Piston, PC: Laptop computer, S1: Sample, S2: Separator, V1: Intake valve, V2: Exhaust valve, V3: Exhaust valve (optional), and V4: Metering valve (optional).

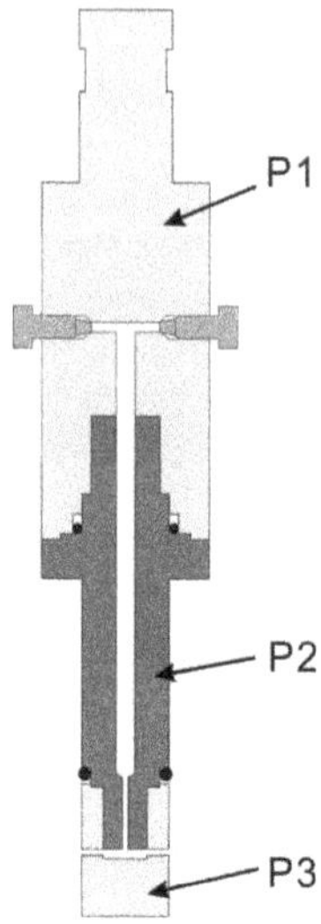

Figure 2. Cross-sectional diagram of the piston components. The piston comprises three main components, P1–P3.

The liquid CO_2 cylinder was connected to the CO_2 inlet of the piston through the introduction valve by a stainless-steel (SUS316) tube. The tube was slacked between the introduction valve and the piston so that the upward and downward movement of the piston was not affected. The stainless-steel tube connected to the outlet of the piston was also connected to the exhaust valve with slack. After the exhaust valve, a T-type branch was provided; one was connected to an additional exhaust valve (V3), and the other was connected to a metering valve (V4). The flow line after the exhaust valve (V2) (dotted red frame in Figure 1) was not provided in the first experiment, and the exhaust was directly released to the atmosphere. Although the introduction valve (V1) was closed during operation, it was temporarily opened when CO_2 was introduced. V2 was normally open, but it was closed when CO_2 was to be retained in the pressure vessel. In operating V4 and V3, when V2 was open and V3 was closed, CO_2 was slowly released through the metering valve. A metering valve (Model number: SS-SS1, Swagelok Inc., Solon, Ohio, United States) was set to the minimum flow rate and used as V4. When V3 was open, CO_2 was instantaneously exhausted. When the experiment is performed without the process outlined in the dotted red frame (see Figure 1), it appears that the same result will be achieved as that when V3 is open. After performing the experiment without the dotted red frame line, the flow line in the red frame was added. This has been depicted as the experiment without a dotted red frame line such that the performed experiment is represented accurately.

The high-pressure container is a container with a hole at the top. The structure was made so that the body part (B1) and the base part (B2) could be separated if the system became clogged. By fixing the container filled with the sample just beneath the piston, the setup of the device is complete.

The laptop computer (PC) controlled each valve and piston such that the timing of the valves opening and closing was consistent for all experiments.

2.3. Experimental Procedure

Five sets of 32 laminated samples (S1) were placed in the high-pressure vessel. The laminated samples were separated by stainless-steel spacers (S2), each having a diameter of 19.5 mm and a thickness of 1 mm. As the inner diameter of the pressure vessel is 20.0 mm, there is enough clearance for CO_2 to flow between the pressure vessel and the spacers. The inner diameter of the stainless tube from the CO_2 cylinder to the piston is 0.8 mm, and the clearance between the pressure vessel and the spacers is much larger than that. The cross-sectional view of the samples when set in the high-pressure vessel is shown in Figure 1. The measured thickness of the center portion of each spacer was, starting from the top spacer, 1.024 mm, 1.009 mm, 1.010 mm, and 1.035 mm. Operation was started when V1 closed and V2 opened. Then, the piston was lowered to the CO_2 introduction position, and V2 was closed. The introduction valve and the exhaust valve were alternately opened three times to replace the air in the pressure vessel with CO_2. As the internal pressure of the liquid CO_2 cylinder is about 6 MPa, once CO_2 is introduced at this pressure and released into the atmosphere (0.1 MPa), the amount of existing air is reduced to 1/60, and the remaining air is replaced with CO_2. The volume of air is estimated to become $(1/60)^3$ following three trials of the procedure. The CO_2 introduction valve was opened again to introduce CO_2, and the introduction valve was closed. Then, the piston was lowered to the press position and set to press for 10 s, after which an exhaustion procedure was performed. The exhaustion operation was conducted in two ways depending on the experiment. One exhaust procedure was an experiment without the flow line enclosed by the dotted red line indicated in Figure 1. In this case, CO_2 was exhausted instantaneously (<1 s). In the other procedure, the flow line enclosed by the dotted red line was attached; V2 was opened when V3 was closed, and CO_2 was slowly exhausted for 30 s by V4. Then, V3 was opened to exhaust CO_2 completely. The distances (when the thickness of the spacer is excluded) between the CO_2 introduction position and the bottom of the pressure vessel and that between the press position and the bottom of the pressure vessel were 9.75 mm and 6.50 mm, 9.00 mm and 6.00 mm, 8.25 mm and 5.50 mm, 7.50 mm and 5.00 mm, and 6.75 mm and 4.50 mm. The experimental results correspond to press positions from 4.50 mm to 6.50 mm. For evaluating the sample, the thickness of the center was measured with a micrometer and the variation due to the sample position was assessed.

Porosity was evaluated from the polymer density, weight of the fabric sheet, and sample thickness. The datasheet for the PET pellet indicated that the density of the used polymer is 1.34 g/mL. For a solid without any voids, the thickness can be easily calculated using the weight, density, and diameter of the fabric sheet. In this experiment, laminated 0.257 g of circular sheets with an 18.0-mm diameter were used, such that the thickness of the solid (L_{solid}) was calculated to be 0.754 mm. The difference between L_{solid} and the actual thickness (L_{sample}) was considered as the total area of the pore, and the porosity α, was calculated as $\alpha = (L_{sample} - L_{solid})/L_{sample}$.

The experiment was performed seven times under each condition and the average value was used for analysis. Error bars are not shown in the plot of the evaluated porosity because the standard deviation of the data was small (maximum 4.9 %); the standard deviation is provided in Tables 1 and 2.

The fiber surface of the central layer of the compressed sample was observed using a scanning electron microscope (SEM; Model number: TM-1000, Hitachi High-Technologies Co., Minato-ku, Japan). The fibrous sheet before compression was also measured using an SEM for comparison

3. Results and Discussion

First, the design of the system, specifically the effect of connecting the CO_2 lines to the piston, is discussed. When they were connected to the high-pressure vessel, the vessel could only be moved within the range that the tubes could extend. To pull the pressure vessel out completely, it was necessary to disconnect the tubes from the pressure vessel, resulting in reduced usability. By connecting the tubes to the piston, the high-pressure vessel became free of this limitation. Moreover, the initial setting (corresponding to the beginning of the operation) was complete by locating the high-pressure vessel including the samples under the piston. If a plurality of high-pressure vessels were set up, it would be possible to prepare and remove samples from the system during press operation, which would greatly improve productivity.

Despite reducing the time for setting and removing the samples, the time required to move the piston, replace the air with CO_2, and press remains absolutely necessary time, which is nearly 30 s. If the processing time per sample should be reduced further, multiple samples must be fabricated simultaneously; at this point, the concept of numbering-up is attractive. Figure 3 schematically presents the preparation of one sample and two kinds of numbering-up. Figure 3b shows the concept of preparing a plurality of pressure vessels and pressing a plurality of samples with one press machine. In this case, numbering-up can be realized under the same conditions as the preparation of one sample and it is possible to increase the production efficiency while suppressing variations in the sample. On the other hand, the maximum load determines the performance of the press machine, and when the press area is increased, the required maximum load for the press machine increases, which increases process cost dramatically and should thus be avoided in the practical process. In Figure 3c, numbering-up by stacking samples without changing the press area is shown. Although the stroke of the piston will be long, if the maximum load does not change, the increased cost for the press machine-related owing to the longer stroke will be small; this will be very advantageous. Porosity is an important characteristic of porous materials. When a sample is compressed to fabricate a porous material, the thickness after compression determines the porosity. Thus, to produce samples of the same porosity, the compression rates must be uniform. In the method shown in Figure 3c, the separators are sandwiched between the samples, but the position of the separator is not controlled. Thus, the compression rates of the samples separated by the separators could differ. Owing to low or no uniformity of the raw material sample, the flow of CO_2 inside the pressure vessel, and the difference of the distance between the piston (moving device) and the samples, the products may vary. Therefore, product variation was evaluated using the porosity of the product as an index. As the objective here is to increase productivity, first, the experiment wherein CO_2 is instantly discarded by opening V2, i.e., without flow line shown within the dotted red line in Figure 1, was conducted.

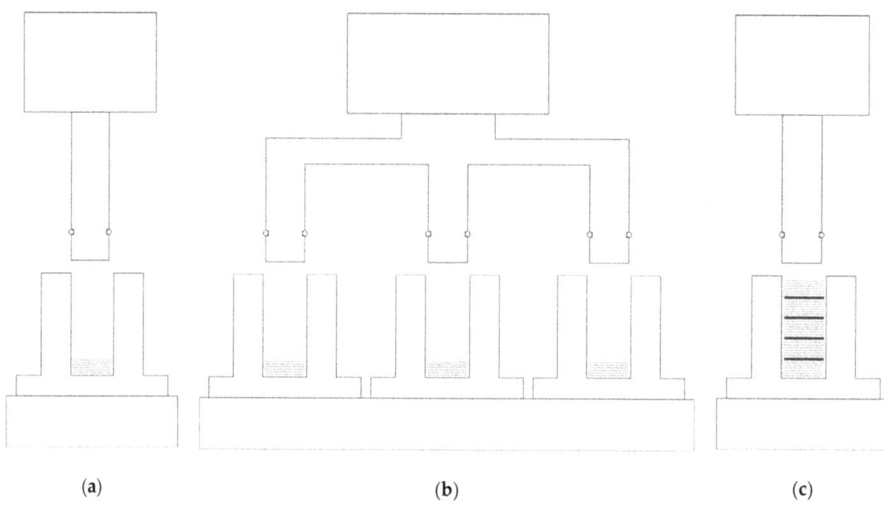

(a) (b) (c)

Figure 3. Schematic drawings showing the (**a**) basic process, (**b**) parallel numbering-up, and (**c**) serial numbering-up.

Figure 4 shows the SEM images of the raw material sheets and CAPC samples. In the CAPC samples, dents, which are adhesion traces are observed on the fibers. The high compression sample (Figure 4b) has more dents than the low compression sample (Figure 4c), which is caused by strong compression. The origin of the pores is the void between the fibers, implying that the porosity depends on the strength of the compression. As the mechanism of the CAPC process has been previously described [5,23], only a brief explanation is provided herein. The shape of a thermoplastic polymer is maintained by the friction between the polymer chains. The fluidity of the polymer increases when the polymer is impregnated with CO_2 as the CO_2 molecules reduce the friction between the polymer chains. In the case of PET, CO_2 is considered to impregnate the amorphous part of the polymer. By compressing the polymer fibers in a plasticized state, bonding points are formed at the overlapping portions of the fibers. The adhesion of the sample is completed by removing the CO_2 and the friction of the polymer chains is recovered.

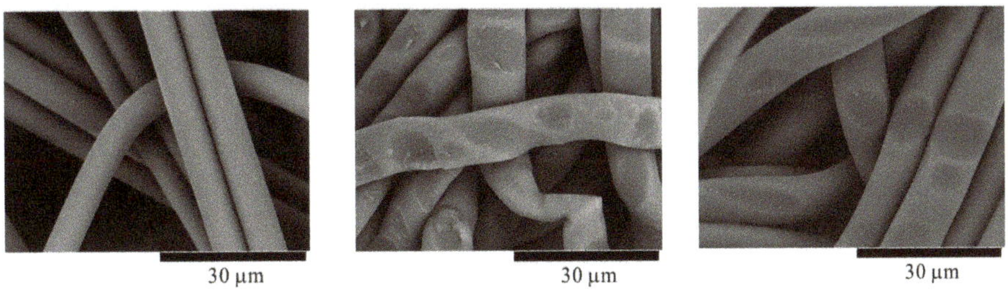

(**a**) (**b**) (**c**)

Figure 4. Scanning electron microscope images of (**a**) raw material, (**b**) sample at 4.5 mm press position, and (**c**) sample at 6.5 mm press position.

Figure 5 shows the product porosity for different press positions. When the press position was small (high compression), the variation in the sample porosity was small. For larger press positions, larger variation was observed. The top and second samples have higher porosity than the lower samples. The experiments were performed seven times under each condition, and the average thicknesses and porosity are provided along with the standard deviation in Table 1. As seen in Figure 4 and Table 1, the differences in the sample porosity were significant.

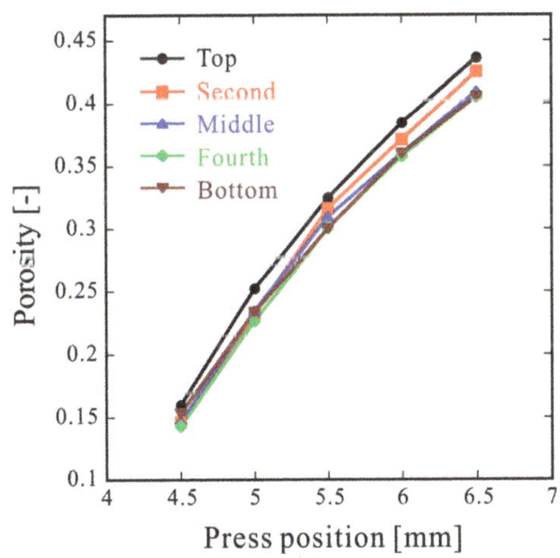

Figure 5. Product porosity at each position for different press positions. Each value is the average thickness based on seven experiments.

I considered the time when variation occurs: When CO_2 is introduced, when it is pressed, or when the CO_2 exhaust is vented from the system. Introduced CO_2 may apply pressure throughout the samples from above by transmitting through the gap between the piston and the high-pressure vessel. However, there is a gap between the spacer and the high-pressure vessel, and CO_2 should flow downward instantaneously through this gap. In addition, as the polymer is plasticized only when CO_2 is dissolved in it, the effect on the polymer before CO_2 is dissolved is expected to be extremely low. When the piston is lowered after introducing the CO_2, compression occurs. Considering that the piston moves suddenly when the sample and the spacer are stationary, the upper sample must be compressed more than the lower sample. Another explanation is that the lower sample may be more pressurized by the weight of upper samples and spacers. However, as the pressure of the piston used to compress the sample is about 1 MPa (1 MPa at 4.5 mm press position, 0.9 MPa at 5.0 mm press position, and 0.8 MPa at 5.5 to 6.5 mm press position), the weight of samples and spacers should be negligibly small. The last possibility is to consider how CO_2 venting affects the system. When CO_2 exhaust is vented by releasing it to the atmosphere, the CO_2 trapped in the pores of the porous material and the CO_2 dissolved in the polymer are vigorously removed. If this effect differs for each sample layer, the samples may vary. Specifically, CO_2 explosively blows out from the upper sample near the outlet, which compresses the lower sample. Because the lower sample, in which CO_2 is sufficiently dissolved in the polymer, is still plasticized, the sample is likely to compress. When CO_2 is removed from the lower sample, which might compress the upper sample, it seems that the CO_2 concentration in the upper sample had already decreased; as decreased CO_2 would prohibit plasticization, the upper sample would be difficult to compress, thus leading to it being thicker.

Table 1. Product thickness and porosity obtained under rapid CO_2 exhaustion.

Press Position [mm]	Sample Position	Thickness [mm]	Standard Deviation [mm]	Porosity [-]	Standard Deviation [-]
4.5	Top	0.897	0.004	0.160	0.004
	Second	0.884	0.006	0.147	0.006
	Middle	0.884	0.005	0.147	0.005
	Fourth	0.880	0.007	0.143	0.007
	Bottom	0.891	0.004	0.154	0.004
5.0	Top	1.009	0.006	0.252	0.004
	Second	0.983	0.007	0.233	0.005
	Middle	0.986	0.010	0.235	0.008
	Fourth	0.976	0.009	0.227	0.007
	Bottom	0.985	0.009	0.234	0.007
5.5	Top	1.117	0.013	0.325	0.008
	Second	1.104	0.009	0.317	0.006
	Middle	1.093	0.009	0.310	0.005
	Fourth	1.079	0.010	0.301	0.006
	Bottom	1.078	0.015	0.300	0.010
6.0	Top	1.225	0.012	0.384	0.006
	Second	1.200	0.014	0.371	0.008
	Middle	1.179	0.006	0.360	0.003
	Fourth	1.175	0.016	0.358	0.009
	Bottom	1.179	0.014	0.360	0.008
6.5	Top	1.338	0.009	0.436	0.004
	Second	1.312	0.012	0.425	0.005
	Middle	1.278	0.011	0.410	0.005
	Fourth	1.269	0.019	0.406	0.009
	Bottom	1.269	0.010	0.406	0.005

Assuming that the sample variation was most likely due to how the CO_2 exhaust was vented from the system, the experiment was conducted with an added metering valve; the metering value should enable even CO_2 discharge from each layer by limiting the exhaustion speed. The experiment was performed under the condition of the press position 6.5 mm, for which the dispersion had been largest. As described in the experimental section (Section 2.3), the evacuation operation was performed in a sequence of exhausting the residual pressure instantaneously after exhausting slowly for 30 s. According to the change in the load applied to the piston, the drop in pressure due to CO_2 exhaustion for 30 s was about 4 MPa. The measured sample thickness and evaluated porosity versus sample position are listed in Table 2. Although the porosity of the top sample was slightly larger, the variation among the samples was greatly reduced. Therefore, these results indicate that the sample variation was caused by the CO_2 venting process. The time required for slow venting doubled the processing time (from 30 s to 60 s), which halves the effect of the 5-fold improvement in the productivity achieved via numbering-up. As the slow venting procedure performed to decrease the sample variation reduces the productivity improvement due to numbering-up, allowable product variation should be considered carefully before adapting slow venting.

Table 2. Product thickness and porosity obtained under slow CO_2 venting conditions.

Press Position [mm]	Sample Position	Thickness [mm]	Standard Deviation [mm]	Porosity [-]	Standard Deviation [-]
6.5	Top	1.311	0.010	0.425	0.004
	Second	1.293	0.019	0.417	0.009
	Middle	1.300	0.008	0.420	0.003
	Fourth	1.294	0.011	0.417	0.005
	Bottom	1.301	0.007	0.420	0.003

4. Conclusions

Two strategies were conducted to improve the productivity of the CAPC method. The CO_2 introduction and exhaustion tubes were connected to the piston instead of the pressure vessel, enabling the free movement of the pressure vessel, which removed the processing time otherwise required for sample introduction and removal. The production per unit time was increased by numbering-up vertically and setting multiple samples without changing the load of the piston. This method is very effective because productivity can be improved without a significant increase in equipment cost. However, sample variation in porosity was detected in samples prepared in the numbering-up scheme, which was found to be caused by the CO_2 venting process. Although controlling the venting rate was one of the solutions, it was accompanied by a decrease in productivity. Therefore, carefully judging the allowable product variation together with the benefits of numbering-up is necessary. Although not discussed in this article, in the future, the flow of CO_2 at the time of exhaustion should be simulated to optimize the high-pressure vessel, spacer, and CO_2 flow path, such that CO_2 is removed at equivalent levels from all samples. This will potentially allow the sample variations to be lowered without reducing the CO_2 venting speed.

Author Contributions: T.A. conceived and designed the experiments, performed the experiments, analyzed the data, and wrote the paper.

Funding: This research received no external funding.

Conflicts of Interest: The author declares no conflict of interest.

References

1. Young, R.J.; Lovell, P.A. *Introduction to Polymers*, 3rd ed.; Routledge: Abingdon, UK, 2011; ISBN 978-0849339295.
2. Fried, J.R. *Polymer Science and Technology*, 3rd ed.; Prentice Hall: Upper Saddle River, NJ, USA, 2014; ISBN 978-0137039555.
3. Mills, N.J. *Polymer Foams Handbook: Engineering and Biomechanics Applications and Design Guide*; Butterworth-Heinemann: Oxford, UK, 2007; ISBN 978-0750680691.
4. Ishizaki, K.; Komarneni, S.; Nanko, M. *Porous Materials—Process technology and applications*; Springer Science+Business Media: Dordrecht, NL, USA, 2014; ISBN 978-1461376637.
5. Aizawa, T. A new method for producing porous polymer materials using carbon dioxide and a piston. *J. Supercrit. Fluids* **2018**, *133*, 38–41. [CrossRef]
6. Kerton, F.; Marriott, R. *Alternative Solvents for Green Chemistry*, 2nd ed.; RSC Publishing: Cambridge, UK, 2013. [CrossRef]
7. Eldik, R.; Klärner, F.-G. *High Pressure Chemistry–Synthetic, Mechanistic, and Supercritical Applications*; Wiley-VCH: Weinheim, Germany, 2002; ISBN 978-3527612628.

8. Jessop, P.G.; Leitner, W. *Chemical Synthesis Using Supercritical Fluids*; Wiley-VCH: Weinheim, Germany, 1999; ISBN 978-3527613687.
9. Taylop, L.T. *Supercritical Fluid Extraction*; Wiley-Interscience: New York, NY, USA, 1996; ISBN 978-0471119906.
10. Kazarian, S.G.; Vincent, M.F.; Bright, F.V.; Liotta, C.L.; Eckert, C.A. Specific intermolecular interaction of carbon dioxide with polymers. *J. Am. Chem. Soc.* **1996**, *118*, 1729–1736. [CrossRef]
11. Li, M.; Huang, X.; Liu, H.; Liu, B.; Wu, Y.; Wang, L. Solubility prediction of supercritical carbon dioxide in 10 polymers using radial basis function artificial neural network based on chaotic self-adaptive particle swarm optimization and K-harmonic means. *RSC Adv.* **2015**, *5*, 45520–45527. [CrossRef]
12. Shieh, Y.-T.; Su, J.-H.; Manivannan, G.; Lee, P.H.C.; Sawan, S.P.; Spall, W.D. Interaction of supercritical carbon dioxide with polymers. I. Crystalline polymers. *J. Appl. Polym. Sci.* **1996**, *59*, 695–705. [CrossRef]
13. Shieh, Y.-T.; Su, J.-H.; Manivannan, G.; Lee, P.H.C.; Sawan, S.P.; Spall, W.D. Interaction of supercritical carbon dioxide with polymers. II. Amorphous polymers. *J. Appl. Polym. Sci.* **1996**, *59*, 707–717. [CrossRef]
14. Nalawade, S.P.; Picchioni, F.; Janssen, L.P.B.M. Supercritical carbon dioxide as a green solvent for processing polymer melts: Processing aspects and applications. *Prog. Polym. Sci.* **2006**, *31*, 19–43. [CrossRef]
15. Lian, Z.; Epstein, S.A.; Blenk, C.W.; Shine, A.D. Carbon dioxide-induced melting point depression of biodegradable semicrystalline polymers. *J. Supercrit. Fluids* **2006**, *39*, 107–117. [CrossRef]
16. Kikic, I. Polymer–supercritical fluid interactions. *J. Supercrit. Fluids* **2009**, *47*, 458–465. [CrossRef]
17. Tomasko, D.L.; Li, H.; Liu, D.; Han, X.; Wingert, M.J.; Lee, L.J.; Koelling, K.W. A review of CO_2 applications in the processing of polymers. *Ind. Eng. Chem. Res.* **2003**, *42*, 6431–6456. [CrossRef]
18. Jacobs, L.J.M.; Kemmere, M.F.; Keurentjes, J.T.F. Sustainable polymer foaming using high pressure carbon dioxide: A review on fundamentals, processes and applications. *Green Chem.* **2008**, *10*, 731–738. [CrossRef]
19. Feng, D.; Liu, P.; Li, L. Fabrication and cell morphology of a microcellular poly(ether imide)-carbon nanotube composite foam with a three-dimensional shape. *J. Appl. Polym. Sci.* **2019**, *136*, 47501. [CrossRef]
20. Heidari, A.; Fasihi, M. Cell structure-impact property relationship of polypropylene/thermoplastic elastomer blend foams. *eXPRESS Polym. Lett.* **2019**, *13*, 429–442. [CrossRef]
21. Viveiros, R.; Rebocho, S.; Casimiro, T. Green strategies for molecularly imprinted polymer development. *Polymers* **2018**, *10*, 306. [CrossRef] [PubMed]
22. Zhu, J.; Yang, C.; Lu, C.; Zhang, F.; Yuan, Z.; Zhuang, X. Two-dimensional porous polymers: From sandwich-like structure to layered skeleton. *Acc. Chem. Res.* **2018**, *51*, 3191–3202. [CrossRef] [PubMed]
23. Aizawa, T. Fabrication of porosity-controlled polyethylene terephthalate porous materials using a CO_2-assisted polymer compression method. *RSC Adv.* **2018**, *8*, 3061–3068. [CrossRef]
24. Aizawa, T. Peel and penetration resistance of porous polyethylene terephthalate material produced by CO_2-assisted polymer compression. *Molecules* **2019**, *24*, 1384. [CrossRef] [PubMed]
25. Wakui, Y.; Aizawa, T. Analysis of sustained release behavior of drug-containing tablet prepared by CO_2-assisted polymer compression. *Polymers* **2018**, *10*, 1405. [CrossRef] [PubMed]
26. Al-Rawashdeh, M.; Yu, F.; Nijhuis, T.A.; Rebrov, E.V.; Hessel, V.; Schouten, J.C. Numbered-up gas–liquid micro/milli channels reactor with modular flow distributor. *Chem. Eng. J.* **2012**, *207–208*, 645–655. [CrossRef]
27. Kashid, M.N.; Gupta, A.; Renken, A.; Kiwi-Minsker, L. Numbering-up and mass transfer studies of liquid–liquid two-phase microstructured reactors. *Chem. Eng. J.* **2010**, *158*, 233–240. [CrossRef]
28. Tonomura, O.; Tominari, Y.; Kano, M.; Hasebe, S. Operation policy for micro chemical plants with external numbering-up structure. *Chem. Eng. J.* **2008**, *135S*, S131–S137. [CrossRef]

© 2019 by the author. Licensee MDPI, Basel, Switzerland. This article is an open access article distributed under the terms and conditions of the Creative Commons Attribution (CC BY) license (http://creativecommons.org/licenses/by/4.0/).

technologies

MDPI

Article

Effects of the Infill Density on the Mechanical Properties of Nylon Specimens Made by Filament Fused Fabrication

Svetlana Terekhina [1], Innokentiy Skornyakov [2], Tatiana Tarasova [2] and Sergei Egorov [2,*]

1 Laboratoire Angevin de Mécanique, Procèdes et innovAtion, Ecole Nationale Supérieure d'Art et Métiers, 49100 Angers CEDEX 01, France
2 Laboratory of Innovative Additive Technologies, Moscow State University of Technology "STANKIN", 127055 Moscow, Russia
* Correspondence: sergey_951@mail.ru

Received: 30 June 2019; Accepted: 13 August 2019; Published: 16 August 2019

Abstract: Additive manufacturing of polymer products over the past decade has become widespread in various areas of industry. Using the fused filament fabrication (FFF) method, one of the most technologically simple methods of additive manufacturing, it is possible to produce parts from a large number of different materials, including wear-resistant nylon. The novelty of the work is properties investigation of ±45° filling configuration with different filling degree for nylon, as well as calculating the effect of infill on the strength characteristics, excluding the shell. This article reflects the process of manufacturing samples from nylon using FFF technology with various internal topologies, as well as tensile tests. The analysis of the obtained results is performed and the relationship between the structure of the sample and the limit of its strength is established. To calculate real filling degree and the effect of internal filling on the strength characteristics of the specimen, it is proposed to use a method based on the geometric and mass parameters. The FFF method is promising for developing methods for producing a composite material. The results of this article can be useful in choosing the necessary manufacturing parameters.

Keywords: additive technologies; additive manufacturing; FFF; 3D printing; nylon

---

1. Introduction

Rapidly developing for several decades, additive technologies are gradually replacing the classic ways of making products in many industries. This fact is due to the main principle of new methods, the layer-by-layer creation of objects based on digital three-dimensional model controlled by a computer. This approach minimizes the number of equipment and the number of technological operations.

A wide range of materials available for additive production (from metallic powders to polymer filaments) makes it possible to produce analogs of products obtained by classical methods, often not inferior to them by mechanical characteristics. A bright representative of polymer materials used in this kind of production is nylon. Nylon is a thermoplastic polymer belongs to the group of polyamides. These plastics are characterized by high wear resistance but increased hygroscopicity (the ability to absorb moisture). Nylon is used in the production of elements of friction pairs and in the medical industry for the manufacturing of prostheses. Nylon is used to make some parts of machines because it is inexpensive and durable. It is often used in the electronics industry as a nonconductive and heat-resistant material. This polymer is one of the main materials for additive plants using fused filament fabrication (FFF). FFF technology refers to the simplest methods of 3D printing. However, this technology allows to produce a wide variety of products, from mock-ups and prototypes to robust functional elements.

One of the ways to increase the competitiveness of the product is to reduce its weight and the cost of its production. In relation to additive production, both objectives can be achieved by replacing the solid product with a shell with the same geometry and dimensions. To ensure its strength, a structure of the same material is formed inside the shell. The volume fraction of this structure inside the shell can vary from 0% to 100%. Obviously, with increasing this value, the strength of the structure will increase. The purpose of this study is to identify the relationship between the volume fraction of this structure inside the shell (20–100%) and the strength of the sample. It should be noted that in this article the filling structure is considered as a composite material, one of the components of which is nylon and the other is air.

At present, all technologies of additive production are fully reflected in the literature [1–3]. The main attention is paid to the manufacturing of products from metal powders [4,5] using selective laser melting technology in view of the increased interest in obtaining functional parts of various machines and mechanisms. However, polymer composite materials, filled with fibers of various types and composition [6,7], already constitute a serious competition to metals. In this connection, on the wave of general interest in additive technologies, attempts are made to directly produce composite materials using 3D printing [8].

Nowadays, experimental and theoretical studies of the effect of the filling structure of parts printed using additive technologies on their properties are being conducted worldwide [9]. The materials chosen are metal alloys (steel [10], aluminum [11], and copper alloys, etc.) as well as polymers (ABS (Acrylonitrile butadiene styrene) [12], PLA (polylactide) [13], polyamides [14,15], etc.). In a number of works, the influence on the mechanical properties of various parameters of the printing process (speed, temperature, layer thickness [16–18], etc.) and the internal architecture (filling geometry [19], directionality [20], and density [21]) is investigated. As the mechanical properties under investigation, tensile [22] and flexural [23] ultimate strength and the modulus of elasticity are most often chosen. Often, a comparative analysis of various materials [24,25], technologies [26], structures [27,28], and filling directions [29] is carried out. In some works, the authors conducted a comparative analysis of experimental and theoretical data [30–32].

To understand the rationality of the transition from traditional manufacturing methods to new ones, the accumulation of experimental data is required. The world scientific community studies the properties of samples from unfilled polymers made by additive technology methods (particularly by the FFF method [33]), and mathematical models of their formation are created [34]. Much attention is paid to the most widely used plastics in FFF installations, ABS and PLA [35,36]. Less common nylon is used in the manufacturing of gears and friction pairs, where it experiences mainly cyclic loads. In this regard, it is more often subjected to fatigue testing [37]. However, the prospects for using nylon as a matrix of composite material are growing because of its high compatibility with biodegradable natural fibers. This can significantly expand the scope of the material. Tests of tensile strength of nylon, useful in further studies, are described in this article.

2. Materials and Methods

Nylon filament (NYLON, manufacturer-Print Product, Saint Petersburg, Russia) was used as a material for the production of samples (Table 1). Due to the high hygroscopicity of the material, the drying of the filament was carried out immediately prior to manufacturing. The drying of the filaments before printing was carried out at 60 °C in a vacuum oven for 6 h. All the samples were then stored in the dry atmosphere of a desiccator prior to testing.

Table 1. Properties of NYLON material.

Mark	Chemical Formula of Monomer	Filament Diameter, mm	Melting Point, °C	Density, g/sm^3	Tensile Strength, MPa	Elastic Modulus, MPa	Percentage Elongation, %
NYLON	$C_{12}H_{22}N_2O_2$	1.75	260 (acceptable softening at 215)	1.14	~80	1700	60

To determine the required dependence, strength tests of a number of samples having different volume fraction of the filling structure were carried out.

Samples were manufactured on equipment developed at MSTU "STANKIN" using elements of the Prusa Mendel project. Manufacturing of products on this equipment is carried out using FFF technology (fused filament fabrication). FFF is one of the methods of additive manufacturing, which consists of feeding a threadlike material into the heated chamber, melting it, squeezing it out through the nozzle, and depositing it onto the working surface. Simultaneous processes of extruding the filament, moving the extruder, and/or the working surface provide a layer-by-layer forming of the product. On the equipment of MSTU "STANKIN", these processes are controlled by computer using the software of Repetier Host. Moving of mobile elements is carried out according to the algorithm, previously formed by the program based on the three-dimensional model of the product and the characteristics set by the operator (Table 2).

Table 2. Characteristics of the fused filament fabrication (FFF) process.

Characteristic	Value
Chamber type	open
Nozzle diameter, mm	0.3
Extruder temperature, °C	240
Bed temperature, °C	80
Layer height, mm	0.15
Extruder movement speed, mm/s	40
Extruder movement speed at first layer, mm/s	12

The shape and dimensions of the samples were selected in accordance with GOST 11262-80 "Plastics. Method of tensile testing" (is an analog of ISO 527-2:2012) (Figure 1a). As a filling scheme, a perpendicular scheme was set. Each layer is filled with parallel tracks (strips of cooled polymer) positioned at an angle of 45° to the axis of the sample and perpendicular to the direction of the tracks of the previous layer (Figure 1b). Varying the distance between adjacent tracks results in a change in the volume fraction of the fill pattern. Several groups of samples were made, for each of which a theoretical volume fraction of the filling structure was set: 20%, 40%, 60%, 80%, and 100%.

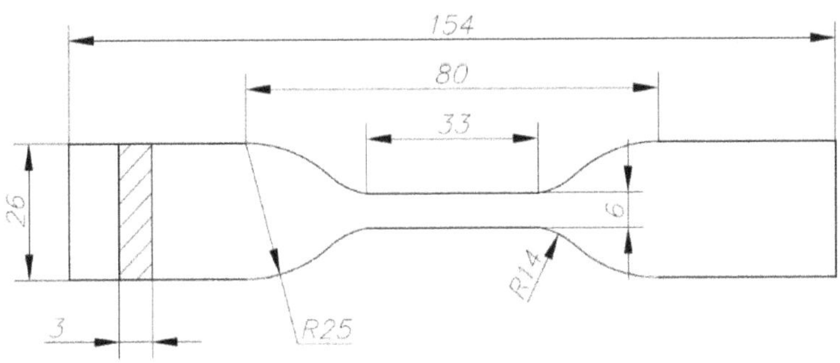

(a)

Figure 1. *Cont.*

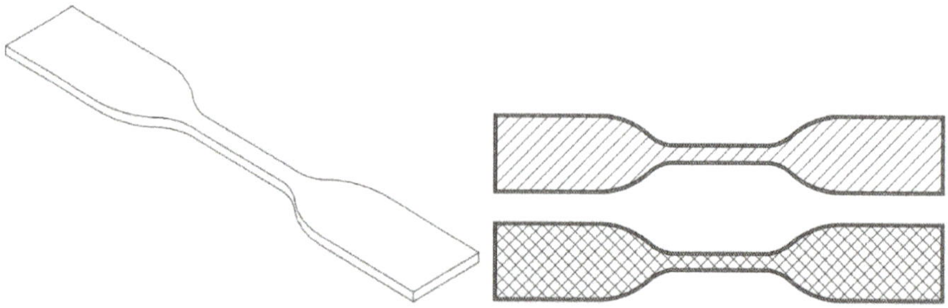

(**b**)

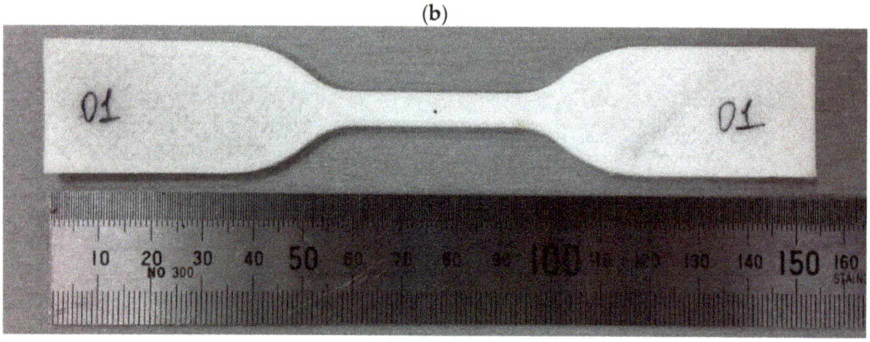

(**c**)

Figure 1. Sample in accordance with GOST (**a**), the layer infill scheme (**b**), real specimen (**c**).

The actual geometric and mass characteristics of the samples were previously determined by caliper measurement (accurate to 0.1 mm) and weighing on the Mettler Toledo XPE analytical scales (accurate to 0.001 g).

To determine the tensile strength of the manufactured samples, the electrodynamic testing system ElectroPuls Model E10000 for axial loading with twisting was used.

The tests of each sample were carried out in accordance with GOST 11262-80 "Plastics. The tensile test method" at a speed of 25 mm/min. Specimen after test is shown in Figure 2.

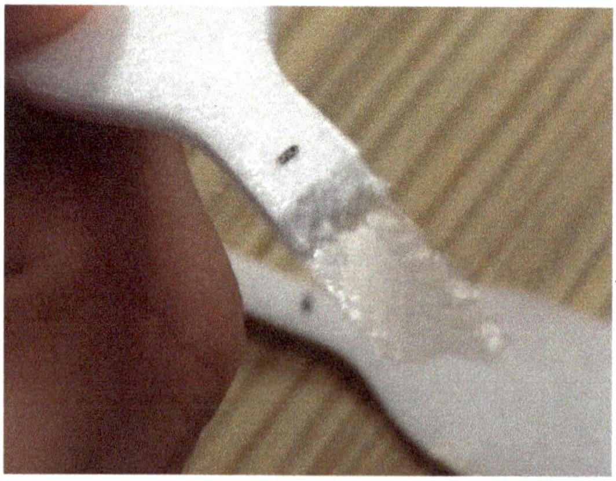

Figure 2. Specimen after test.

3. Results

After the tests, stretch diagrams of the samples were obtained (some of them are shown in Figure 3) and the strength limits were calculated.

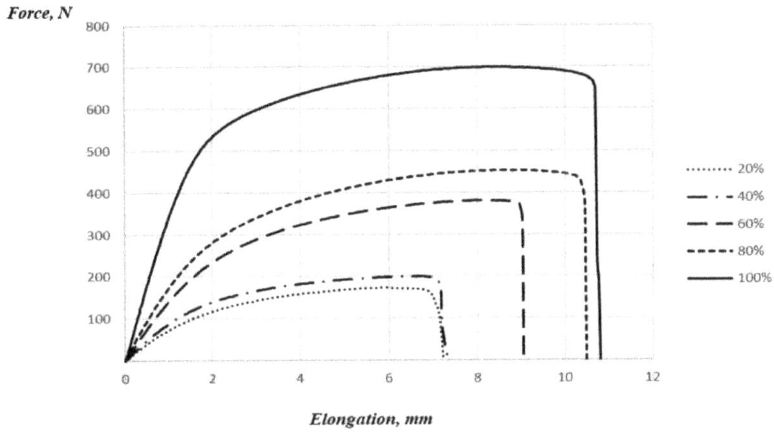

Figure 3. Tensile diagrams of some samples.

As the degree of filling increases, the ability of the sample to plastic deformation increases.

The obtained data cannot fully characterize the relationship between strength and volume fraction of the sample filling pattern. To identify this relationship, the actual fraction of filling of each sample and the strength of the filling (without taking into account the strength of the shell) were calculated. Calculation of the actual fraction of sample filling was made on the basis of geometric and mass parameters.

The fabricated sample consists of N_Σ layers of thickness t, of which N_L lower ones and N_U upper ones have a continuous filling (Figures 4 and 5a). The number of lower layers varies from sample to sample due to different conditions for separating samples from the bed surface after manufacturing. Each of the N_{INT} intermediate layers consists of a solid contour and filling structure with a theoretical volume fraction $A_{T\%}$ (Figure 5b). Thus, the total number of layers:

$$N_\Sigma = N_L + N_U + N_{INT} \tag{1}$$

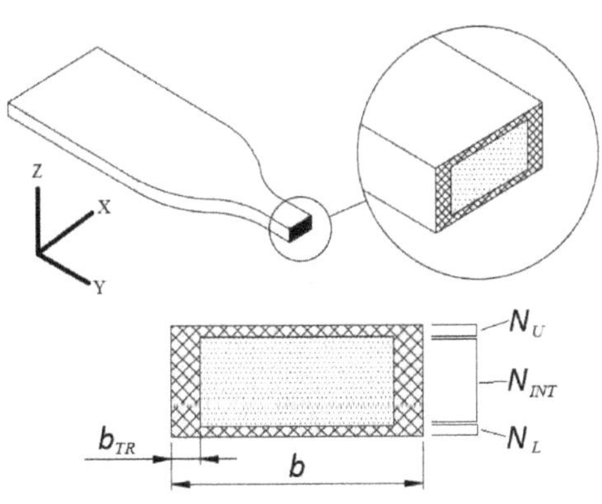

Figure 4. Working cross-section of the sample.

(**a**)

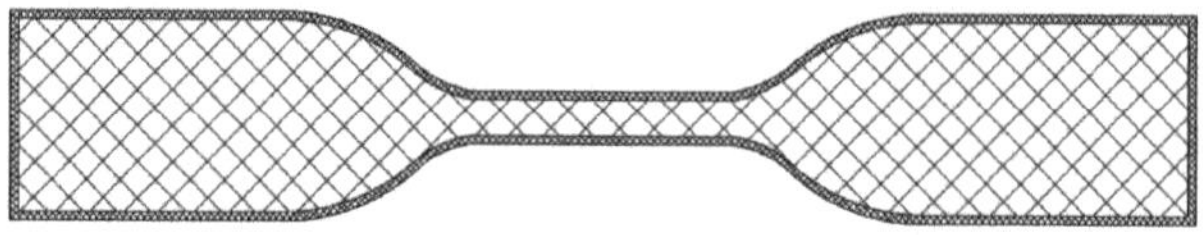

(**b**)

Figure 5. Infill scheme of upper and lower layers (**a**); infill scheme of intermediate layers (**b**).

Taking into account the actual dimensions of each manufactured sample using the AutoCAD 2016 software, the contour was drawn and the values of the areas of the different zones of the layer were determined (Figure 6):

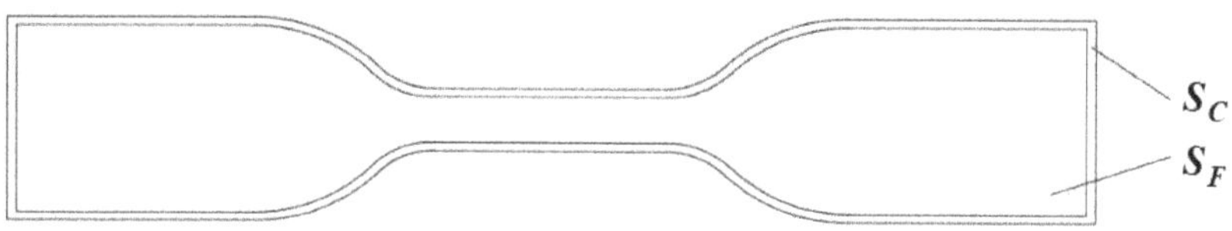

Figure 6. Layer zones.

S_C—solid contour area,
S_F—filling area,

$$S_{\sum} = S_C + S_F \tag{2}$$

—total area of the continuous layer.

The difference in these values in the samples is due to changes in the temperature state of the working zone during manufacturing.

According to obtained values, the volumes of solid material and filling were calculated:
$V_{SM} = (N_L + N_U) \cdot t \cdot S_{\sum} + N_{INT} \cdot t \cdot S_C$—volume of solid material,
$V_F = N_{INT} \cdot t \cdot S_F$—volume of filling.
The measured mass m of the sample:

$$m = V_{SM} \cdot \rho_N + V_F \cdot \rho_A$$

ρ_N—density of solid material,
ρ_A—actual filling density.
Thus, the actual filling density:

$$\rho_A = \frac{m - V_{SM} \cdot \rho_N}{V_F} = \frac{m - [(N_L + N_U) \cdot t \cdot S_{\sum} + N_{INT} \cdot t \cdot S_C] \cdot \rho_N}{N_{INT} \cdot t \cdot S_F} \tag{3}$$

The mass of the sample is composed of the weight of solid material and the weight of air:

$$m_F = m_N + m_{AIR}$$

$$\rho_A \cdot V_F = \rho_N \cdot V_N + \rho_{AIR} \cdot V_{AIR}$$

ρ_{AIR}—air density,

V_N—volume of solid material in the filling,

V_{AIR}—volume of air in the filling.

Taking into account $V_F = V_N + V_{AIR}$ and $V_N = V_F \cdot \frac{A_{ACT\%}}{100\%}$, the formula was obtained:

$$\rho_A \cdot V_F = \rho_N \cdot V_F \cdot \frac{A_{ACT\%}}{100\%} + \rho_{AIR} \cdot V_F \cdot \left(1 - \frac{A_{ACT\%}}{100\%}\right)$$

$A_{ACT\%} = \frac{\rho_A - \rho_{AIR}}{\rho_N - \rho_{AIR}} \cdot 100\%$—actual volume fraction of the filling structure.

Neglecting the density of air, the following dependence was obtained:

$$A_{ACT\%} = \frac{m - [(N_L + N_U) \cdot t \cdot S_\Sigma + N_{INT} \cdot t \cdot S_C] \cdot \rho_N}{N_{INT} \cdot t \cdot S_F \cdot \rho_N} \cdot 100\% \quad (4)$$

Calculation of the tensile strength of the filling structure (without taking into account the strength of the shell) was done on the basis of the following formula:

$$\sigma_B = \sigma_{BN} \cdot A_N + \sigma_{BF} \cdot A_F \quad (5)$$

σ_B—tensile strength of the entire sample,

σ_{BN}—tensile strength of solid nylon (due to the change in material properties because of thermal stresses, the maximum tensile strength of 100%-filled sample was used),

σ_{BF}—tensile strength of the filling structure,

A_N—percentage of solid nylon contour in cross-section,

A_F—percentage of filling structure in cross-section.

Taking into account the geometrical parameters of each sample (number of upper, intermediate, and lower layers, working cross-section width, track width (Figure 3)), the values of A_N and A_F were calculated:

$$A_N = \frac{(N_U + N_L) \cdot b + 2b_{TR} \cdot N_{INT}}{(N_U + N_L + N_{INT}) \cdot b} \quad (6)$$

$$A_F = \frac{N_{INT} \cdot (b - 2b_{TR})}{(N_U + N_L + N_{INT}) \cdot b} \quad (7)$$

The obtained dependence for the tensile strength of the filling structure:

$$\sigma_{BF} = \left(\sigma_B - \sigma_{BN} \cdot \frac{(N_U + N_L) \cdot b + 2b_{TR} \cdot N_{INT}}{(N_U + N_L + N_{INT}) \cdot b}\right) \cdot \frac{(N_U + N_L + N_{INT}) \cdot b}{N_{INT} \cdot (b - 2b_{TR})} \quad (8)$$

Based on the obtained values of the volume fractions (theoretical and actual) and the strength of the filling structure of each sample, the arithmetic mean values were determined (Table 3) and the dependence was obtained (Figure 7).

Table 3. Average arithmetic values of the volume fraction and the tensile strength of the infill structure.

Average Theoretical Infill Volume Fraction $A_{T\%}$, %	Average Actual Infill Volume fraction $A_{ACT\%}$, %	Average Infill Tensile Strength σ_{BF}, MPa	Average Weight in Relation to $A_{T\%}$	Shell/Infill Ratio, %
20	23.71	0.83	0.13717	5.73
40	41.74	6.15	0.09901	5.77
60	54.35	14.20	0.090386667	5.71
80	70.23	28.20	0.0759925	5.72
100	72.49	32.70	0.062868	5.59

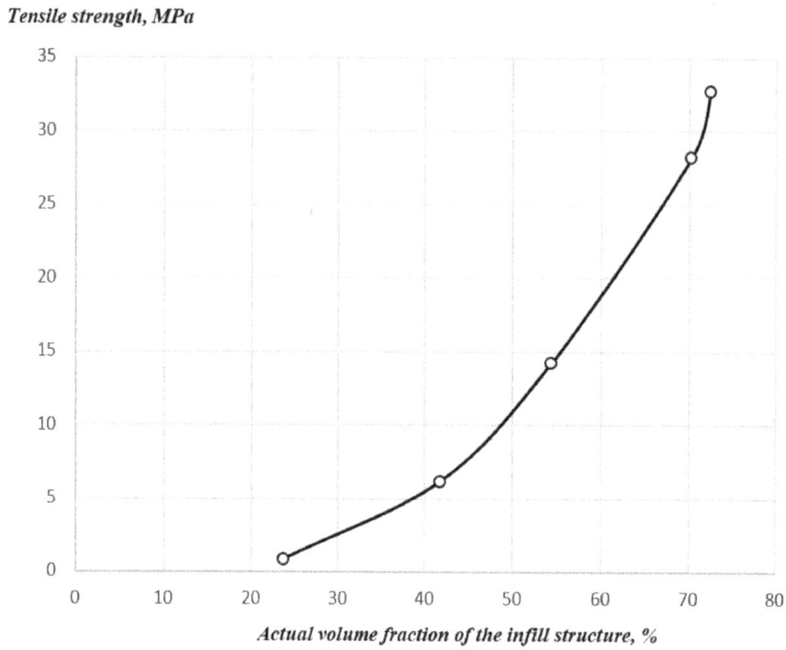

Figure 7. Dependence of the tensile strength on the actual volume fraction of the infill structure.

4. Discussion

The obtained results show that with the increase of the volume fraction of the infill structure, the discrepancy between the preproduction (theoretical) and actual values is growing. This can be explained by the fact that the material has time to partially cool before reaching the bed, and is stacked as an oval-shaped track (Figure 8). As a result, voids are formed, leading to a decrease in the volume fraction of plastic. The increase of the volume fraction and, as a consequence, the supply of material, leads not so much to a decrease of these voids as to an increase of the cross-section of the sample. Thus, the conducted experiment did not allow to establish a dependence in the area from 72% to 100% of the infill, which does not make it possible to compare samples fabricated by the FFF technology and cast samples.

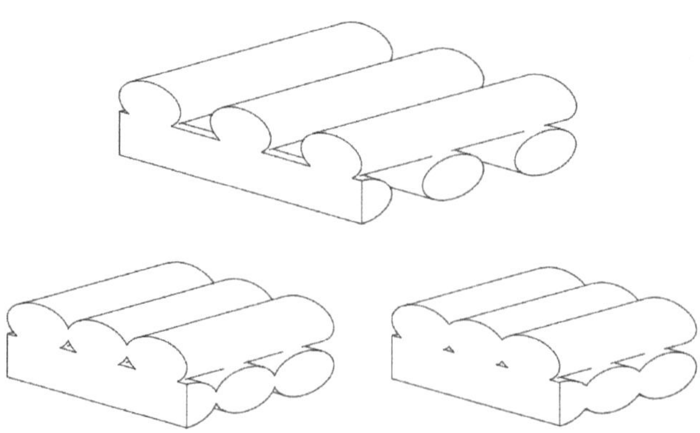

Figure 8. Tracks interaction of different theoretical volume fractions of the infill structure: 40% (**top**), 80% (**left**), 100% (**right**).

Nevertheless, this dependence can be useful in choosing the volume fraction of the infill structure in the obtained range of values depending on the required strength characteristics.

Carrying out a qualitative assessment of the results, it can be noted that after increasing the volume fraction of the infill structure above 60%, a significant increase in strength occurs (Figure 7). The ultimate strength of the fabricated samples is determined not only by the amount of material but also by the contact between the parallel tracks. Observation of the manufacturing process showed that when setting a theoretical volume fraction of filling in the range of 20–40%, neighboring tracks of the same layer do not touch each other. When the parameter is increased to 60% (which corresponds to the actual value of 54%), the parallel tracks contact (Figure 8), which leads to the formation of a continuous layer and increases the strength of the entire sample.

Thus, the observation of the production of samples from nylon and the tensile tests made it possible to establish a relation between the internal structure and the strength properties of the samples, and to characterize the effects that occur during the printing process.

Author Contributions: Conceptualization, S.T.; data curation, S.T.; formal analysis, S.T.; funding acquisition, T.T.; investigation, I.S. and S.E.; methodology, S.T. and I.S.; project administration, T.T.; resources, I.S. and S.E.; software, I.S.; supervision, T.T.; validation, S.T. and T.T.; visualization, S.E.; writing—original draft, I.S.; writing—review and editing, T.T.

Funding: The work was carried out with the financial support of the Ministry of Education and Science of the Russian Federation in the framework of state task No. 11.1267.2017/4.6.

Acknowledgments: The studies were carried out using the equipment of the Center for Collective Use of MSTU "STANKIN". The authors express their gratitude to the anonymous reviewers, whose remarks made it possible to improve the article.

Conflicts of Interest: The authors declare no conflict of interest. The funders had no role in the design of the study; in the collection, analyses, or interpretation of data; in the writing of the manuscript; or in the decision to publish the results.

References

1. Bourell, D.L.; Beaman, J.J.; Leu, M.C.; Rosen, D.W. *A Brief History of Additive Manufacturing and the 2009 Roadmap for Additive Manufacturing: Looking Back and Looking Ahead*; RapidTeach: Istanbul, Turkey, 2009; Available online: https://pdfs.semanticscholar.org/4716/c69f0b90a158589e54248a524a57ad78f4a3.pdf (accessed on 16 August 2009).
2. Gibson, I.; Rosen, D.; Stucker, B. *Additive Manufacturing Technologies: 3D Printing, Rapid Prototyping, and Direct Digital Manufacturing*, 2nd ed.; Springer Publ.: New York, NY, USA, 2015; p. 498.
3. ISO/ASTM. *52900:2015 Additive Manufacturing General Principles—Terminology*; ISO/ASTM International: Geneva, Switzerland, 2015; p. 19. Available online: https://www.sis.se/api/document/preview/919975/ (accessed on 15 August 2015).
4. Grigoriev, S.N.; Tarasova, T.V. Possibilities of additive manufacturing technology for the manufacturing of complex profile parts and the production of functional coatings for the production of powders. Metallovedenie i termicheskaya obrabotka metallov = Metal Science and Heat Treatment of Metals. 2015. 10:51–1. Available online: http://mitom.folium.ru/contents/2015/20151--0.php (accessed on 15 August 2015). (In Russian)
5. Zlenko, M.A.; Nagaytsev, M.V.; Dovbysh, V.M. *Additivnye Tekhnologii v Mashinostroyenii: Posobie dlya Inzhenerov*; NAMI Publ.: Moscow, Russia, 2015; p. 220. Available online: http://vneshtechnika.ru/rus/books/123pd.pdf (accessed on 15 August 2015).
6. Mashkov, Y.K.; Baibaratskaya, M.Y.; Grigorevsky, B.V. *Konstruktsionnye Plastmassy i Polimernye Kompozitsionnye Materialy: Ucheb. Posobie*; OmGTU Publ.: Omsk, Russia, 2002; p. 129. Available online: https://lib-bkm.ru/11434 (accessed on 15 August 2002).
7. Perepelkin, K.E. *Armiruyuschie Volokna i Voloknistye Polimernye Kompozity. Monografiya*; Scientific Basics and Technologies Publ.: St. Petersburg, Russia, 2009; p. 380. Available online: https://elibrary.ru/item.asp?id=20246555 (accessed on 15 August 2009).
8. Skornyakov, I.A.; Tarasova, T.V. [Possibilities of additive technologies in the production of polymer composite materials]. Mashinostroenie: traditsii i innovatsii: sbornik trudov nauchno-tekhnicheskoy konferentsii

technologies

MDPI

Article

Compensation for Geometrical Deviations in Additive Manufacturing

Christoph Hartmann [1,*], Philipp Lechner [1], Benjamin Himmel [1], Yannick Krieger [2], Tim C. Lueth [2] and Wolfram Volk [1]

1 Chair of Metal Forming and Casting, Technical University of Munich, Walther-Meissner-Strasse 4, 85748 Garching, Germany; philipp.lechner@utg.de (P.L.); benjamin.himmel@outlook.de (B.H.); wolfram.volk@utg.de (W.V.)

2 Institute of Micro Technology and Medical Device Technology, Technical University of Munich, Boltzmannstr. 15, 85748 Garching, Germany; yannick.krieger@tum.de (Y.K.); tim.lueth@tum.de (T.C.L.)

* Correspondence: christoph.hartmann@utg.de; Tel.: +49-89-289-13769

Received: 28 October 2019; Accepted: 28 November 2019; Published: 2 December 2019

Abstract: The design of additive manufacturing processes, especially for batch production in industrial practice, is of high importance for the propagation of new additive manufacturing technology. Manual redesign procedures of the additive manufactured parts based on discrete measurement data or numerical meshes are error prone and hardly automatable. To achieve the required final accuracy of the parts, often, various iterations are necessary. To address these issues, a data-driven geometrical compensation approach is proposed that adapts concepts from forming technology. The measurement information of a first calibration cycle of manufactured parts is the basis of the approach. Through non-rigid transformations of the part geometry, a new shape for the subsequent additive manufacturing process was derived in a systematic way. Based on a purely geometrical approach, the systematic portion of part deviations can be compensated. The proposed concept is presented first and was applied to a sample fin-shaped part. The deviation data of three manufacturing cycles was utilised for validation and verification.

Keywords: additive manufacturing; 3D-printing; compensation; accuracy; precision

1. Introduction

A central, key technology in the future will be additive manufacturing (AM) [1]. Opportunities are arising in combination with ongoing digitisation and cross-linking between production and design in line with industry 4.0. For certain components this will allow one to avoid costly and time-consuming production processes, such as the manufacturing of moulds and tools. An increase in flexibility and the production of individualised products containing complex structures without additional effort go hand in hand. In this way, AM can make a significant contribution to mastering the complexity in the production process associated with the rapidly increasing number of variants [2].

However, along with all of the opportunities AM processes offer, a comprehensive transfer to mass production is nowhere in sight. Central hurdles for the application in serial production are above all, the cost-related general conditions (high material costs in combination with an insufficient ratio of plant costs to productivity) and the high manual effort in the process chain due to the lack of physical and digital line integration. This current lack of the integration of AM processes in conventional production environments is due to the specifics of AM processes, such as batch production, and the currently low degree of automation in the peripheral systems, such as component handling, quality measurement and transport chains. On the component level, one major critical point in mass production is accuracy and reproducibility. In AM processes, deviations between the nominal and the actual parts are primarily caused by the highly complex thermo-mechanical conditions leading to high gradients and high temperatures, for example. This significantly limits the use and the associated potential of AM processes for in series production [3].

2. State of the Art

When considering the dimensional quality of manufactured components, the systematic and stochastic portions of the deviations have to be carefully separated. Varying environmental conditions, material fluctuations, tribology, and changing thermo-mechanical boundary conditions are typical sources of stochastic errors. It is impossible to completely eliminate the stochastic portion in real applications and direct elimination is hardly possible. To cope with stochastic phenomena, process robustness, reasonable component design and the suitable definitions of tolerances are necessary. The systematic deviations can be related to identifiable causes and possess unique, unidirectional characteristics. For that reason, systematic deviations can be addressed by geometric compensation. For a separation of both deviation types, a sufficient number of measurements is necessary.

2.1. *Causes for Geometrical Deviations in Additive Manufacturing*

Due to the complexity of the process chain from the CAD model to the finished component, geometric deviations can occur at different points in the manufacturing process. These can be divided into the following error categories [4]:

- Deviations in data preprocessing;
- Errors during material processing;
- Machine errors;
- Stochastic errors.

In laser powder bed fusion (L-PBF), each layer is two-dimensional. The three-dimensional component is, thus, created step-by-step by joining individual layers of the same thickness along the Z coordinate. For this purpose, mathematical layer information is extracted from the digital 3D data set during pre-processing. This results in deviations during the creation of a discrete mesh (usually .STL). In addition, layer-by-layer construction can result in steps in a component, since the smallest possible resolution in the Z-direction is the layer thickness [4,5].

During processing, the powder is heated by a laser to a temperature above its melting temperature in order to bond the particles through phase transformation. There is a risk of thermal stresses due to an inhomogeneous temperature distribution in the build chamber. As a result, distortions occur in the component, as the material shrinks inhomogeneously during cooling, resulting in thermally induced stresses. Dimensional deviations resulting from material shrinkage are dependent on the geometry of the respective component [6,7].

Based on temperature measurements in an SLS machine, a significant temperature gradient in the x-y plane of a layer was determined. From this it can be concluded that the material shrinkage of a geometry additionally depends on its x and y-positions in the build chamber [6]. The dimensional accuracy in the x-y plane of a layer depends directly on the scanning system used by the AM system and its ability to move the laser beam along the desired path [8]. Such machine errors, on the other hand, are measurable and can be compensated to a certain extent in order to minimise their effect on the end product to an acceptable minimum. In addition to the factors already mentioned, despite an identical processing environment, quality fluctuations can occur between the end products, the exact cause of which cannot be defined, so they are, therefore, regarded as stochastic errors. Of all these factors, the temperature-induced shrinkage of the material is the main cause of inaccuracies in the end product [4,8].

2.2. Classical Methods to Compensate for Deviations in Additive Manufacturing

In the literature, different methods have been found to counter geometry deviations in the component by conventional means. However, these are usually limited to one specific error. In their work, Pham et al. analysed various factors that influence the accuracy of additive manufactured components. This began with the generation of the STL file. The chord height and the angle control have a big influence on the accuracy. Some of these error sources can also be avoided by using alternative data formats [4,8,9].

During further processing from 3D to layer information (slicing), discretisation errors occur, which are mainly due to the z-direction resolution, which is limited to one layer in thickness [10]. These effects can, however, be reduced by various algorithms; for example, by adjusting the layer thickness [11]. In addition, the orientation of the parts in the build chamber can be adjusted. Masood et al. describe in their research, the possibility of minimising the error caused by the step effect by adjusting the orientation in the build chamber. The best possible orientation can be calculated on the basis of the minimum volumetric error. This leads to an improvement of the accuracy and the surface quality of the component by minimising the effects of the stair tread effect [12].

In order to prevent geometric deviations due to residual stresses in the component, heat treatments are used to reduce mechanical stresses in the component. This prevents stresses from being released and dimensional deviations from occurring in one of the subsequent processing stages, such as separation from the construction platform. Reference [13] In addition, thermal distortion is influenced by the scanning strategy and scanning time and can be improved by adjusting them. References [14–16] Kamat and Pei analytically calculated deformations due to residual stresses in parts with internal channels, which can be used to compensate for deformations [17]. To compensate for shrinkage effects, different scaling factors are applied to the individual 2D layers. Furthermore, a compensation of the laser diameter can be achieved by an offset in the 2D planes [4,18,19].

2.3. Geometrical Deviation Compensation in Additive Manufacturing

Due to interacting factors, such as material shrinkage, geometric complexity, quality fluctuations and machine faults and inaccuracies, it is challenging to predict and control the extent of dimensional deviations. The straightforward way to compensate for deviations is currently to use a constant scaling factor across a part to calculate material shrinkage. Alternatively, different scaling factors can be applied to individual sections in the CAD model. It is assumed that the dimensional deviations occur evenly in the individual sections. However, this assumption is often not legitimate due to the complexity of some geometries, and this can lead to transfer effects, among other things [20].

Huang et al. deal with the compensation of dimensional deviations in the x-y plane and define various quality standards that minimise volume and surface deviations [20,21].

In addition, initial work has been done to use pre-deformation to compensate for dimensional deviations in selective laser beam melting. This involves simulating thermal distortion in the component using a finite element calculation, and then pre-deforming the CAD model against the expected deviations [22–24]. Zhu et al. proposed a statistical shape analysis, which is trained with data obtained by simulation and can predict geometrical deviations for similar parts [25]. Furthermore, they proposed a compensation method for the layers of the sliced part [26]. Schmutzler et al. compared a simulation-based pre-deformation to a deviation compensation in theory, which is based on previous manufactured iterations, for a cantilever geometry. They showed that the simulation-based compensation converged to the desired geometry within three iterations [27].

3. Hypothesis

Geometrical accuracy has always been one of the main objectives in component manufacturing. The established manufacturing processes, therefore, have a variety of approaches that address this problem. In forming technology, two systematic approaches to deviation compensation are the displacement adjustment method and the stress-based adjustment method [28,29]. The stress-based method uses measured deviations to calculate a certain stress state on the target geometry. This stress state is reversed and is used as a boundary condition to deform the forming tool. The result of the deformation is the compensated forming tool. The displacement adjustment method works similarly. The measured deviations are directly reversed and used for the geometrical adaptation of the forming tool. Both methods have been used successfully in different applications. The stress-based adjustment methods seem to perform better for processes dominated by a global deformation pattern—sheet metal forming, for example, where the displacement adjustment method tends to be suitable for local compensation, as with bulk forming.

In AM, no tools are needed for manufacturing, but a manufacturing geometry has to be provided to the AM system. We proposed that this manufacturing geometry could be seen as a forming tool in a geometrical compensation sense. We further proposed that the displacement adjustment method, which directly incorporates the deviation data, is a suitable method for AM, because of its incremental characteristics.

In this paper, the objective was to prove those hypotheses. Therefore, a general geometrical compensation procedure based on the displacement adjustment method is suggested for AM to capture the problem of accuracy and reproducibility also with regard to an implementation in production lines. The compensation scheme is shown using a SLS process.

4. Materials and Methods

For the investigation in this paper, a SLS process is used to apply the conducted general geometrical compensation procedure. In the following subsections, the parts created, manufacturing steps and further processing are presented in detail.

4.1. Specimen and Manufacturing

A sample part was designed to show the effectiveness of the compensation algorithm. Figure 1 depicts the whole job. The fin-shaped geometry of the component was chosen specifically to boost thermal deviations. A total of 27 parts were positioned in the build chamber in three rows with nine parts each. We measured and compensated the blue coloured parts a_i–e_i. The light blue parts were measured but not compensated and served as a reference to their compensated direct neighbour to study the reproducibility of the process. For the digitisation of the components, the structured white light scanning device GOM ATOS II 400 (GOM, Braunschweig, Germany) was utilised. For each component, approximately 15 frames had to be acquired. With the setup we used, a resolution of 0.05 mm was achieved. The test specimens were fabricated using an EOS Formiga P100 (EOS, Krailing, Germany) in combination with polyamide 12 (EOS PA2200, Krailing, Germany). The powder we used had a 50/50 ratio of new and recycled powder. The chosen layer thickness was 0.1 mm. The job was positioned in the centre of the build platform with the standard offset of six millimetres in the z plane. The temperature of the process chamber was set to 172 degrees Celsius. The material-dependent scaling for the jobs was (X:3.086%; Y:3.019%; Z(0):2.2%; Z(300):1.6%). These machine parameters were kept constant for all jobs.

4.2. Compensation Algorithm

The compensation strategy was based on a framework proposed for bulk forming processes [30,31]. The compensation approach was fully driven by geometrical data, which, in principle, ensures general validity and scale invariance. In Figure 2 the flow chart for the compensation procedure is shown (forming in the upper half and the adaptation of AM in the lower half). Each process originates from the part design and the target geometry, which could be provided in any form; for example, as a CAD-design or reverse engineering measurement object. In case of forming processes a forming tool geometry is derived, which is close to the negative of the target geometry, but usually is divided into an upper and a lower tool part due to the uni-axial working direction of the presses. The active surfaces of the forming tool in the closed state show a certain transformation of the target geometry's surface with the same topology. In the surface generation process (centre of Figure 2), this transformation, also called compensation, is deduced through the iterative tool redesign. Based on the target geometry for AM manufacturing, a discrete AM manufacturing geometry was determined based on the part design. This step corresponds to the generation of active tool surfaces in forming processes. Since AM is an incremental process, no physical tool is necessary, but an active closed surface is required, which is a building instruction for the AM system. It is transferred to the AM system for initial manufacturing. As an alternative, the framework is also capable of handling data in a numerical simulation framework. Whether part manufacturing, numerical analysis or both should used, depends on different process and part specific characteristics and attendant circumstances. Both from real manufacturing or numerical results, a final workpiece geometry was achieved and evaluated. The deviation data of the component compared to the target geometry served as input data for the systematic compensation. As a result, an updated AM manufacturing geometry is provided. This surface, in the sense of forming, "forms" the part, in the AM manufacturing environment. More abstract, in both manufacturing processes, surfaces have to be generated based on the target geometry, from which the actual manufacturing process result is controlled.

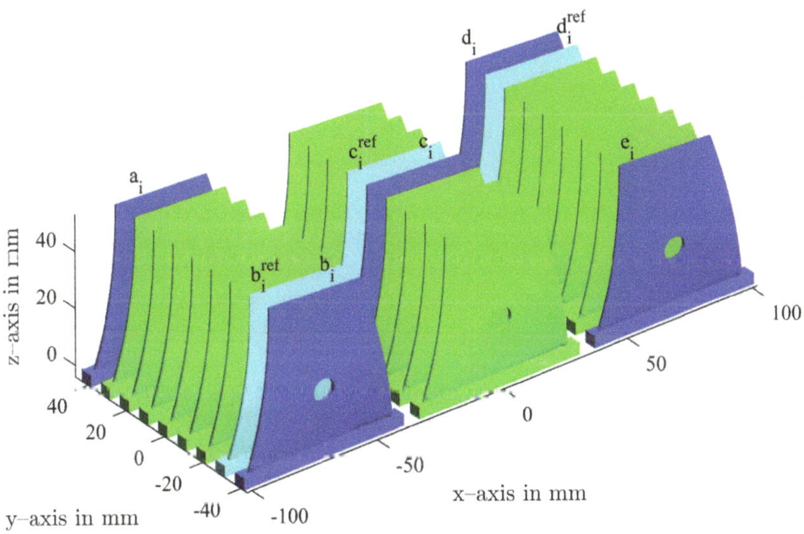

Figure 1. Stacking of the components in the build chamber. Blue components were measured and compensated; cyan components were measured for reference and not compensated.

The purely geometric approach assumes that all physical causes for deviations occurring during the manufacturing process are included in the systematic portion of the component shape. Hence, different physical boundary conditions have no effects on the general procedure. For example, different materials or processing temperatures can be treated in the same way. The approach also allows one, for example, to incorporate additional processing steps, such as heat treatments or machining. Material properties and environmental conditions are implicitly contained in the approach through the measurement data. The three key aspects pointed out in [31] in the context of bulk forming, reduce to two key aspects that also arise in AM for an efficient and effective implementation of the compensation procedure:

- Identification of the stochastic portion of the deviations;
- Automated and structured modification of manufacturing geometries.

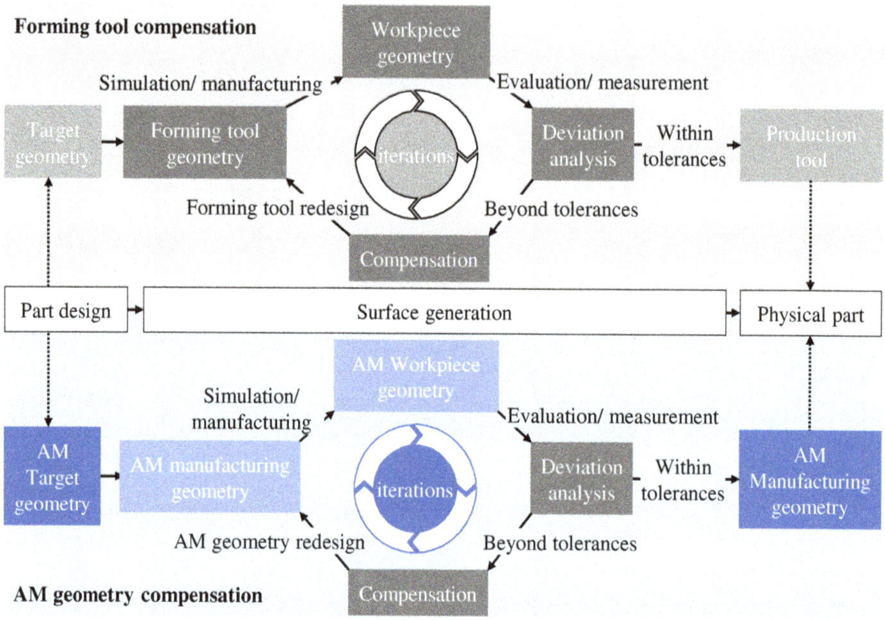

Figure 2. Flow chart of the compensation process for determining a suitable manufacturing geometry for AM in comparison to the forming tool compensation approach.

The problem of finding a suitable manufacturing geometry presents as an inverse problem under stochastic conditions. Hence, the design process is built in an iterative manner. A certain number of iterations i need to be passed, as shown in Figure 2. The manufacturing geometry M_i leads to a workpiece geometry W_i. With a predefined reference coordinate system and the target geometry W_{target}, the deviations D_i can be derived. It is assumed that the deviation values D_i are small compared to the component dimension. Hence, a non-rigid transformation $\phi_{W_i} : W_{target} \rightarrow W_i$ between the target shape W_{target} and the workpiece shape W_i can be deduced. Additionally the manufacturing geometry is supposed to be a non-rigid transformation $\phi_{M_i} : W_{target} \rightarrow M_i$ of the the target geometry W_{target}. The idea of the geometrical compensation is to use the information of the non-rigid transformations to derive a new manufacturing geometry M_{i+1} in a recursive fashion by using

$$M_{i+1} = M_i - \alpha_{i+1}(W_i - W_{target}) = \phi_{M_i}(W_{target}) - \alpha_{i+1}(\phi_{W_i}(W_{target}) - W_{target}). \quad (1)$$

The factor α_{i+1} weighs the compensation term. In its easiest form, α_{i+1} is scalar, but in principle, depending on the topological structure of W_{target}, tensor-valued α_{i+1} values are possible, leading to anisotropic compensations. Working with transformations of the target geometry is beneficial for computational reasons, since the same topology is ensured for straight forward calculations. In the present paper, the transformations are based on the target surface normals, which means that the measurement information can be directly used to build the transformations. The transformation ϕ_{W_i} in this case reads

$$\phi_{W_i}(W_{target}) = W_{target} + D_i = W_i. \quad (2)$$

At the beginning of the compensation process, an initial manufacturing geometry has to be defined, either based on experience, numerical analysis or simply the target geometry. After the first run, the geometric adjustment and stochastic phenomena in turn influence the entire AM process again, in the sense of the inverse problem, which could call for another iteration to achieve the desired tolerances.

In this paper, the target geometry W_{target} is represented by a surface triangulation, as in common in AM. Based on the node coordinates and the connectivity list, a straightforward application of the transformations ϕW_i and ϕM_i is possible.

5. Results and Discussion

5.1. Geometrical Deviation in the Calibration and Compensation Cycles

In the calibration cycle (index 0), all 27 components in the build chamber were produced with the target geometry. In Figures 3 and 4, the first rows with index 0 show seven measured parts with their respective positions in the build chamber in Figure 1. Two main geometrical errors can be observed in the measured parts. There is a volume error due to the shrinkage of the material, and the upper right edge of the fin is distorted. It is evident that no optimal, universal compensation for all components is possible, due to the influence of the position in the build chamber. Hence, a unique compensation was calculated for each of the compensated parts a_i–e_i.

The results of the first compensation cycle are shown in the second row of Figure 3 with part numbers a_1–e_1. For part numbers a_1–d_1, the compensation reduced the geometrical deviation significantly. e_1 was not improved. Analysing the calibration cycle ($a_0 - e_0$), e_0 already is an outlier in the area of the upper-right corner. The compensation was calculated to this deviation and did not improve the accuracy. After further investigation, a finishing error was detected. There was still loose powder on the part in this area, which emulated a thicker part to the measuring system. Hence, the compensation led to a thinner part in e_1.

The second compensation cycle (index 2) showed minor geometrical improvements for a_2–c_2. For d_2, a minor deviation increase was measured, while e_2 was improved significantly. This shows that this compensation process is quite robust regarding process outliers, since in the next compensation cycle they can be corrected without manual interaction. The limit of the compensation is reached when the statistical deviations are in the same order of magnitude as the deterministic deviations. The minor increase in deviation for d_2 shows that this limit is possibly reached after two compensation cycles. In Figure 4, the results of the reference components, which were not compensated, can be analysed. The parts show minor changes from one cycle to the next as well. This is due to the statistical deviations.

Comparing cycles 0 and 2, the compensated parts were significantly improved in terms of geometrical accuracy. Furthermore, the parts were more homogeneous and the influence of the position in the build chamber was diminished. In order to quantify this improvement and to analyse the limits of this method, the deviations were numerically analysed; see the next section.

5.2. Evolution of the Part Deviations

In Figure 5, the evolution of the mean absolute error and the evolution of the standard deviation are shown over the three iterations. For the reference parts (b_i^{ref}, c_i^{ref} and d_i^{ref} in Figure 1) the mean absolute error and the standard deviation remains almost constant, which is in accordance with the impression of Figure 4. For all compensated parts, an enhancement in the mean absolute error and the standard deviation was achieved. The process outlier e_i is also clearly visible in the evolution of the mean absolute error, which stayed higher after cycle 1 but improved significantly in cycle 2. In Figure 5, another outlier with respect to standard deviation in cycle 1 can be seen, which is part d_i. The high standard deviation in cycle 1 finally led to the increased mean absolute error of part d_2 in cycle 2.

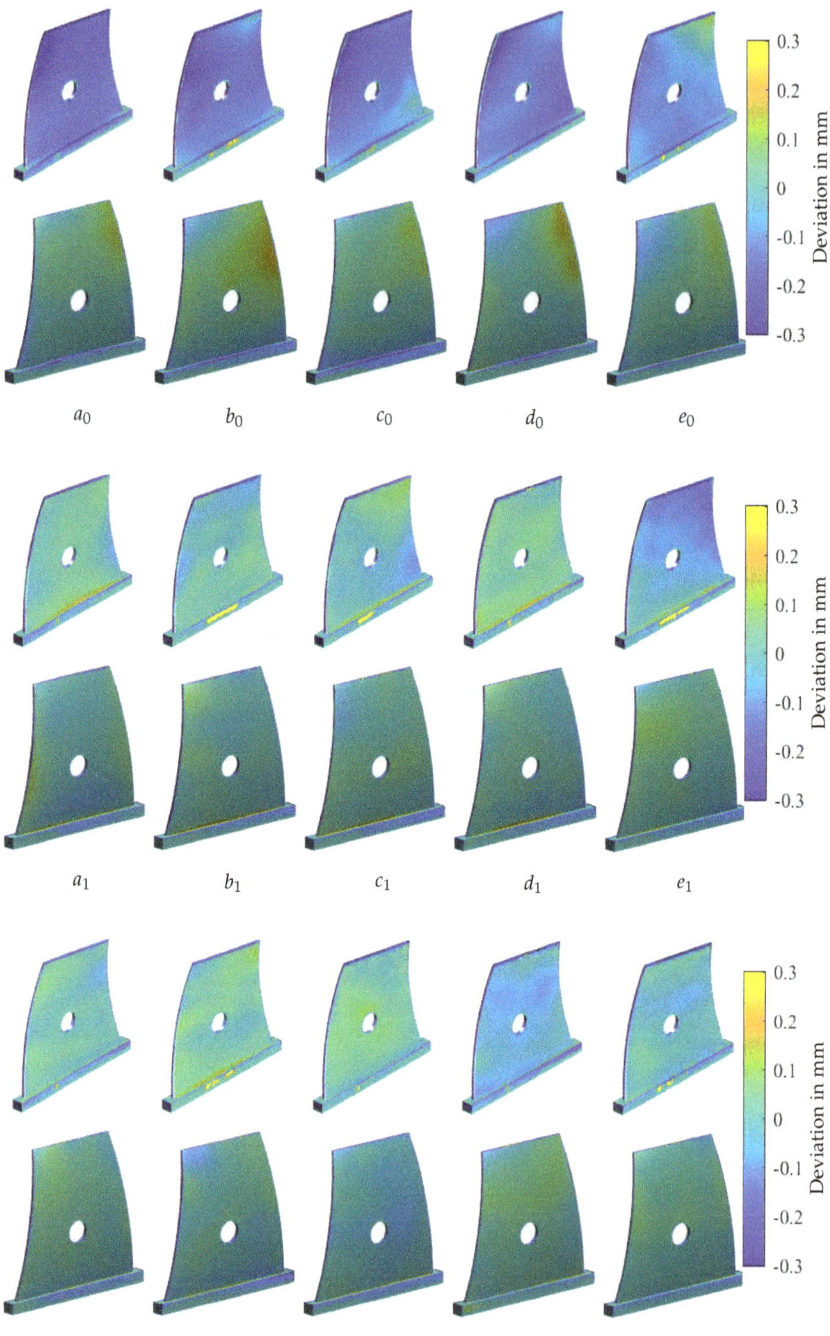

Figure 3. Compensated components for the calibration cycle (0) and two compensation cycles (1 and 2). The positions on the building platform of the parts *a*–*e* are detailed in Figure 1.

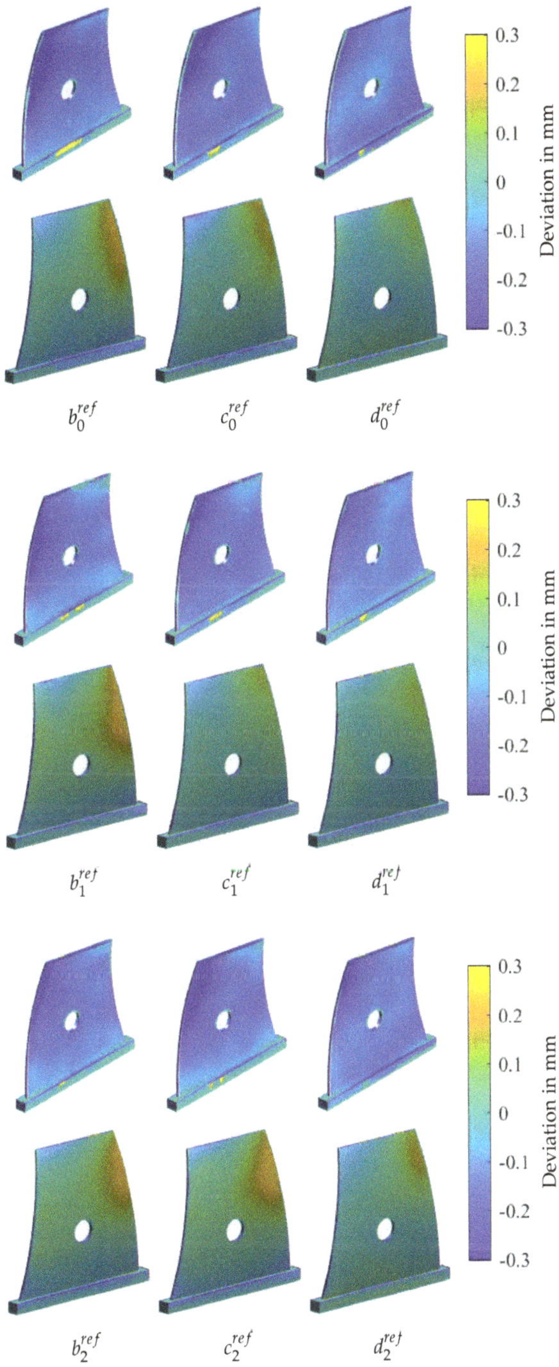

Figure 4. Reference components without compensation for the calibration cycle (0) and two compensation cycles (1 and 2). The positions on the building platform of the parts b^{ref}-d^{ref} are detailed in Figure 1.

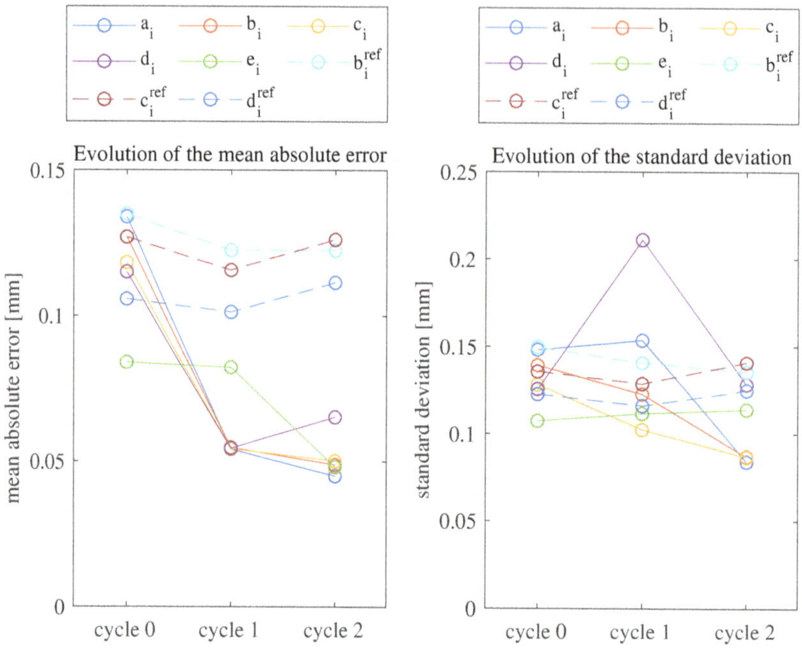

Figure 5. Evolution of the mean absolute error and the standard deviation for the three cycles.

5.3. Stochastic Deviations

For the separation of systematic and stochastic errors, three reference parts b_i^{ref}, c_i^{ref} and d_i^{ref} are considered in each manufacturing job. In order to determine the stochastic portion of the procedure, the deviations are normalised with respect to the mean of the surface deviations according to Equation (3) for each measurement point.

$$x_{normalised} = x_i - \overline{x}. \tag{3}$$

Based on the normalised values, the stochastic mean absolute error and the stochastic standard deviation can be calculated and are summarised in Table 1. The stochastic error measures give the values for the whole procedure shown in Figure 2, including manufacturing (or numerical analysis), measurement (evaluation) and the redesign process of the manufacturing geometry.

The average values of the stochastic errors over all reference parts and all iterations are suitable estimates to characterise the stochastic behaviour of the procedure. With regard to the approach, which is capable of dealing with systematic deviations only, these values could be seen as a process limit for this specific geometry and job in this configuration using the procedure described.

Table 1. Separated stochastic errors of the compensation procedure conducted for the three reference parts. (mae: mean absolute error; std: standard deviation).

Part	Iteration 1		Iteration 2		Iteration 3		Average	
	Mae	**Std**	**Mae**	**Std**	**Mae**	**Std**	**Mae**	**Std**
b_i^{ref}	0.113	0.150	0.111	0.141	0.108	0.135	0.111	0.142
c_i^{ref}	0.115	0.145	0.100	0.132	0.115	0.144	0.110	0.140
d_i^{ref}	0.090	0.123	0.085	0.116	0.097	0.125	0.091	0.121
average	0.106	0.139	0.099	0.130	0.107	0.135	**0.104**	**0.134**

6. Conclusions

In this study, a systematic compensation approach from forming technology was adapted and applied for AM. The Am manufacturing geometry was successfully used as a forming tool in the sense of geometrical compensation. The displacement adjustment method, which directly incorporates deviation data for compensation, presented itself as suitable for AM. Compared to conventional manufacturing technologies, the presented procedure is even more reasonable for AM processes, since no expensive physical forming tools or moulds are necessary.

The presented approach is a data-based method, which relies on an initial calibration job. We think that with regard to high engineering and computational costs of AM numerical analysis, this approach is reasonable, especially with regard to build chamber dependencies. The necessary effort and experiments for a highly complex material model have to be taken into account as well.

By using non-rigid transformations of the initial geometry topology, a direct application of measurement data for the compensation is realised. This way, each part is treated individually, which ensures a compensation of spatial dependencies in the build chamber. The systematic procedure is suitable for series production especially, where the first iteration can be seen as a kind of calibration step. We conclude that the proposed hypothesis could be proven for the setup conducted. Of course, future work has to be done to further verify the approach.

During production, the approach can also be used for process control, in the sense of a closed-loop control, and, for example, to compensate for other raw materials, system characteristics or changed environmental conditions. In future work, the approach will also be tested for other AM processes and different materials. Furthermore, the incorporation of simulation results for the initial geometry is a promising way to further enhance the procedure.

Author Contributions: Conceptualisation, C.H., P.L. and B.H.; methodology, C.H., B.H. and P.L.; investigation, C.H., P.L. and Y.K.; writing—original draft preparation, C.H. and P.L.; writing—review and editing, B.H., T.C.L. and W.V.; supervision, T.C.L. and W.V.

Funding: This research was funded by the Federal Ministry of Education and Research of Germany (BMBF) under grant number: 13N15085, Industrialisierung und Digitalisierung von Additive Manufacturing (AM) für automobile Serienprozesse, acronym IDAM.

Conflicts of Interest: The authors declare no conflict of interest.

Abbreviations

The following abbreviations are used in this manuscript:

2D	two dimensional
3D	three dimensional
AM	additive manufacturing
CAD	computer aided design
MAE	mean absolute error
SLS	selective laser sintering
STD	standard deviation
STL	stereolithography, standard tessellation language

References

1. Woern, A.L.; Pearce, J.M. Distributed Manufacturing of Flexible Products: Technical Feasibility and Economic Viability. *Technologies* **2017**, 5, 71. [CrossRef]
2. Brischetto, S.; Maggiore, P.; Ferro, G. Special Issue on Additive Manufacturing Technologies and Applications. *Technologies* **2017**, 5, 58. [CrossRef]
3. Prashanth, K.; Scudino, S.; Chatterjee, R.; Salman, O.; Eckert, J. Additive Manufacturing: Reproducibility of Metallic Parts. *Technologies* **2017**, 5, 8. [CrossRef]

4. Ning, Y.; Wong, Y.S.; Fuh, J.; Loh, H.T. An approach to minimize build errors in direct metal laser sintering. *IEEE Trans. Autom. Sci. Eng.* **2006**, *3*, 73–80. [CrossRef]
5. Gebhardt, A. *Generative Fertigungsverfahren: Additive Manufacturing und 3D Drucken für Prototyping; Tooling; Produktion*, 1st ed.; Carl Hanser Fachbuchverlag: Munich, Germany, 2013.
6. Manetsberger, K.; Shen, J.; Muellers, J. Compensation of Non-Linear Shrinkage of Polymer Materials in Selective Laser Sintering. In Proceedings of the Solid Freeform Fabrication Symposium Proceedings, Austin, TX, USA, 6–8 August 2001; pp. 346–356.
7. Yang, H.J.; Hwang, P.J.; Lee, S.H. A study on shrinkage compensation of the SLS process by using the Taguchi method. *Int. J. Mach. Tools Manuf.* **2002**, *42*, 1203–1212. [CrossRef]
8. Pham, D.T.; Dimov, S.; Lacan, F. Selective laser sintering: Applications and technological capabilities. *Proc. Inst. Mech. Eng. Part B J. Eng. Manuf.* **1999**, *213*, 435–449. [CrossRef]
9. Ma, W.; But, W.C.; He, P. NURBS-based adaptive slicing for efficient rapid prototyping. *Comput. Aided Des.* **2004**, *36*, 1309–1325. [CrossRef]
10. Zhu, Z.; Keimasi, S.; Anwer, N.; Mathieu, L.; Qiao, L. Review of Shape Deviation Modeling for Additive Manufacturing. In *Advances on Mechanics, Design Engineering and Manufacturing*; Lecture Notes in Mechanical Engineering; Eynard, B., Nigrelli, V., Oliveri, S.M., Peris-Fajarnes, G., Rizzuti, S., Eds.; Springer International Publishing: Cham, Switzerland, 2017; pp. 241–250. doi:10.1007/978-3-319-45781-9_25. [CrossRef]
11. Tata, K.; Fadel, G.; Bagchi, A.; Aziz, N. Efficient Slicing for layered manufacturing. *Rapid Prototyp. J.* **1998**, *4*, 151–167. [CrossRef]
12. Masood, S.H.; Rattanawong, W.; Iovenitti, P. A generic algorithm for a best part orientation system for complex parts in rapid prototyping. *J. Mater. Process. Technol.* **2003**, *139*, 110–116. [CrossRef]
13. Khorasani, A.; Gibson, I.; Goldberg, M.; Littlefair, G. On the role of different annealing heat treatments on mechanical properties and microstructure of selective laser melted and conventional wrought Ti-6Al-4V. *Rapid Prototyp. J.* **2017**, *23*, 295–304. [CrossRef]
14. Boillat, E.; Kolossov, S.; Glardon, R.; Loher, M.; Saladin, D.; Levy, G. Finite element and neural network models for process optimization in selective laser sintering. *Proc. Inst. Mech. Eng. Part B J. Eng. Manuf.* **2004**, *218*, 607–614. [CrossRef]
15. Wang, R.J.; Wang, L.; Zhao, L.; Liu, Z. Influence of process parameters on part shrinkage in SLS. *Int. J. Adv. Manuf. Technol.* **2007**, *33*, 498–504. [CrossRef]
16. Wang, X.C.; Laoui, T.; Bonse, J.; Kruth, J.P.; Lauwers, B.; Froyen, L. Direct Selective Laser Sintering of Hard Metal Powders: Experimental Study and Simulation. *Int. J. Adv. Manuf. Technol.* **2002**, *19*, 351–357. [CrossRef]
17. Kamat, A.M.; Pei, Y. An analytical method to predict and compensate for residual stress-induced deformation in overhanging regions of internal channels fabricated using powder bed fusion. *Addit. Manuf.* **2019**, *29*, 100796. [CrossRef]
18. Moesen, M.; Craeghs, T.; Kruth, J.P.; Schrooten, J. Robust beam compensation for laser-based additive manufacturing. *Comput. Aided Des.* **2011**, *43*, 876–888. [CrossRef]
19. Nelson, C.; McAlea, K.; Gray, D. *Improvements in SLS Part Accuracy*; University of Texas: Austin, TX, USA, 1995.
20. Huang, Q. An Analytical Foundation for Optimal Compensation of Three-Dimensional Shape Deformation in Additive Manufacturing. *J. Manuf. Sci. Eng.* **2016**, *138*, 061010. [CrossRef]
21. Huang, Q.; Zhang, J.; Sabbaghi, A.; Dasgupta, T. Optimal offline compensation of shape shrinkage for three-dimensional printing processes. *IIE Trans.* **2015**, *47*, 431–441. [CrossRef]
22. Bayerlein, F.; Zeller, C.; Zaeh, M.F.; Weirather, J.; Wunderer, M.; Seidel, C. Improving cost effectiveness in additive manufacturing—Increasing dimensional accuracy in laser beam melting by means of a simulationsupported process chain. In Proceedings of the CADFEM Users' Meeting 2015, Bremen, Germany, 24–26 June 2015.
23. Seidel, C. Finite-Elemente_Simulation des Aufbauprozesses Beim Laserstrahlschmelzen. Ph.D. Thesis, Technische Universität München, Munich, Germany, 2016.
24. Afazov, S.; Denmark, W.A.; Lazaro Toralles, B.; Holloway, A.; Yaghi, A. Distortion prediction and compensation in selective laser melting. *Addit. Manuf.* **2017**, *17*, 15–22. [CrossRef]
25. Zhu, Z.; Anwer, N.; Mathieu, L. Geometric deviation modeling with Statistical Shape Analysis in Design for Additive Manufacturing. *Procedia CIRP* **2019**, *84*, 496–501. [CrossRef]

26. Zhu, Z.; Anwer, N.; Mathieu, L. Deviation Modeling and Shape Transformation in Design for Additive Manufacturing. *Procedia CIRP* **2017**, *60*, 211–216. [CrossRef]
27. Schmutzler, C.; Bayerlein, F.; Janson, S.; Seidel, C.; Zaeh, M.F. Pre-compensation of Warpage for Additive Manufacturing. *J. Mech. Eng. Autom.* **2016**, *6*. [CrossRef]
28. Fourment, L.; Chenot, J. Optimal Design for Non-Steady-State Metal Forming Processes-II. Application of Shape Optimization in Forging. *Int. J. Numer. Methods Eng.* **1996**, *39*, 51–65. [CrossRef]
29. Ponthot, J.P.; Kleinermann, J.P. A cascade optimization methodology for automatic parameter identification and shape/process optimization in metal forming simulation. *Comput. Methods Appl. Mech. Eng.* **2006**, *195*, 5472–5508. [CrossRef]
30. Hartmann, C.; Eder, M.; Opritescu, D.; Volk, W. Process-integrated Compensation of Geometrical Deviations for Bulk Forming. *Procedia Eng.* **2017**, *207*, 466–471. [CrossRef]
31. Hartmann, C.; Eder, M.; Opritescu, D.; Maier, D.; Santaella, M.; Volk, W. Geometrical compensation of deterministic deviations for part finishing in bulk forming. *J. Mater. Process. Technol.* **2018**, *261*, 140–148. [CrossRef]

© 2019 by the authors. Licensee MDPI, Basel, Switzerland. This article is an open access article distributed under the terms and conditions of the Creative Commons Attribution (CC BY) license (http://creativecommons.org/licenses/by/4.0/).

technologies

MDPI

Article

Analysis, Optimization, and Characterization of Magnetic Photonic Crystal Structures and Thin-Film Material Layers

Mikhail Vasiliev *, Kamal Alameh and Mohammad Nur-E-Alam

Electron Science Research Institute, School of Science, Edith Cowan University, 270 Joondalup Drive, Joondalup, WA 6027, Australia

* Correspondence: m.vasiliev@ecu.edu.au; Tel.: +61-8-6304-5809

Received: 20 May 2019; Accepted: 4 July 2019; Published: 5 July 2019

Abstract: The development of magnetic photonic crystals (MPC) has been a rapidly evolving research area since the late 1990s. Magneto-optic (MO) materials and the techniques for their characterization have also continually undergone functional and property-related improvements. MPC optimization is a feature-rich Windows software application designed to enable researchers to analyze the optical and magneto-optical spectral properties of multilayers containing gyrotropic constituents. We report on a set of computational algorithms which aim to optimize the design and the optical or magneto-optical spectral analysis of 1D MPC, together with a Windows software implementation. Relevant material property datasets (e.g., the spectral dispersion data for the refractive index, absorption, and gyration) of several important optical and MO materials are included, enabling easy reproduction of the previously published results from the field of MPC-based Faraday rotator development, and an effective demonstration-quality introduction of future users to the multiple features of this package. We also report on the methods and algorithms used to obtain the absorption coefficient spectral dispersion datasets for new materials, where the film thickness, transmission spectrum, and refractive index dispersion function are known.

Keywords: 1D magnetic photonic crystals; multilayer film modeling; modeling of Faraday rotation spectra; MPC optimization; exhaustive computation; materials characterization

1. Introduction and Background

In recent years, there has been some resurgence in the research interest dedicated to engineering and characterization of magneto-optic iron garnet materials [1–4]. Thin-film magnetic garnets are semitransparent magnetic dielectrics possessing record specific Faraday rotations of up to several thousand °/cm, in the near-infrared spectral range, if containing bismuth substitution [5]. Generically, the chemical composition of garnet materials of this type is described by the formula $(Bi_xRe_{3-x})Fe_yM_{5-y}O_{12}$, where Re is rare-earth metal (e.g., Dy, Sm, Lu, Nd, or Ce), and M is transition metal such as Ga or Al [5]. The exploration of this important subclass of functional materials has a decades-long history, starting from the days of bubble-domain magnetic memory development, and more recently, continued with renewed research activities, in application areas ranging from on-chip nonreciprocal components (waveguide isolators, [2,6]), to magnetic recording [7]. Magnetic garnet materials synthesized by a range of thin-film deposition techniques and crystal-growth methods have also attracted a significant research and development momentum since the 1990s, and throughout the last two decades, in areas ranging from photonic crystals to spintronics [8–15].

Various approaches to the design and manufacture of magnetic photonic crystals (MPC) with tunable properties, and potentially suitable for the fabrication of nonreciprocal optical components have been reported [8–18]. The one-dimensional MPC are structured (periodic or quasi-periodic) sequences of the magnetic and nonmagnetic material layers deposited on optical substrates, which are defined by the layer order and thickness. Many research groups have focused on optimizing the thickness of one-dimensional (1D) MPC structures to simultaneously achieve 45° of Faraday rotation angle and maximal possible transmission at optical telecommunication wavelengths. The MPC based on quarter-wavelength thin-film stacks, which are sequences of magnetic and nonmagnetic layers with multiple embedded phase shifts (termed defects, or missing layers), have been shown to possess a significant potential for practical implementation of integrated optical isolators. This is due to the necessity of approaching Faraday rotations as large as 45°, which has been shown to be attainable, due to the resonant enhancement of Faraday rotation observed in such structures. Complex 1D MPC designs featuring a "flat-top" spectral response, with almost 100% transmission within a large bandwidth (several nanometers), and close to 45° of Faraday rotation at 1550 nm can contain in excess of 200 layers [10,11], limiting their applications to the near-infrared range, where the magnetic garnets possess very low (negligible) optical absorption. Since the original development of the MPC optimization program in 2005 [18], multiple garnet material development efforts have been undertaken within our group [19–21], all of which have relied substantially on material characterization options featured within the same software package. In particular, the option of deriving the data for the spectral dispersion of the absorption coefficient using the measured transmission spectrum and refractive index dispersion data has been very useful to characterize new nanocomposite-type garnet materials synthesized by cosputtering using an extra oxide material source [19]. The present-day performance limits of 1D MPC in the visible spectral range have been evaluated using the same software [22] and using the available spectral data on the optical properties of the best-performing magnetic garnet compositions. Optical constants data of multiple recently-synthesized magnetic garnet compositions have been evaluated using the measured transmission spectra in conjunction with MPC Optimization software and Swanepoel envelope method [23]. Our group's preferred method for calibrating the quartz microbalance sensor's tooling factors of various deposition sources also involves fitting the actual film layer thickness using MPC Optimization software in conjunction with measured transmission spectra.

A graphical snapshot of the controls and features implemented within MPC Optimization software is shown in Figure 1, which also presents a sample optimization result. The computation time to obtain this result is typically around one min (after analyzing almost 5000 MPC designs out of a possible 10,000 defined by the preset film structure features). Some of calculated designs exceed the maximum limits set for thickness or layer number, and thus are not fully evaluated. The default spectral range of calculations is extremely broad, slowing down the optimization algorithm significantly, since the default wavelength settings and resolution have been entered to assist in fitting film layer thickness conveniently, which is one of the most common everyday applications of this software. Running the actual MPC optimization algorithm is best performed by zooming onto the spectral range of interest, which is usually represented by a wavelength region surrounding a narrow peak of transmission or reflection, where it is possible to engineer the enhancement of Faraday rotation.

The default-entered materials-related data used to optimize MPC designs (within the parameter space and constraints also entered as default values not relevant to any particular application) relate to a MPC design which is based on magnetic garnet composition $Bi_2Dy_1Fe_4Ga_1O_{12}$ and SiO_2 L-type layers, deposited on a glass substrate. The optical constants and gyration data for wavelengths near 630 nm have been obtained from thick garnet layers of this composition, and the corresponding index dispersion dataset is also preloaded into the menu item "Extra data | Account for refractive index dispersion".

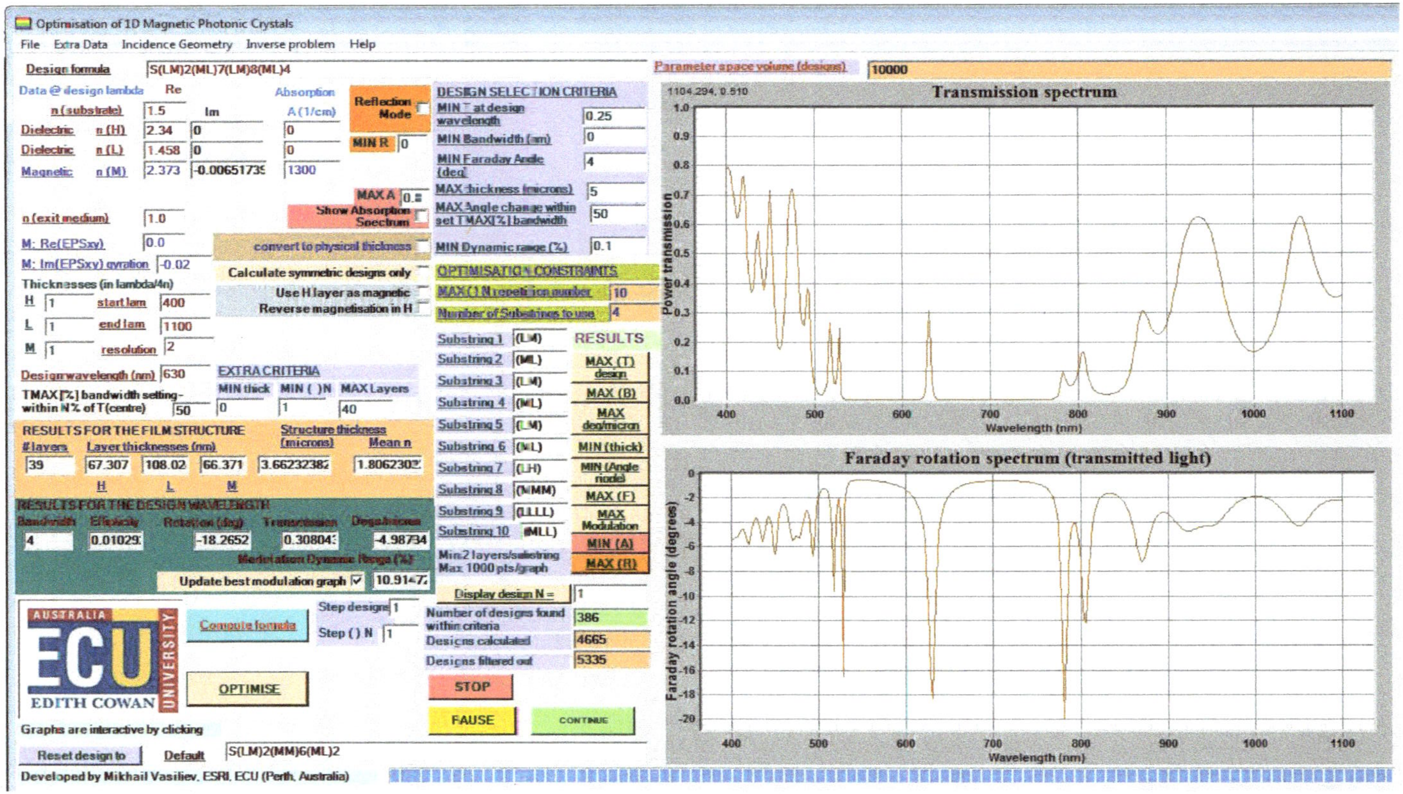

Figure 1. Front-panel controls of MPC Optimization software and a sample optimized MPC design obtained by clicking the "optimize" button without changing the default-entered data. The result is a MPC structure which is designed to operate at 630 nm, reliant on a magnetic garnet material of specific Faraday rotation near 2°/µm at that wavelength, however, the structure enhances the Faraday rotation to 4.98°/µm within the spectral transmission peak.

The contents of the compiled html (.chm) help file accessible from the "Help" menu are sufficient to enable most beginner-level MPC (or thin-film) designers to quickly learn the main features of program and its data representation formats. In the following sections of this article, these main features are described in detail, with examples provided to enable the productive and convenient use of this feature-rich software package. The main aim of this present work has been to provide a set of computational tools and algorithms for use by the developers of MPC and other nanostructured material systems (e.g., thin-films and multilayers containing MO materials). These tools will enable both the characterization of functional material layers and the application-specific design of Faraday rotators. Other contributions of this work include a number of experimentally-validated optical and MO property datasets of several garnet material types possessing giant Faraday rotation and good MO quality, synthesized by our group using RF sputtering followed by high-temperature annealing crystallization processes. Examples of particularly important MPC-based Faraday rotator design types are reviewed, focusing on the ways that strong peaks of Faraday rotation are engineered to spectrally coincide with the peaks of transmission, while paying special attention to the role of limiting factors such as absorption.

2. Overview of Package Operation and Key Examples

Since the structure of 1D MPC is essentially represented by a sequence of magnetic and nonmagnetic material layers on a substrate, most of the core terminology, design structure description conventions, and analysis techniques are derived from the field of multilayer thin-film design. The input datasets necessary to define the layer sequence and the optical properties of each material type and individual layer are entered into the relevant text boxes, starting from the top-left corner of the Windows Form. The design formula field defines the layer sequence, starting necessarily from the capital letter S defining the substrate. The layer sequence must be entered in an alphanumeric string format containing special symbols such as round, square, or curled brackets, corresponding to one of the three main design-string representation types. These types are: (i) quarter-wave stack-based notation, e.g., SLH(ML)2 which is perhaps the most common system of layer structure abbreviation in thin-film design; (ii) physical thickness-based notation, e.g., SM [1000]L[5 0], and (iii) advanced designs notation, e.g., S{1.0}(L/2HL/2)1{1.06}(L/2HL/2)3 further described within the help documentation. The optical properties of the substrate material (which is presumed to be dielectric, semi-infinite, and non-absorbing) are defined using only the real part of refractive index. Up to three different optical materials are allowed, denoted by the letters H, L, and M, however, H-type layers are not restricted to mean "high refractive index". The layer-naming conventions are derived from thin-film terminology for convenience only. In the preceding thin-film abbreviation example (SLH(ML)2), the symbols mean a substrate (S) with the following sequence of layers deposited, in the following order: L, H, M, L, M, and L.

The M layers can optionally be modeled as magnetic dielectrics, where a gyration value at the design wavelength (the imaginary part of the nondiagonal dielectric tensor component, or Im(ε_{xy})) needs to be entered into its relevant text box. This imaginary component of the off-diagonal dielectric tensor element describes magnetic circular birefringence and manifestes as Faraday rotation of the polarization plane in the transmitted or reflected light waves. The real part of this tensor component is treated as zero, in most applications, unless the experimentally-measured value is known, which is related to characterizing magnetic circular dichroism and polarization state ellipticity effects. For applications requiring good numerical accuracy over broad spectral ranges, the spectral dispersion of both the gyration and the refractive indices of all relevant materials need to be loaded from .txt format data files, using the options within the "Extra Data" menu. Optionally, H-type layers are selected to also represent a magnetic dielectric, with the same off-diagonal dielectric tensor components as in M-type material (but optionally with different refractive index and absorption), in order to model a special physical situation where an MPC has layer-specific magnetization reversal possibilities in some individual magnetic layers within the structure. If M layers are modeled as nonmagnetic (e.g., just implying medium-index dielectric layers), then both the real and imaginary parts of ε_{xy} are entered as zero values, however, if a small gyration is still entered by error, it will not measurably distort the transmission or reflection spectral calculations, and therefore, in this case, the results regarding the Faraday rotation spectra should be ignored.

The details of the physical situation being modeled, in terms of incidence geometry, and the ways in which the transmitted (or reflected) light intensity is normalized with respect to the incident wave intensity, are defined using sets of checkboxes within the menu entry "Incidence Geometry", where the relevant descriptions are given. The incidence geometry settings defined as default are the ones used most often and generic in general, since these enable the convenient and accurate fitting of lab-measured transmission spectra in deposited film-substrate systems, to the corresponding theoretically-modeled spectra, where the effects of each interface (including the back surface of substrate) are correctly accounted for in the model. The only parameter not accounted for is the physical thickness of substrate, which is modeled as non-absorbing. Alternative settings for the incidence geometry are useful for considering more theoretical situations, such as calculations of the optical intensity transmitted into a semi-infinite substrate medium, or in reflection-mode calculations, where it is often necessary to compare reflected-wave intensities, which vary with the direction of incidence.

One of the most important material parameters in all layer types is the optical absorption coefficient at the design wavelength (entered in cm^{-1} into relevant text boxes; the corresponding extinction coefficients will then be displayed after the film design is characterized by pressing the "compute formula" button), especially for materials possessing significant spectral dependency in their optical absorption. For accurate characterization of thin films or MPC, ideally, every material should have its optical constants dataset available for loading from the "Extra Data" menu option "Account for refractive index dispersion" and loaded into the specialized form (shown in Figure 2) prior to calculations. The material-specific optical constants data files are prepared using zeros entered in place of an unknown absorption coefficient, for the purposes of physical layer thickness fitting, based on the accurately measured transmission spectrum data (as discussed more in detail in subsequent sections). The end-of-file (EOF) marker in these .txt material data files prepared using editor applications such as Notepad must be placed immediately after the last figure in the last column, by way of making sure to delete any possible symbols or empty row spaces below.

Once the refractive index dispersion information is loaded from data file(s), the data in text boxes that corresponds to the index and absorption at the design wavelength are no longer used in the main spectral calculations, but used only for calculating the quarter-wave stack physical thickness in nm. For all wavelength points between the 27-point data grid, the values of refractive index and absorption are linearly interpolated from the nearest grid-located points. This can erroneously cause small spectral shifts to appear in the spectral locations of the transmission, reflection, or Faraday rotation peaks, seen away from the precise design wavelength, if the interpolated refractive index at that wavelength does not coincide with the value entered into front-panel text box. This is not a significant issue for the experienced designers of MPC, once the origin of these possible small data errors is known or eliminated by entering precise (same as interpolated from the index dataset for the design wavelength) index data into front-panel text boxes. It is a known a priori that the actual spectral response peak locations in quarter-wave stack-type designs will appear precisely at the design wavelengths, due to the nature of optical interference-related phenomena. In situations when the refractive index dispersion data are only available within a limited spectral interval of interest, rather than for all wavelengths in the 350–1600 nm range, it is recommended that the available refractive index data are entered into the newly-generated data file. The refractive index and absorption coefficient values at other wavelengths still need to be entered. The recommended practice is as follows: for example, if the available data starts from 500 nm, enter the same values as are known at 500 nm into the wavelength grid positions for all shorter wavelengths; alternatively, use any available models to predict the unknown values (e.g., Cauchy dispersion model). If the refractive indices or absorption are only available up to 800 nm, it is best to enter (for all larger wavelengths), the same (n, A) data as at the last data point (800 nm).

Whilst the optical properties cannot be extrapolated outside a known range by the values measured at the range limits, the designers of thin film and MPC structures are encouraged to make numerical experiments with the package to evaluate the errors expected to result from the possible refractive index data inaccuracies, in each particular design case. The refractive index of the exit medium surrounding the substrate-film system is defined only by its real part (typical value is 1.0 for air), and the same exit medium is presumed to precede the (thickness-undefined) substrate, and to exist beyond the last deposited film layer, regardless of the direction of light incidence. The checkbox "reflection mode" sets up the calculations of the reflection spectra, and also the Faraday rotation spectrum for the reflected light, if checked. Optionally, the "show absorption spectrum" checkbox is checked, after which the "compute formula" button will initiate new calculations, resulting in the display of the calculated absorption spectrum.

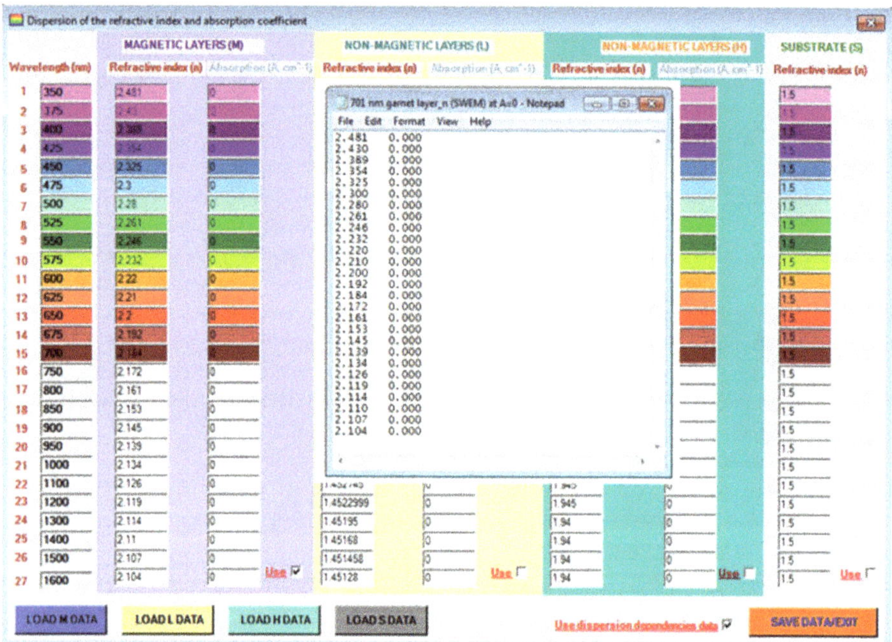

Figure 2. Form dedicated to loading the optical constants data from text files prepared as shown, using the software-specific 27-row wavelength grid and containing the columns data for the refractive index and optical absorption coefficient (in cm^{-1}) at each wavelength point within the spectral grid.

The theoretical analysis of the spectral properties of MPC has been implemented using the 4×4 complex-valued transfer matrix approach [16]. The method is reliant on finding a transfer matrix that relates the complex electric and magnetic field amplitudes of the light in front of the MPC and behind it. The transfer matrix for the whole structure is represented by the product of the transfer matrices calculated for each layer of the MPC. The transfer matrices of layers are determined by the thickness of the layers and their dielectric tensor components. Nonmagnetic layers are treated as isotropic media having a diagonal dielectric tensor, whereas, magnetic layers are described by tensors containing the magnetization-dependent off-diagonal dielectric tensor components containing wavelength-dependent gyration information, which are related to specific Faraday rotation spectra. When running the optimization algorithm, a C++ code implemented using Microsoft Visual Studio.NET 2003 generates a look-up table of all possible design structure variations (within the design constraints specified). From this table, an optimum MPC structure is obtained by successive tightening of the desired spectral response specifications, guided by the user inputs. The optimization approach used in our algorithm is based on the "exhaustive computation" of the entire parameter space defined by an arbitrary selected type of a structural formula composed of "design substrings" representing user-selected elementary building blocks of the MPC design.

2.1. Multi-Defect Multilayer MPC Characterization and Optimisation Examples Suitable for Validating Calculations

To illustrate the suitability of software to correctly calculate the spectral responses of complex, multi-defect MPC designs, it is easiest to use the design or optimization examples described in the published literature, e.g., [11] and [22]. One of the common goals of optimizing the MPC structures has been to achieve strong spectral peaks in either the transmission or reflection (ideally approaching 100%) that coincident spectrally with peaks of enhanced Faraday rotation, in either the transmission or reflection-mode operation, and ideally approach 45°, if efficient modulation of light intensity is required. It is important to note that Faraday rotation in the reflected wave is different in its physical nature from Kerr rotation [5] (although there may exist some terminological misinterpretations, even in the published literature). This program calculates only the angles of polarization-plane rotation due to Faraday effect, in either geometry, and does not account for Kerr effect. Figure 3 presents a graphical summary of the input parameters necessary to enter in the relevant text boxes to evaluate one remarkable MPC design from the published literature [11], as well as the results of calculations. The practical implementation of this MPC design is expected to be difficult, due to factors such as the extremely high number of layers, large total thickness, the expected nonzero (but possibly well below about 0.1 cm^{-1}) optical absorption coefficient in garnets at 1550 nm, as well as the scattering effects expected to occur at multiple layer boundaries. From the theoretical insight perspective, this high-performance MPC design is still an outstanding example of MPC application potential, especially in systems using optimized surrounding media, index-matched to the mean refractive index of structure. A .mpc file (MPC Design from JLT Vol. 19 No. 12 p. 1964.mpc) is included in the subfolder "Optimization results files" of the program installation directory, and can be loaded from the "File" menu, enabling easy recalculation of the contents of Figure 3, using preloaded design data. The way this MPC has been modeled also involves the customized incidence geometry settings, where film-side incidence is modeled, without normalizing the transmitted intensity after the substrate. Rather, transmission into the substrate material is modeled, which explains why the modeled transmission within 1550 nm peak closely approaches 100%. Running the MPC optimization algorithm, on the other hand, requires using substrate-side incidence geometry settings only.

All materials-related data values were used as per the description in [11], and the data in the calculated graphs reproduce the results presented in Figure 3b of [11] with precision. Due to materials-related constraints, such as the spectral dependencies of the absorption coefficient and gyration, achieving strong enhancements in Faraday rotation simultaneously with low optical losses becomes progressively more difficult with the reducing design wavelength. Across the visible spectral range, the optical absorption in bismuth-substituted iron garnet materials becomes the limiting factor, placing stringent limits on the achievable MPC performance characteristics, regardless of their intended application area or the design type. Therefore, the ability to generate and compare multiple and differently-structured optimized MPC designs is crucial for achieving the best possible performance characteristics, limited only by the fundamental, materials-related constraints.

In order to directly reproduce the optimization result reported in [11], by way of running a constrained optimization algorithm using the MPC Optimization software, the materials-related datasets and a set of optimization constraints, as shown in Figure 4, must be used prior to clicking the "optimize" button. It is also necessary to set the substrate-side incidence geometry, and uncheck the second checkbox related to the way the transmitted intensity is normalized. The "flat-top" optimized MPC design will be retrieved from the database of calculated MPC designs within the set of defined criteria (60 different designs will be found to fit these overall criteria and constraints, as set per data of Figure 4), by selecting the design with the maximized spectral response bandwidth. This is done by pressing the "MAX (B)" button. Importantly, all design substrings must be entered as per Figure 4 to reproduce this optimization result, with precisely five substrings and a maximum of 30 entered for the substring-repetition index. Another constraint which needs to be entered relates to the maximum allowed layer number which is 203. The checkbox "calculate symmetric designs only" needs to be checked, and it is best to set the spectral resolution to 1 nm during optimization, followed by changing it to 0.1 nm for calculating the spectral properties more accurately. The optimized design equivalent to the design of Figure 3 (and [11]) is retrieved from 60 possible designs found within the criteria, by pressing the "MAX (B)" button. This retrieves the design with maximized full fidth at half maximum (FWHM) bandwidth, according to the data entered into the TMAX(%) text box.

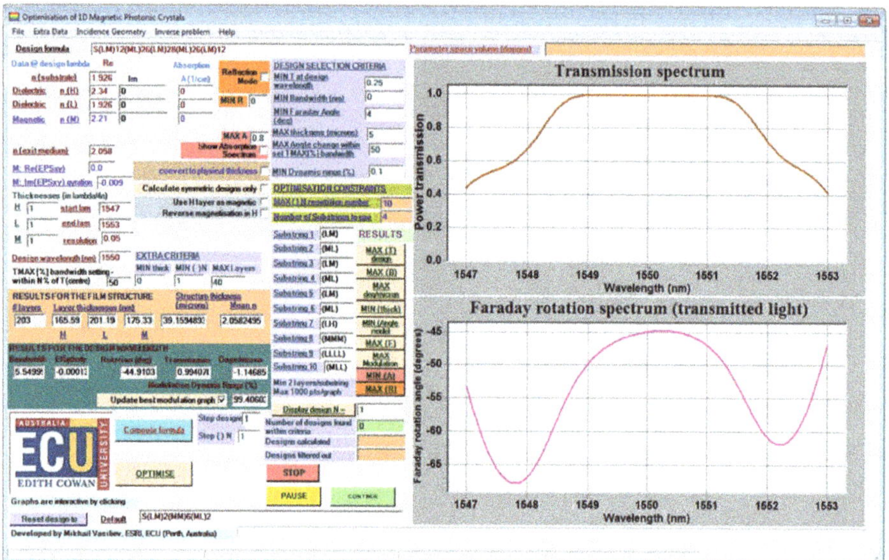

Figure 3. MPC optimization software used to reproduce the flat-top MPC transmission and Faraday rotation spectral properties for a four-defect design reported in [11]. The calculated graphs (obtained after entering the design data and pressing the "compute formula" button) reproduce the data originally presented graphically in Figure 3b of [11].

The settings applied to the normalization of the transmitted intensity within the Incidence Geometry menu corresponded to the physical situation equivalent to that applied during the calculation of the MPC properties in [11], stipulating the transmitted intensity normalization procedure involving transmission "from within" the substrate material, into the index-matched exit medium. Figure 5 shows a graphic summary of the MPC optimization results first reported in [22], which illustrate the current performance limits of MPC designs aimed at developing transmission-mode magneto-optic light intensity modulators working at near 650 nm. The optical and MO material parameters relevant to $Bi_2Dy_1Fe_4Ga_1O_{12}$ and similar highly bismuth-substituted nanocrystalline garnet materials (synthesized by techniques such as RF sputtering) were used in the calculations.

The data of Figure 5 can be reproduced by running optimization of four-defect structures with up to 40 layers and thickness up to 5 μm, composed of five sequenced (LM) and (ML) substrings, as shown, using the default entered n(L) value and n(M) = 2.376. The gyration value corresponding to 2°/μm needs to be entered as −0.02 for wavelengths near 650 nm, accounting for the composition-specific sign of Faraday rotation, which is defined by the convention reported in [5] and other sources. The maximum repetition index value is set to either 10 or, e.g., 12, prior to running the optimization with either absorption coefficient.

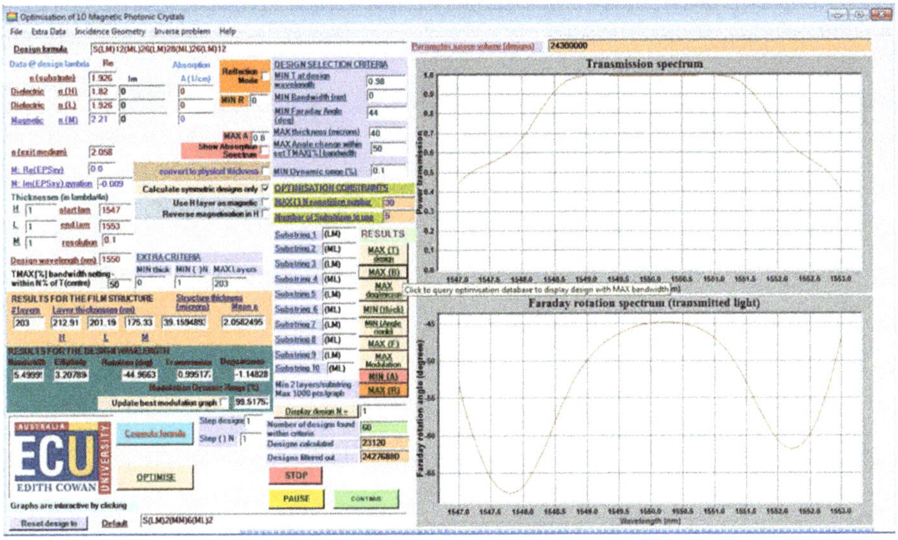

Figure 4. Optimized "flat-top" high-performance MPC design reproduced by running the MPC optimization algorithm, using checkbox "Calculate symmetric designs only" for speeding up calculations. The exact four-defect MPC design reported in [11] (Figure 3b of [11]) is shown after selecting the design with maximum spectral response bandwidth from the 60 possible MPC designs found to be within the optimization criteria and constrains, as shown.

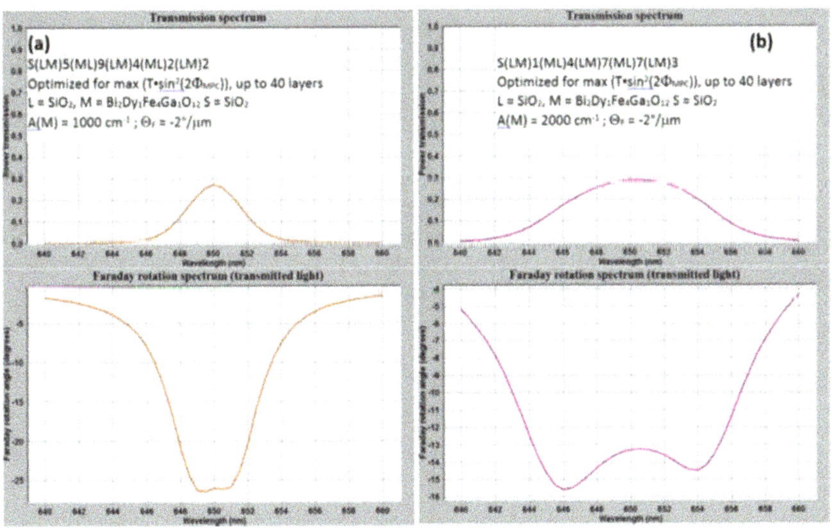

Figure 5. Calculated spectral performance parameters for multi-defect (four-defect structures, having up to 40 total layers) MPC optimized by exhaustive computation to achieve a maximum light intensity modulation capability (maximized value of parameter (T·sin^2(2Φ_{MPC})), for n(M) = 2.376, n(L) = 1.458, and using two different optical absorption coefficients for magnetic material at 650 nm (a) α = 1000 cm^{-1} and (b) α = 2000 cm^{-1}. These refractive index and absorption coefficients were considered constant within the design-specific wavelength region, as well as gyration (−0.02), corresponding to approximately 2°/μm near 650 nm. The graphical information is reproduced from [22].

The results shown in Figure 5 illustrate clearly that the optical absorption is the limiting factor in the visible-range MPC design, even at long-wavelength red wavelengths, where the absorption is still moderate, and the thin constituent garnet layers used within MPC would have been almost visually clear. This is demonstrated by entering a design string, such as SM[68.39], into design formula text box, and running the absorption-mode calculation (e.g., using $\alpha = 2000\ cm^{-1}$) to reveal the graph area by zooming in with mouse, that individual MO layers absorb only about 1.2% of the incident power on each single-pass transmission).

2.2. Generating Optimized Antireflector Film Designs Using Spectral Target Points

It is possible to apply additional optimization constraints at up to three selected wavelength points, to force the algorithm to output the designs with specific spectral features, in either the transmission or reflection mode. An example of obtaining the optimized five-layer thin-film broadband antireflector coating designs is stored in the file Try_optimise_5-layer antireflector.mpc which is placed into the subdirectory of "Optimization results files" in the installation directory. The menu item "Extra Data | Multi-wavelength spectral targets" is used to enter the additional optimization constraints, regardless of whether the Faraday rotation features are being optimized or not. Figure 6 shows the required inputs within the two submenus related to the multi-wavelength spectral targets and the incidence geometry options, which will result in reproducing the five-layer antireflector design shown also in Figure 6. Selecting the design with maximum transmission at the design wavelength (either after running the optimization, or simply after loading the relevant example file) will reveal the reflection spectrum as shown.

This example also demonstrates the use of scaled QWOT layers for use in thin substrings, the thickness of which is then being optimized by the algorithm by adjusting the repetition indices. The calculation of more than 750,000 designs should still take only a few minutes. A number of antireflector-type designs are revealed by using the button "display design N=" with any corresponding number not exceeding the number of designs found within criteria.

Since the optimization algorithm presumes the substrate-side incidence, it is convenient to define air as the subtrate, and set the exit medium to glass. The obtained design S(LL)10(HH)14(LL)4(HH)1(LL)14 needs to have its deposition sequence reversed and be re-evaluated for the film-side incidence case from air. Using the physical thickness notation is preferable in this case, i.e., the design needs to be changed into S(LL)14(HH)1(LL)4(HH)14(LL)10. In physical thickness notation, for a practical deposition-ready design description, this is equivalent to SL[139.49]H[8.19]L[39.85]H[114.68]L[99.64]. It is important to not forget to set n(S) to 1.5 and n(exit) to 1.0(air) in this case. Now, the incidence geometry settings are checked to correspond to the film-side incidence. Note, the way the reflectivity of the back side of the substrate is accounted for in the detailed incidence geometry settings.

2.3. Fitting of the Measured Thin-Film Transmission Spectra to Theoretical Models

One of the most frequently used applications of MPC Optimization software, apart from optimizing the MPC structures, is expected to be the fitting of actual deposited layer thickness, for thin-film materials with known refractive index dispersion function. The option of loading the measured transmission spectrum for immediate comparison with the modeled transmission spectrum of the same substrate-film system is available from the menu item "Inverse problem | Load T spectrum for fitting". To enable accurate modeling, both of the two checkboxes that correspond to the substrate-side incidence within the menu "Incidence Geometry | Define geometry" must be checked, which is done by default. A measured reflection spectrum (if available for the normal-incidence reflection, which is rare with most instruments) can also be fitted in the same way as transmission, by running the calculations in reflection mode, with the corresponding checkbox checked. During the calibration of in-situ thickness control systems, e.g., quartz crystal microbalances or reflectometer-type systems, the task of determining the actual physical thickness of deposited thin-film layers is very common, where MPC Optimization software can be used effectively, in conjunction with other methods such as SEM or profilometry characterization.

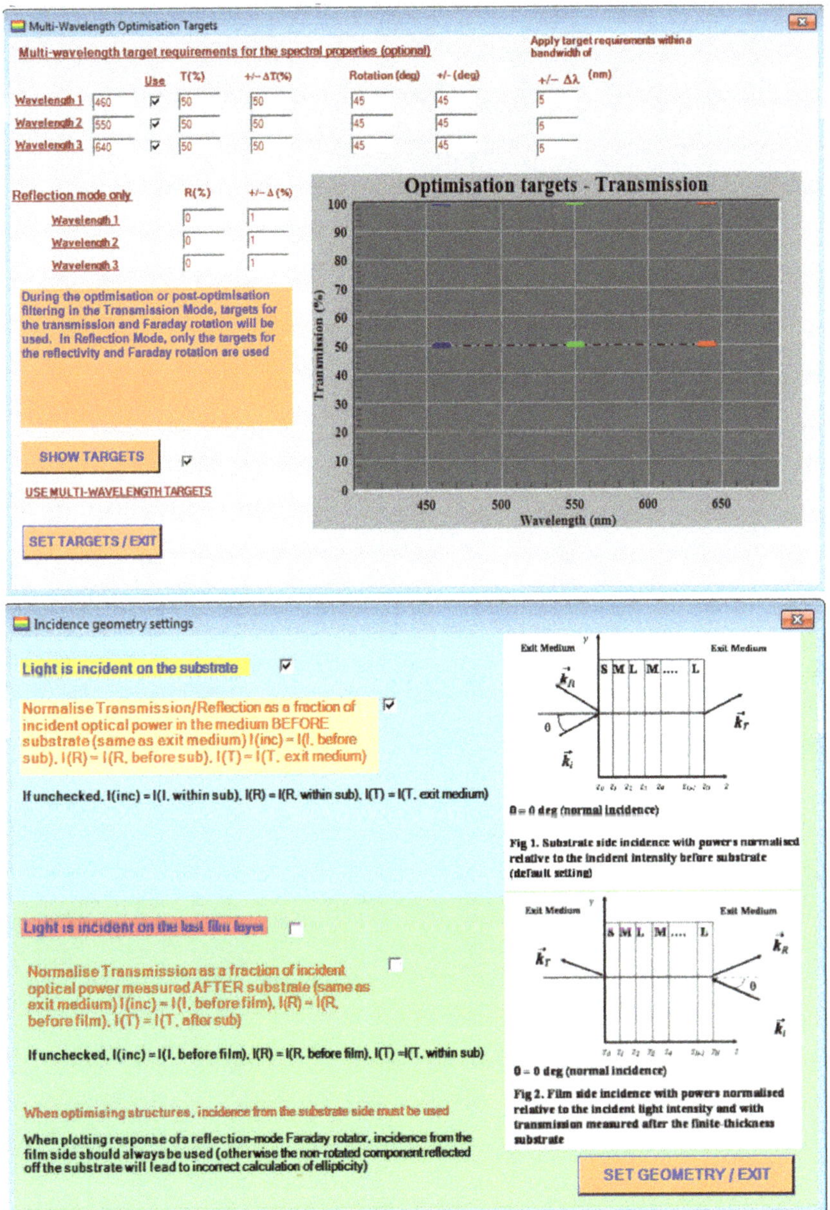

Figure 6. *Cont.*

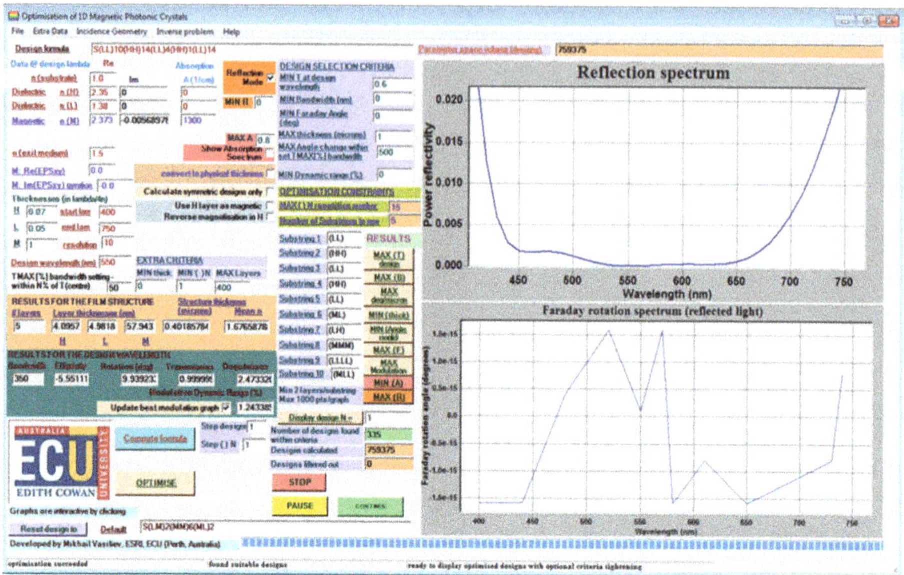

Figure 6. Example of menu and algorithm settings required to generate a number of optimized five-layer antireflector film designs on a glass substrate. In this example, the optical materials (ZnS and MgF_2) are presumed to possess constant refractive indices and zero absorption across the entire visible spectrum.

Figure 6 illustrates graphically the results of fitting the loaded (measured) transmission spectra of two thin films of composition type $Bi_{0.9}Lu_{1.85}Y_{0.25}Fe_{4.0}Ga_1O_{12}$ (the refractive index dispersion data file for this composition is supplied within the appropriate sub-folder in the program installation directory of the target machines). A thinner (684 nm) film modeled as deposited on a glass substrate (n(S) = 1.5) was fitted using the refractive index and absorption coefficient data file related to the as-deposited (amorphous-phase) garnet films of this composition. Since the transmission of an annealed (garnet-phase) film was actually loaded, its transmission at shorter wavelengths was in excess of that modeled. The absorption of garnet-phase films is practically always less in the crystallized state as compared with amorphous garnet-precursor films. A thicker film of fitted thickness (transmission shown in Figure 7b) was measured in the amorphous phase, and therefore the quality of fit is better. Some systematic transmission discrepancies across a wide spectral range can be attributed to a combination of possible light scattering on the film surface features and film layer refractive-index nonuniformity.

If the spectral dependency of the film material absorption coefficient is completely unknown (or the data are not reliable), but the dispersion of its refractive index is well known (e.g., from variable-angle spectroscopic ellipsometry data), then the index dispersion data files need to be prepared with zero values entered for all absorption coefficients. These data files still enable, in many cases, very reliable fitting of the physical thickness. It is highly recommended, then, to back up these thickness fitting results by also applying the Swanepoel method-based techniques, e.g., methods reported in [23].

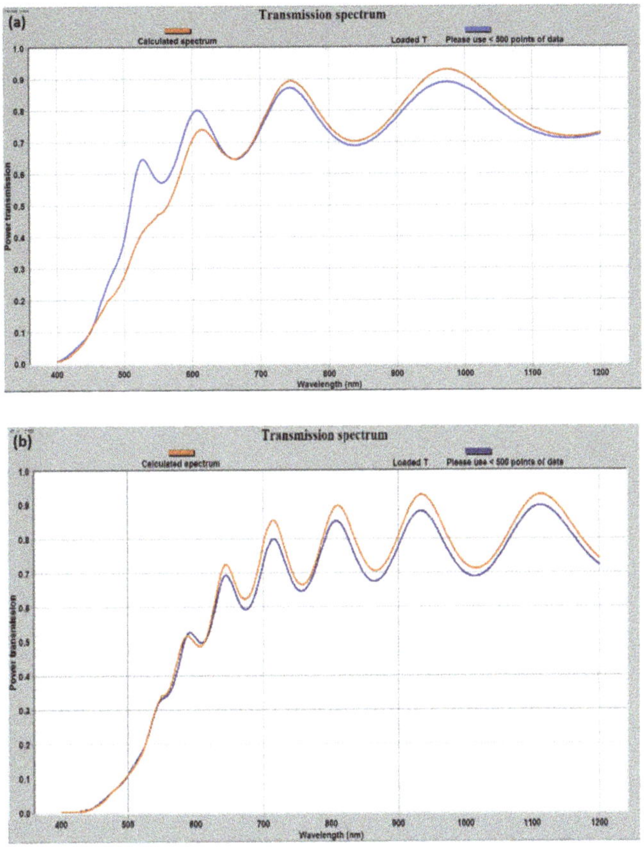

Figure 7. Magneto-photonic crystal (MPC) software fitted transmission spectra of different thin film garnet layers. (**a**) Annealed garnet sample of composition type $Bi_{0.9}Lu_{1.85}Y_{0.25}Fe_{4.0}Ga_{1}O_{12}$ and thickness 684 nm and (**b**) as-deposited garnet sample of the same composition but from another batch and thickness 1310 nm

2.4. Fitting of the Absorption Coefficient Spectra in Single-Layer Films of Known Refractive Index Dispersion Function, Transmission Spectrum, and Thickness

In situations when only the refractive index spectral distribution and the measured transmission spectrum of a semitransparent material layer are available, it is possible to use the custom-prepared "zero absorption" refractive index dispersion data file, and then first fit the physical thickness (typical results are shown in Figure 8a), followed by a derivation of fitted absorption coefficient spectrum using a dedicated algorithm from the program menu item "Inverse problem | Derive absorption spectrum". In order to minimize errors, this combination of fitting procedures is only recommended if several deposited films of the same material are available which have a significantly different physical thickness, and the fitting procedures applied to more than one film consistently lead to obtaining well correlated datasets as a result of the absorption fitting. It is also recommended to apply the Swaneopol method-based fitting procedures [23], to reconfirm the validity and accuracy of the physical thickness data.

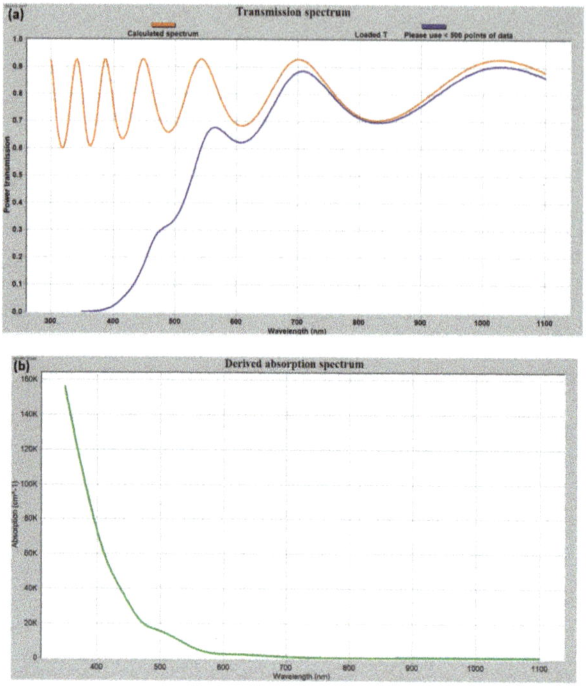

Figure 8. Iterative (bisection algorithm-assisted) fitting of the absorption coefficient spectral dependency and the required pre-fitting of film thickness through matching transmission peak features. (**a**) Peak-aligned transmission spectrum pre-fitting result, from which the physical film thickness information and the measured dataset on the refractive index dispersion are then used, within the option available in the "Inverse Problem" menu, to derive the absorption coefficient data; (**b**) the algorithm-derived (fitted) absorption coefficient spectrum (α, cm^{-1}) for 481 nm thick as-deposited $Bi_{0.9}Lu_{1.85}Y_{0.25}Fe_{4.0}Ga_{1}O_{12}$ garnet-precursor film sample on a glass substrate (its measured transmission spectrum is shown as blue curve in a).

Figure 8 illustrates the results of physical thickness and absorption fitting procedures obtained using a transmission file of a 481 nm thick $Bi_{0.9}Lu_{1.85}Y_{0.25}Fe_{4.0}Ga_{1}O_{12}$ as-deposited film on a glass substrate, and the corresponding material's "zero-absorption" refractive index data file, both of which are included with program distribution in the corresponding subdirectories of the installation folder. The data file named 701 nm $Bi_{0.9}Lu_{1.85}Y_{0.25}Fe_{4}Ga_{1}O_{12}$ garnet layer_n (SWEM) at A = 0 .txt must be loaded using the "load M data" button and the subfolder corresponding to this material type (we previously characterized the refractive index of a 701 nm thick film of this material). After the best fitted thickness value (481 nm) is obtained by comparing different models of design string, such as SM[450], … , SM [481], the same design string must be re-entered using a quarter-wave thickness notation, e.g., S(MM)3, where the QWOT multiplier for M thickness is set to 1.125 (for n(M) = 2.21, corresponding to the nearest index data point to the default 630 nm), to match the physical thickness of 481 nm in this notation. The text box "structure thickness (microns)" must be used to check the physical thickness changes in response to changing either the (MM)N repetition index, or the QWOT multiplier for the M layers. The index n(M) = 2.21 should be entered into its corresponding text box, after looking up the 630 nm (default design wavelength) data for the actual refractive index, to avoid possible misrepresentations of the QWOT data. This notation conversion is required for running the absorption coefficient fitting algorithm, as well as using M-type layers only, regardless of whether the material possesses any magnetic properties or not.

Once within the "derive absorption spectrum" submenu, the same measured transmission file should again be reloaded on the graph from file, followed by a relatively self-explanatory procedure for deriving the graph of absorption spectrum. No changes are usually required in any other text boxes. The result of the fitting procedure with the data files (as described above) is shown in Figure 8b.

In the cases where the fitting algorithm produces exceptions such as described within a dialog text box, e.g., "at some or all wavelengths, even at zero absorption in M layers, the transmission of this structure should be less than specified", adjustments are made to either check that the incidence geometry settings are correct, or the refractive index of the substrate is increased in the model, removing these fitting procedure errors. In cases when these issues persist in the low-absorption wavelength ranges, the wavelength range of the model is reduced to include only the regions where the fit results can be obtained. Nonuniformities in real thin films can lead to reduced refractive index values, leading then to reductions in the modeled reflection, thus showing increased transmission at some wavelengths as compared with theory models. After the absorption spectrum fitting procedure has been completed, the plotted data points are exported into other formats or saved in the data files using the options provided.

3. Installation and Code-Related Information

The installation of MPC Optimization software is easy, and it is enabled by running the installer (.msi) file supplied within the .zip archived folder used for program redistribution. A necessary prerequisite to program installation is the Microsoft .NET Framework 1.1, which must be installed on any Windows machine prior to running the MPC Optimization installer. The .NET 1.1 Framework installation file (dotnetfx.exe) is supplied within the archived folder file used to redistribute MPC optimization. So far, no known problems have been identified to exist in relation to installing this older version of .NET Framework on modern computer systems. A number of example data files containing the samples of measured garnet thin film transmission spectra and files containing the data on the refractive index and absorption coefficient spectra of various garnets and other optical materials are placed into the selected program installation folder on installation. These data enable the users to recreate the example calculations presented within this article, and therefore are useful in mastering the software operation.

The program has been written as Microsoft Visual Studio .NET 2003 Solution and it is composed of three projects: (i) class library project written using managed C++ code, and built as .dll class library providing matrixrelated computation functions; (ii) main program project, written using visual C# and implemented as Windows forms project; and (iii) deployment project, used for including the necessary reference assemblies, class libraries, license files, organizing the file system structure within the installation directory, and generating the Installer files for program distribution. The installed MPC optimization program runs on any Windows system architecture, whether 32-bit or 64-bit, regardless of the processor type.

Two main third-party software components have been embedded in the program structure and the necessary developer licenses have been purchased commercially. Bluebit Matrix Library 2.2 has been used to provide classes that enable efficient complex-valued 4 × 4 matrix operations functionality. The class library .dll assembly referencing the Bluebit Matrix Library assembly has been required, through the terms of a developer-grade matrix software license, to be compiled as a signed assembly, using a strong-name assembly key file, which is not allowed for redistribution by the developers. The second embedded component was GigaSoft ProEssentials 5 .NET package, which has been installed on the developer's computer to provide ActiveX controls that enable scientific-type graphing output functions. The assemblies GigaSoft.ProEssentials.dll and Pegrpcl.dll are referenced by the C# project, embedding the scientific graphing controls as Microsoft component object model (COM) objects into the structure of Windows forms project-related software assemblies. The use of licensed third-party software components, with the associated restrictions, is the primary reason why the MPC optimization program code is not intended for open-source distribution. Additionally, newer versions of Bluebit Matrix Library designed to work with present-day versions of .NET Framework, as well as significant code syntax changes applied throughout the projects, would have been required to successfully port the solution into current versions of Visual Studio, e.g., VS2015 or VS2017. The program installation pre-requisites, installation files, and related documentation are available for download from the Supplementary Material section. Appendix A provides a concise summary of the program functionality, code-related technical details, and outlines some limitations in relation to using the embedded graphing controls.

4. Conclusions

In summary, a set of computational approaches, and a custom software package have been described, designed to enable the design and optimization of 1D magnetic photonic crystals in terms of the achievable combinations of Faraday rotation, transmission, and reflection spectra. A number of important MPC design examples have been introduced, illustrating both the desirable optical property combinations, and the materials-related performance limits of MPC-based Faraday rotators. The same package allows computational modeling of the optical spectral properties of various dielectrics-based generic single- and multilayer thin films. Additional program features include the tools for fitting of the experimentally-measured transmission or reflection spectra to theoretical models, allowing the film physical thickness data recovery, if detailed refractive index information is available. Fitting of the absorption coefficient spectra in absorbing material layers is possible, using an automated algorithm reliant on the data for the measured transmission spectrum, refractive index spectrum, and physical thickness. A number of experimentally- and computationally-derived optical constant-related datasets of different magneto-optic garnet compositions possessing giant Faraday rotation have also been reported.

Supplementary Materials: The following are available online at http://www.mdpi.com/2227-7080/7/3/49/s1, Zip archive containing program installation files. A number of MO garnet material-related datasets are available from the program installation directory after installation.

Author Contributions: M.V. wrote the solution code for the program using Microsoft Visual Studio 2003. NET Professional and multiple code changes and feature additions have been applied between 2005–2019. During this time frame, the development of the algorithm and code features of the program have been greatly influenced by the academic discussions within our group (K.A. and M.N.-E.-A.), and on-going work on multiple materials-related projects. Both K.A. and M.N.-E.A. contributed to the preparation of the final manuscript and its pre-submission review. All co-authors have performed multiple laboratory and computational experiments with a range of magneto-optic material types using the reported software tools.

Funding: This research received no external funding.

Acknowledgments: This work was supported by the Edith Cowan University.

Conflicts of Interest: The authors declare no conflict of interest.

Appendix A : Program Summary and Technical Details

Program title: Optimization of 1D Magnetic Photonic Crystals (alternatively, MPC Optimization)

Program installation files doi: available from the Supplementary section of manuscript.

Licensing provisions: Creative Commons Attribution-NonCommercial-3.0 Unported (CC BY-NC-3.0)

Programming language: Visual C#, compiled using Microsoft Visual Studio .NET 2003

Nature of computational problem: Calculation of the optical transmission, reflection, and Faraday rotation spectra in multilayer thin films containing gyrotropic constituents (magnetized material layers possessing magneto-optic properties); optimization of magnetic photonic crystal (MPC) designs aimed at achieving maximized transmission or reflection coincident with maximized Faraday rotation, according to sets of defined criteria; fitting of the experimentally measured transmission or reflection spectra to theory models; and fitting of the absorption coefficient spectra of single-layer thin film materials using the data for optical transmission, film thickness, and refractive index spectra.

Implementation: The program exhaustively calculates multiple possible multilayer structure designs, based on the design structure type(s) defined prior to running optimization. complex-valued 4 × 4 transfer matrix method (accounting for all dielectric tensor components, including the non-diagonal terms responsible for gyrotropic effects) is implemented to compute the complex field amplitudes and optical intensities in either the transmitted or reflected left-hand and right-hand circularly-polarized eigenwaves propagated through the thin-film substrate structure.

Restrictions: The program is designed for use in conjunction with reliable optical constant datasets for up to three component dielectric materials, one of which can be modeled as magnetic dielectric possessing Faraday rotation; metallic layers are not implemented. The embedded ActiveX controls, which enable graphical data output, accept only up to 1000 data points per graphing control, whether plotting a single curve or several.

References

1. Kharratian, S.; Urey, H.; Onbaşlı, M.C. RGB magnetophotonic crystals for high-contrast magnetooptical spatial light modulators. *Sci. Rep.* **2019**, *9*, 644. [CrossRef] [PubMed]
2. Du, Q.; Fakhrul, T.; Zhang, Y.; Hu, J.; Ross, C.A. Monolithic magneto-optical oxide thin films for on-chip optical isolation. *MRS Bull.* **2018**, *43*, 413–418. [CrossRef]
3. Shoji, Y.; Mizumoto, T. Silicon waveguide optical isolator with directly bonded magneto-optical garnet. *Appl. Sci.* **2019**, *9*, 609. [CrossRef]
4. Jakubisova-Liskova, E.; Visnovsky, S.; Chang, H.; Wu, M. Optical spectroscopy of sputtered nanometer-thick yttrium iron garnet films. *J. Appl. Phys.* **2015**, *117*, 17B702. [CrossRef]
5. Zvezdin, A.K.; Kotov, V.A. *Modern Magnetooptics and Magnetooptical Materials*; Institute of Physics Publishing: Bristol, Philadelphia, 1997; ISBN 075030362X.
6. Drezdzon, S.M.; Yoshie, T. On-chip waveguide isolator based on bismuth iron garnet operating via nonreciprocal single-mode cutoff. *Opt. Express* **2009**, *17*, 9276–9281. [CrossRef] [PubMed]
7. Stupakiewicz, A.; Szerenos, K.; Davydova, M.D.; Zvezdin, K.A.; Zvezdin, A.K.; Kirilyuk, A.; Kimel, A.V. Selection rules for all-optical magnetic recording in iron garnet. *Nat. Commun.* **2019**, *10*, 612. [CrossRef]
8. Inoue, M.; Fujii, T. A theoretical analysis of magneto optical Faraday effect of YIG films with random multilayer structures. *J. Appl. Phys.* **1997**, *81*, 5659–5661. [CrossRef]
9. Sakaguchi, S.; Sugimoto, N. Transmission properties of multilayer films composed of magneto-optical and dielectric materials. *J. Light. Technol.* **1999**, *17*, 1087–1092. [CrossRef]
10. Steel, M.J.; Levy, M., Osgood, R.M. High transmission enhanced Faraday rotation in one-dimensional photonic crystals with defects. *IEEE Photon. Technol. Lett.* **2000**, *12*, 1171–1173. [CrossRef]
11. Levy, A.; Yang, H.; Steel, M.; Fujita, J. Flat-top response in one-dimensional magnetic photonic bandgap structures with Faraday rotation enhancement. *J. Light. Technol.* **2001**, *19*, 1964–1969. [CrossRef]
12. Kato, H.; Matsushita, T.; Takayama, A.; Egawa, M.; Nishimura, K.; Inoue, M. Properties of one-dimensional magnetophotonic crystals for use in optical isolator devices. *IEEE Trans. Magn.* **2002**, *38*, 3246–3248. [CrossRef]
13. Kato, H.; Matsushita, T.; Takayama, A.; Egawa, M.; Nishimura, K.; Inoue, M. Effect of optical losses on optical and magneto-optical properties of one-dimensional magnetophotonic crystals for use in optical isolator devices. *Opt. Commun.* **2003**, *219*, 271–276. [CrossRef]
14. Zvezdin, A.K.; Belotelov, V.I. Magnetooptical properties of photonic crystals. *Eur. Phys. J. B* **2004**, *37*, 479–487. [CrossRef]
15. Brataas, A.; Kent, A.D.; Ohno, H. Current-induced torques in magnetic materials. *Nat. Mater.* **2012**, *11*, 372–381. [CrossRef] [PubMed]
16. Kahl, S. Bismuth Iron Garnet Films for Magneto-Optical Photonic Crystals. Ph.D. Thesis, Department Condensed Matter Physics/KTH, Royal Institute Technology, Stockholm, Sweden, 2004.
17. Belotelov, V.I.; Zvezdin, A.K. Magneto-optical properties of photonic crystals. *J. Opt. Soc. Am. B* **2005**, *22*, 286–292. [CrossRef]

18. Vasiliev, M.; Alameh, K.; Belotelov, V.; Kotov, V.; Zvezdin, A. Magnetic photonic crystals: 1-D optimization and applications for the integrated optics devices. *J. Light. Technol.* **2006**, *24*, 2156–2162. [CrossRef]
19. Vasiliev, M.; Alam, M.N.E.; Kotov, V.A.; Alameh, K.; Belotelov, V.I.; Burkov, V.I.; Zvezdin, A.K. RF magnetron sputtered $(BiDy)_3(FeGa)_5O_{12}$:Bi_2O_3 composite materials possessing record magneto-optic quality in the visible spectral region. *Opt. Express* **2009**, *17*, 19519–19535. [CrossRef]
20. Nur-E-Alam, M.; Vasiliev, M.; Alameh, K. Nano-structured magnetic photonic crystals for magneto-optic polarization controllers at the communication-band wavelengths. *Opt. Quantum Electron.* **2009**, *41*, 661–669. [CrossRef]
21. Nur-E-Alam, M.; Vasiliev, M.; Alameh, K. High-performance RF-sputtered Bi-substituted iron garnet thin films with almost in-plane magnetization. *Opt. Mater. Express* **2017**, *7*, 676–686. [CrossRef]
22. Kotov, V.; Shavrov, V.; Vasiliev, M.; Alameh, K.; Nur-E-Alam, M.; Balabanov, D. Properties of magnetic photonic crystals in the visible spectral region and their performance limitations. *Photonics Nanostruct.-Fundam. Appl.* **2018**, *28*, 12–19. [CrossRef]
23. Nur-E-Alam, M.; Vasiliev, M.; Belotelov, V.; Alameh, K. Properties of Ferrite Garnet (Bi, Lu, Y)3(Fe, Ga)5O12 Thin Film Materials Prepared by RF Magnetron Sputtering. *Nanomaterials* **2018**, *8*, 355. [CrossRef] [PubMed]

© 2019 by the authors. Licensee MDPI, Basel, Switzerland. This article is an open access article distributed under the terms and conditions of the Creative Commons Attribution (CC BY) license (http://creativecommons.org/licenses/by/4.0/).

technologies

MDPI

Article

Microwave Assisted Green Synthesis of Silver Nanoparticles Using Mulberry Leaves Extract and Silver Nitrate Solution

Le Ngoc Liem [1,*], Nguyen Phuoc The [2] and Dieu Nguyen [2]

1 Office of the Board of Trustees, Duy Tan University, 03 Quang Trung, Danang 550000, Vietnam

2 Faculty of Natural Sciences, Duy Tan University, 03 Quang Trung, Danang 550000, Vietnam; nguyenphuocthe@dtu.edu.vn (N.P.T.); nguyentxuandieu@dtu.edu.vn (D.N.)

* Correspondence: lengocliem@dtu.edu.vn; Tel.: +84-915-995-376

Received: 20 December 2018; Accepted: 31 December 2018; Published: 5 January 2019

Abstract: In this work, silver nanoparticles (AgNPs) were synthesized quickly and in an eco-friendly manner using the extract of Mulberry leaves and aqueous solution of silver nitrate without any toxic chemicals (Yuet et al. *Int. J. Nanomed.* **2012**, *7*, 4263–4267; Krishnakuma and Adavallan. *Adv. Nat. Sci. Nanosci. Nanotechnol.* **2014**, *5*, 025018). The Mulberry leaves extract functions as both a stabilizing and reducing agent. The UV-Vis spectroscopy shows a peak maximum at 430 nm. The transmission electron microscopy (TEM) image illustrated of synthesized AgNPs were nearly spherical-shaped particles whose sizes range from 15 to 20 nm. The TEM image of Nano Silver solution sample synthesized by the microwave assisted method shows nearly spherical particles, with an average particle size estimated at 10 nm. The absorption UV-vis spectrum of silver nanoparticles synthesized by the microwave assisted method (AgNPsmw) shows a sharp absorption band around 415 nm. The UV-Vis spectrum of AgNPsmw after two months of storage shows negligible peak changes of silver nanoparticles.

Keywords: silver nanoparticles; mulberry leaves extract

1. Introduction

Nanoparticles (NPs) are defined as small particles sized between 1 and 100 nm. Compared with the material bulk states, NPs have gained prominence in recent technological advancements due to their tunable physicochemical characteristics such as their melting point, wettability, electrical and thermal conductivity, catalytic activity, light absorption, better tunable optical properties, and higher reactivity. Various chemical and physical methods have been employed to prepare silver nanoparticles, including chemical reduction electrochemical techniques, and photochemical reduction. Among all the synthetic methods, chemical reduction is most commonly used. However, the chemical synthesis of nanoparticles may lead to the presence of some toxic chemicals. Several studies have focused on green synthesis approaches to avoid using hazardous materials. Synthesizing silver nanoparticles using mulberry leaves extract as a reducing agent not only offers many advantages but also uses less chemicals, thus reducing the pollution caused to the environment. In recent years, nanostructured materials have obtained many applications relevant to daily life. Silver nanoparticles (AgNPs) are used in a wide range of applications, including pharmaceuticals, cosmetics, medical devices, foodware, clothing and water purification, agriculture and in wastewater treatment, etc. due to their antimicrobial properties [1–7]. In this study, mulberry leaves extract had been used as a reduction agent and stabilizing agent. Synthesizing silver nanoparticles using mulberry leaves extract as a reducing agent not only offers many advantages but also uses less chemicals. The green synthesis combined with the microwave assistance make up a highly effective and eco-friendly method.

The mulberry (Figure 1) is a woody plant that grows quickly and within a short proliferation period. There are about 10 to 16 species of genus Morus that are found in the subtropical climates and the warm and temperate regions of Africa and North America [8]. Those species have been cultivated in many Asian countries such as China, India, Korea, Japan, Thailand and Vietnam, where the leaves have been used as food for silkworms [9]. Mulberry is known to be used in some traditional Chinese medicinal formulas. There are several studies have shown that it may provide health benefits. There are many biologically active compounds and many phytochemicals in mulberry leaves. There are also many important pharmacological properties such as antibacterial, antiviral [5–9], antitussive, hypoglycemic, antiatherogenic [10], hypotensive [11], diuretic, astringent, antioxidant [12,13] and α-amylase inhibitory effects [10].

Figure 1. Picture of Mulberry trees leaves.

2. Materials and Methods

2.1. Preparation of Mulberry Leaves Extract and Silver Nitrate Solution

Mulberry leaves were collected from a residential garden house in Hoi An, Quang Nam province, Vietnam. The leaves collected must be intact and at their prime (neither be too young or too old). Those fresh leaves were then cleaned with fresh water and let air dried by laying them out evenly. 10 g of fresh leaves was obtained, cut into thin strips, then placed into a 200 mL heat-resistant glass flask. Then, they were boiled with distilled water in 5 min, cooled and the mixture was filtered with Whatman filter paper using a vacuum filter [4–7]. The mulberry leaves extract had a light yellow color. The extracted solution was stored in a fridge for further use. Dissolved silver nitrate ($AgNO_3$) from Sigma Aldrich was mixed with distilled water to get 4.10^{-3} M aqueous $AgNO_3$ solution.

2.2. Synthesis of Nano Silver Material

2.2.1. Non-Microwave Assisted Synthesis of Nano Silver

Visual Observation and UV-vis Spectral

$AgNO_3$ 4.10^{-3} M solution and mulberry leaves extract were mixed at different ratios as shown in Table 1. These mixtures were placed on a shaker and stirred for 30 min, 150 rpm at room temperature. After 30 min, the color of all the mixtures changed from light brown to dark red, except for mixtures M_0 and M_1. The color of each mixture varies depending on the ratio of $AgNO_3$ concentration and mulberry leaves extract that were used (Figure 2). This observation implied that Nano Silver particles were formed.

Table 1. Synthesized samples.

$AgNO_3$ Solution (mL)	Mulberry Leaves Extract (mL)	Samples
0	10	M_0
40	0	M_1
40	1	M_2
40	3	M_3
40	5	M_4
40	6	M_5
40	7	M_6
40	8	M_7

Characterization Techniques: In order to investigate the optical properties, we measured the UV-vis absorption spectra using GE Ultrospec 7000 UV-vis spectrophotometer (GE Lifesciences, Freiburg, Germany). Transmission Electron Microscope (TEM) analysis of silver nanoparticles was done using JEOL JEM 1010 (JEOL, Tokyo, Japan). X-ray diffraction spectra were measured by the diffractometer Bruker D8-Advance (Bruker, Karlsruhe, Germany), Fourier transform infrared (FTIR) Spectra. For mulberry leaves, extract was obtained in the range 400–4000 cm^{-1} with IRAffinity-1S Shimadzu FTIR spectrophotometer (Shimadzu, Tokyo, Japan).

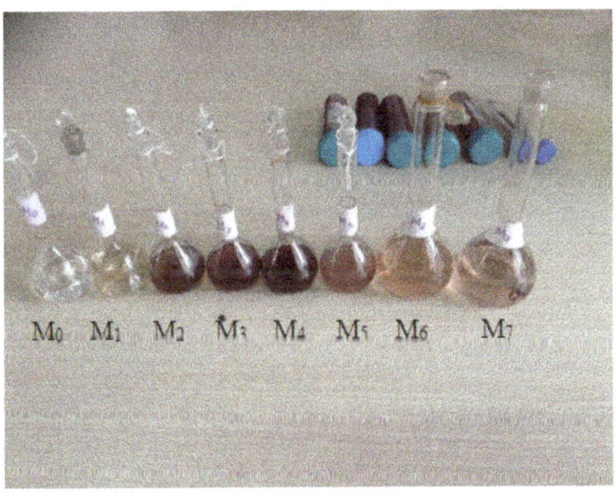

Figure 2. Photograph color change of colloids.

The intensity of absorption within the wavelength of 200 to 250 nm is very strong. Therefore, before measuring UV-vis, samples were diluted 40 times. The UV-vis spectrum of the material shows a strong surface plasmonic resonance band centered at 430 nm. For the samples M_2, M_3 and M_4, the absorbance intensity increasesd as we increased the volume of mulberry leaf extract from 1 mL to 5 mL. For samples whose volume of mulberry leaves extract were more than 5 mL, the intensity decreased as the volume of mulberry leaves extract increased. Combining spectral information with qualitative observation, we could conclude that AgNPs has been formed when mixing $AgNO_3$ solution with mulberry leaves extract. Among our samples, the sample M_4 not only showed the highest peak intensity value but also had the darkest red color; the observations indicated that it had the most AgNPs particles. We can also infer that, at high concentration of mulberry leaves extract, the rate of AgNPs production is so rapid that it prevents the formation of protective layer between particles. For this to happen, the aggregation phenomenon between particles occurred and increased the particle size. The increased particle size reduced peak intensity and shifted the peak to a longer wavelength (as in M_5, M_6 and M_7 samples). The absorption spectrum of $AgNO_3$ aqueous solution, shown in the olive colored line in Figure 3, had no peak in the measuring range. The absorption spectrum of mulberry leaves extract, shown in orange line in Figure 3, had absorption peak in the short wavelength region due to an organic substance found in the extract solution and had no peak at wavelengths longer than 400 nm. The spectral information indicated that AgNPs particles were formed only when $AgNO_3$ was mixed with mulberry extract. When magnified from 380 to 700 nm of the absorption spectrum (Figure 3), discrete lines were shown.

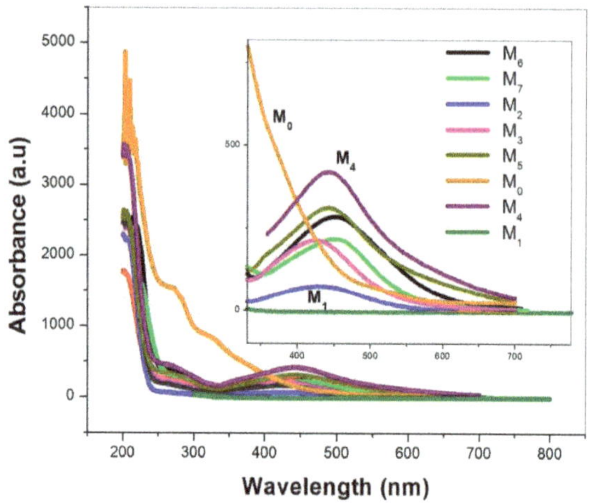

Figure 3. UV-Vis spectra of $AgNO_3$ solution (M1), Mulberry extract (M0) and AgNps prepared at different Mulberry leaves extract.

2.2.2. Microwave Asissted Systhesis of Nano Silver

Visual Observation and UV-vis Spectral

We mixed 50 mL of $AgNO_3$ solution with 6 mL of mulberry leaves extract and divided it into 2 equal portions. Then, we put the first portion (M_8) on a shaker machine and stirred for 30 min, 150 rpm at room temperature. The other portion (M_9) was heated in a microwave for one minute. The changed colors of M_9 from light yellow to dark red within one minute in the microwave indicated that the efficiency of (AgNPsmw) synthesis using the microwave assisted method was higher than the non-microwave assisted method. The sample colors are shown in Figure 4.

Figure 4. Photograph of silver nanoparticles (AgNPs) (M_8) and of AgNPsmw (M_9).

The UV-vis spectra (Figure 5) shows that absorption intensity at around 415 nm of M_9 is higher than that of M_8 and the maximum wavelength of M_9 is smaller than the M_8. After two months of storage, the M_9 sample was renamed M_{10}. Compared to the spectrum of M_9, the UV-vis spectrum taken for M_{10} (the blue line of Figure 5) shows a slight decrease in the peak intensity and the peak shift of 5 nm towards a longer wavelength. The above observations indicated that the synthesized colloid solution is highly stable. Compared with the work of other authors, our study resulted in a smaller particle size, short-time synthesis, and a higher absorption peak intensity.

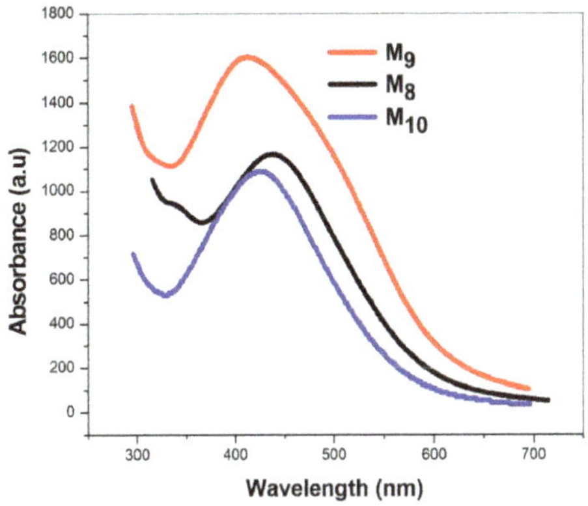

Figure 5. UV-Vis spectra of AgNPs (M_8), of AgNPsmw (M_9) and of AgNPsmw after two months of storage (M_{10}).

2.3. X-ray Diffraction (XRD) Studies

Cloth was cut into 25 cm^2 pieces and dipped into a colloids solution of AgNPs for 10 min, and another piece of cloth was not dipped into a colloids solution of AgNPs. The XRD pattern of the cloth dipped into a colloids solution AgNPs and the cloth not dipped into a colloids solution AgNPs are shown in Figure 6. The XRD pattern of the undipped cloth showed no characteristics of peak silver. On the other hand, the XRD pattern of the cloth that was dipped in the colloids AgNPs showed the characteristic of peak silver. Four main characteristics of the diffraction peak for Ag were observed at 2θ values of 38.2°, 44.1°, 64.5°, and 77.6°, which correspond to the (111), (200), (220) and (311) crystallographic planes of face-centered cubic (fcc) Ag crystals, respectively (JCPDS 00-004-0783) [14,15]. Because the AgNps concentration was low and the cloth's internal structure had lots of empty spaces, not many AgNPs particles were deposited onto the cloth surface, thus the peaks intensity was low.

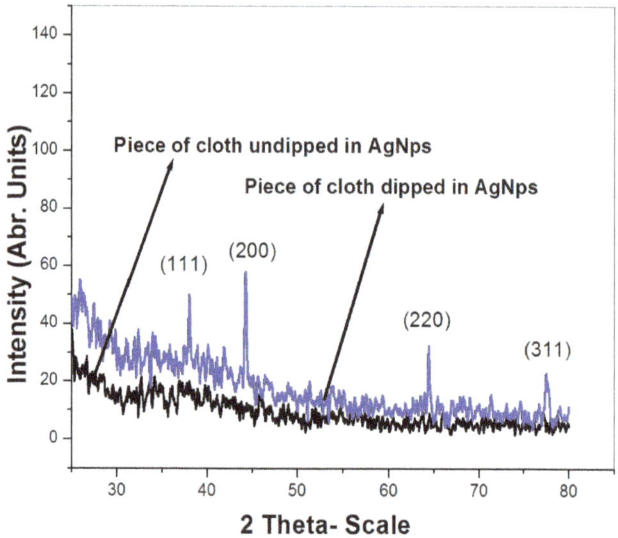

Figure 6. X-ray diffraction of a pieces of cloth dipped into a colloids solution AgNPs (blue line), and a pieces of cloth not dipped into a colloids solution AgNPs (black line).

2.4. Transmission Electron Microscope (TEM) Analysis

The size and shape of the AgNPs was further confirmed by TEM analysis, is shown in Figure 7. The TEM image of M_8 showed a relatively uniform spherical particle size, ranging from 15–20 nm. The TEM image of M_9 showed particle size about 10 nm. The particle sizes were more uniform and no sign of nanoparticle clustering was observed.

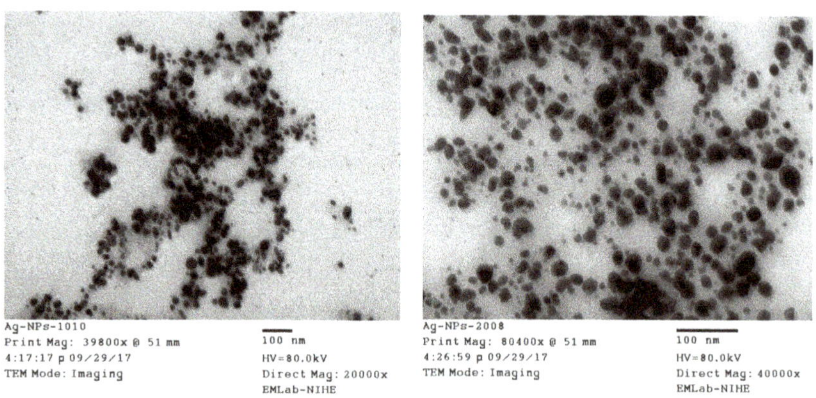

(a) (b)

Figure 7. Transmission Electron Microscope (TEM) images of silver nanoparticles (M_8 and M_9).

Figure 8 shows the histogram of size distribution of silver nano particles. The average particle size measured from the TEM image is 10 nm. This large variation in particle size was due to the presence of a few irregular shaped particles.

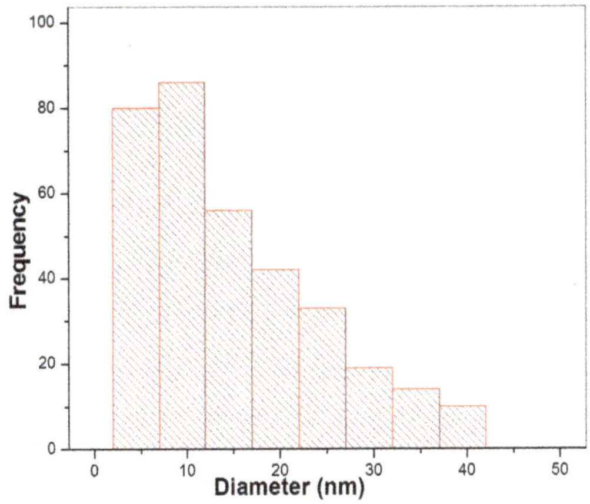

Figure 8. Histogram showing the particle sizes of AgNPs corresponding to TEM images M_9.

2.5. FT-IR Spectrum

0.5 mL of the $AgNO_3$ solution was dropped into 50 mL of mulberry leaves extract to prevent the extracted solution from rotting before taking the spectrum for mulberry leaf extract (From the sample synthesis to the FT-IR spectrum takes about 5 days). An FT-IR spectrum of silver nanoparticles synthesized by this green method is shown in Figure 9. A number of absorption peaks at 3261 cm^{-1} and 1637 cm^{-1}. The peaks at 3261 cm^{-1} corresponds to O-H and N-H bonds, the peak at 1637 cm^{-1} corresponds to the C=O bond, indicating the biomaterial bind to the silver nanoparticles through amine and C=O of amide I and amid II of the protein [1,6,7,16]. These results indicate that mulberry leaves extract acts as a reducing and stabilizing agent for silver particles. According to studies by A.K. Mittal et al., the extract of Dhatura metel contained alkaloids, proteins, enzymes, amino acids, alcoholic compounds, and polysaccharides, which were said to be responsible for the reduction of the silver ions to nanoparticles. Quinol and chlorophyll pigments present in the extract also contributed to the reduction and stabilization of the nanoparticles [17]. Leaf extracts of Mulberry also contain similar leaf extracts of Dhatura metel [18]. Therefore, mulberry leaf extract also played a role in the reduction and stabilization of the nanoparticles.

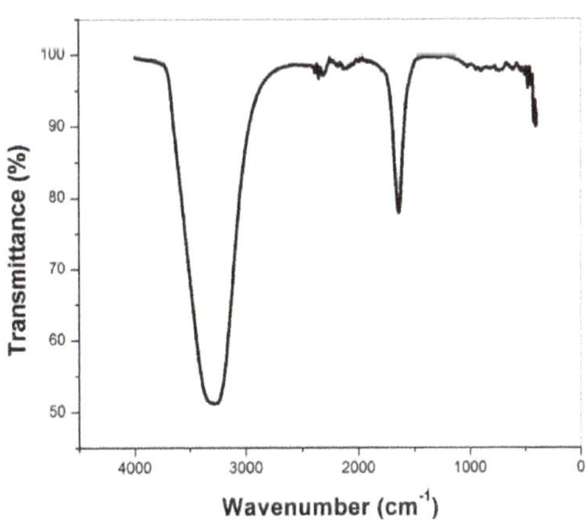

Figure 9. FTIR spectrum of mulberry leaves extract.

3. Conclusions

The silver nanoparticles (AgNPs) colloid solution has been successfully synthesized, using a green and eco-friendly method. This is a quick, highly effective and less chemical-consuming method. The Microwave assisted green synthesis method is more effective than the non-microwave assisted method. The UV-Vis spectrum of AgNPs has an absorbance peak ranging from 425 nm–435 nm. Nano-silver particles are spherical shaped with size ranging from 15 nm to 20 nm. UV-vis spectrum of AgNPsmw shows peak at 415 nm, with an average particle size of 10 nm. The results observed after two months of storage of AgNPsmw are quite stable.

Author Contributions: Conceptualization, L.N.L.; Methodology, L.N.L. and N.P.T.; Validation, L.N.L. and N.P.T.; Formal analysis, L.N.L.; Investigation, L.N.L., D.N. and N.P.T.; Writing—Original draft preparation, L.N.L.; Writing—Review and Editing, L.N.L., D.N. and N.P.T; supervision, L.N.L.

Funding: This research received no external funding.

Conflicts of Interest: The authors declare no conflict of interest.

References

1. Yuet, Y.-L.; Buong, W.-C.; Mitsuaki, N.; Son, R. Synthesis of Silver Nanoparticles by Using Tea Leaf Extract from Camellia Sinensi. *Int. J. Nanomed.* **2012**, *7*, 4263–4267.
2. Krishnakuma, N.; Adavallan, K. Mulberry leaf extract mediated synthesis of gold nano particles and its anti-bacterial activity against human pathogens. *Adv. Nat. Sci. Nanosci. Nanotechnol.* **2014**, *5*, 025018.
3. Jaiswal, S.; Duffy, B.; Jaiswal, A.-K.; Stobie, N.; McHale, P. Enhancement of the antibacterial properties of silver nanoparticles using β-cyclodextrin as a capping agent. *Int. J. Antimicrob. Agents* **2010**, *36*, 280–283. [CrossRef] [PubMed]
4. Sun, Q.; Cai, X.; Li, J.; Zheng, M.; Chen, Z.; Yu, C.-P. Green synthesis of silver nanoparticles using tea leaf extract and evaluation of their stability and antibacterial activity. *Colloids Surf. Physicochem. Eng. Asp.* **2014**, *444*, 226–231. [CrossRef]
5. Sajeshkumar, N.-K.; Prem, J.-V.; Jiby, J.-M.; Anupa, S. Synthesis of silver nano particles from curry leaf (murraya koenigii) extract and its antibacterialactivity. *CIBTech J. Pharm. Sci.* **2015**, *4*, 15–25.
6. Awwad, A.-M.; Salem, N.-M. Green synthesis of silver nanoparticles by Mulberry leavesextrac. *Nanosci. Nanotechnol.* **2012**, *2*, 125–128. [CrossRef]
7. Awwad, A.-M.; Salem, N.-M.; Abdeen, A.-O. Biosynthesis of silver nanoparticles using Loquat leaf extract and its antibacterial activity. *Adv. Mater. Lett.* **2013**, *4*, 338–342. [CrossRef]
8. Shahid, L.; Umer, Y.; Sirajuddin; Kim, W.-C.; Raja, A.-S.; Uddin, M.-K. Proximate Composition and Antioxidant Potential of Leaves from Three Varieties of Mulberry (*Morus* sp.): A Comparative Study. *Int. J. Mol. Sci.* **2012**, *13*, 6651–6664.
9. Nuengchamnong, N.; Ingkaninan, K.; Kaewruang, W.; Wongareonwanakij, S.; Hongthongdaeng, B. Quantitative determination of 1-deoxynojirimycin in mulberry leaves using liquid chromatography-tandem mass spectrometry. *J. Pharm. Biomed. Anal.* **2007**, *44*, 853–858. [CrossRef] [PubMed]
10. Bahman, N.; Golboo, M. Influence of Three Morus Species Extracts on α-Amylase Activity. *Iran. J. Pharm. Res.* **2009**, *8*, 115–119.
11. Chirino, Y.-I.; Pedraza, C.-J. Role of oxidative and nitrosative stress in cisplatin-induced nephrotoxicity. *Exp. Toxicol. Pathol.* **2009**, *61*, 223–242. [CrossRef] [PubMed]
12. Yang, Y.; Gong, T.; Liu, C.-H.; Chen, R.-Y. Four new 2-arylbenzofuran derivatives from leaves of *Morus alba* L. *Chem. Pharm. Bull.* **2010**, *58*, 257–260. [CrossRef] [PubMed]
13. Kobayashi, Y.; Miyazawa, M.; Kojima, T. The use of Morus alba L (mulberry): And Eucommia ulmoides (Tochu): Leaves as functional foods: A promising approach in the management of hyperlipidemia. *J. Tradit. Med.* **2010**, *27*, 227–230.
14. Jagpreet, S.; Navalpreet, S.; Aditi, R.; Deepak, K.; Mohit, R. Facile Approach to Synthesize and Characterization of Silver Nanoparticles by Using Mulberry Leaves Extract in Aqueous Medium and its Application in Antimicrobial Activity. *J. Nanostruct.* **2017**, *7*, 134–140.

15. Siby, J.; Beena, M. Synthesis of silver nanoparticles by microwave irradiation and investigation of their catalytic activity. *Res. J. Recent Sci.* **2014**, *3*, 185–191.
16. Dhivya, G.; Rajasimman, M. Synthesis of silver nanoparticles using Momordica charantia and its applications. *J. Chem. Pharm. Res.* **2015**, *7*, 107–113.
17. Amit, K.-M.; Yusuf, C.; Banerjee, U.C. Synthesis of metallic nanoparticles using plant extract. *Biotechnol. Adv.* **2013**, *31*, 346–356.
18. Olfa, R.; Manel, B.; Sami, F.; Mohamed, A. Phytochemicals from Mulberry extract (Morus sp.): Antioxidant and neuroprotective potentials. *J. Appl. Pharm. Sci.* **2017**, *7*, 217–222.

© 2019 by the authors. Licensee MDPI, Basel, Switzerland. This article is an open access article distributed under the terms and conditions of the Creative Commons Attribution (CC BY) license (http://creativecommons.org/licenses/by/4.0/).

technologies

Letter

Facile Fabrication of Macroscopic Self-Standing Ni or Co-doped MnO_2 Architectures as Catalysts for Propane Oxidation

Long Chen [1,2] and Xiping Song [1,*]

1 State Key Laboratory for Advanced Metals and Materials, University of Science and Technology Beijing, Beijing 100083, China; clong1113@gmail.com

2 Beijing Institute of Metrology, Beijing 100029, China

* Correspondence: xpsong@skl.ustb.edu.cn; Tel.: +86-10-6233-3213

Received: 17 October 2019; Accepted: 8 November 2019; Published: 11 November 2019

Abstract: The fabrication of macroscopic self-standing architectures plays a key role in the practical applications of nanomaterials. A facile strategy to assemble MnO_2 nanowires into macroscopic self-standing architectures via hydrothermal reaction followed by ambient pressure drying was developed. The obtained sample was robust and showed excellent mechanical strength with a Young's modulus of 127 MPa, which had the possibility for practical applications. In order to promote the catalytic activity for propane oxidation, Ni or Co doping into MnO_2 was studied. The results showed that the obtained macroscopic self-standing Ni-MnO_2 and Co-MnO_2 architectures exhibited enhanced catalytic activities for propane oxidation. Specifically, the conversions of propane over Co-MnO_2 and Ni-MnO_2 samples at 400 °C were 27.3% and 25.7% higher than that over pristine MnO_2 sample.

Keywords: macroscopic self-standing architectures; Ni-doped MnO_2; Co-doped MnO_2; propane oxidation; mechanical properties

1. Introduction

Amongst various transition metal oxides, manganese dioxide has been considered as one of the most potential low-temperature catalysts due to its environmental friendliness as well as its low cost [1]. In order to further improve the catalytic activity of MnO_2, one of the most common strategies is doping with a different cation, such as Ni [2,3] and Co [4,5]. On the other hand, self-assembly of nanomaterials into macroscopic architectures provides the possibility for exploring the practical applications. However, the method to assemble nanomaterials into macroscopic architectures remains a challenge [6,7].

Long et al. [8] obtained a macro-assembly with MnO_2 nanowires via a hydrothermal method followed by a freeze-drying process, which showed selective adsorption of cationic dyes. Jung et al. [9] constructed MnO_2 nanowire hydrogel/aerogels via hydrothermal synthesis (over four days) and supercritical drying, which could be used to remove heavy metal ions and toxic organic contents in water. Suib et al. [10] constructed macroscopic free-standing OMS-2 sponges through hydrothermal reaction (250 °C for four days) and freeze-drying process, which could be used to separate oil and water. Rong et al. [11] fabricated a three-dimensional manganese dioxide framework combining δ-MnO_2 nanosheets and α-MnO_2 nanowires, which had interconnected network structures and showed excellent oxidation activity for ppm-level HCHO to CO_2 at low temperatures (≤120 °C). However, the reported process was time-consuming and high cost, and the catalytic application at higher temperature was rarely concerned. Therefore, it is necessary to explore simpler preparation methods and study the catalytic performance at higher reaction temperatures.

This research aims to fabricate macroscopic self-standing architectures with metal doped MnO_2 nanomaterials via facile hydrothermal reaction followed by ambient pressure drying and study the catalytic activity for propane oxidation. To the best of our knowledge, the study of macroscopic architectures with metal doped MnO_2 nanowires as catalysts for propane oxidation has been scarcely reported.

2. Materials and Methods

The macroscopic self-standing MnO_2 architectures were prepared by a modified hydrothermal process followed by ambient pressure drying. Typically, the A aqueous solution consisting of manganese acetate ($Mn(CH_3COO)_2{\cdot}4H_2O$, 1.18 g) and ammonium sulfate ($(NH_4)_2SO_4$, 3 g) was added slowly into the B aqueous solution consisting of potassium permanganate ($KMnO_4$, 0.506 g) and cetyl trimethylammonium bromide ($C_{19}H_{42}BrN$, 0.09 g) with continuous stirring. Afterwards, the resultant slurry was treated under hydrothermal conditions at 140 °C and kept for 6 h. Then the produced wet gel was washed in distilled water at 50 °C repeatedly. After ambient pressure drying, the macroscopic free-standing MnO_2 with the specific shape of the drying vessel was obtained. The schematic diagram of the preparation process for macroscopic self-standing architectures was shown in Figure 1.

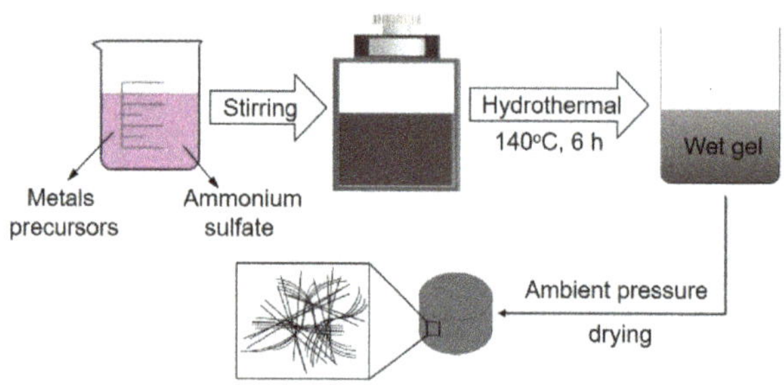

Figure 1. The schematic diagram of the preparation process for macroscopic self-standing architectures.

The macroscopic self-standing Ni- or Co-doped MnO_2 architectures were prepared by a similar procedure including the addition of nickel nitrate or cobalt nitrate precursors. The doping amounts of Ni (Ni/Mn molar ratio of 3/5) or Co (Co/Mn molar ratio of 1/5) were optimized in previous research [12,13] and the obtained samples were designated as Ni-MnO_2 and Co-MnO_2, respectively.

An X-ray diffractometer (XRD, D8 Advance A25, Bruker, Germany) was employed to identify the phase structure. The morphologies were investigated with Scanning Electron Microscopy (SEM, SUPRATM 55, Carl Zeiss, Germany). An Instron 5940 universal testing machine (Shanghai, China) was employed for compression testing of pristine MnO_2, Ni- or Co-doped MnO_2 samples.

The catalytic activities of propane oxidation over pristine MnO_2, Ni- or Co-doped MnO_2 samples were evaluated in a fixed-bed reactor with continuous flow. The feed gas consisted of 1000 ppm propane, 5% oxygen in a nitrogen balance gas, and total flow rate was 300 mL/min. A mixture of 0.4 g sample and quartz sands was used, and the heating rate of the oxidation reaction was 5 °C/min. The maximum pressure reached inside the reactor was 103 kPa. The propane concentration in the outlet gas was on-line monitored with a MultiGas analyzer (MKS MultiGas ™ 2030, USA). The propane conversion was defined according to the following equation:

$$\text{Propane conversion (\%)} = (1 - [C_3H_8]_{outlet}/[C_3H_8]_{inlet}) \times 100\%, \quad (1)$$

3. Results and Discussion

The appearance of prepared macroscopic self-standing MnO_2 architectures with cylindrical shapes was shown in the inset of Figure 2. The appearance of Ni-MnO_2 and Co-MnO_2 were similar to that of MnO_2. It demonstrates that the Ni or Co doping did not interrupt the assembling process of macroscopic self-standing MnO_2 architectures. The XRD patterns (Figure 1) suggested that the crystal phases of pristine MnO_2, Ni-MnO_2 and Co-MnO_2 samples corresponded well to the tetragonal MnO_2 (JCPDS no. 44-0141). No impurity phase was observed, which validated the high dispersion state of Ni or Co within the MnO_2 framework. The statistical average bulk densities of MnO_2, Ni-MnO_2 and Co-MnO_2 samples were 0.72, 0.69 and 0.70 g cm^{-3}, respectively. The morphologies of pristine MnO_2, Ni-MnO_2 and Co-MnO_2 samples were shown in Figure 3. The basic component unit of pristine MnO_2 was nanowire, with a length of several hundred micrometers. The bundles of ultra-long nanowires intertwined and assembled into a network structure of macroscopic self-standing MnO_2 architecture. After Ni or Co doping, the morphology of nanowires did not change distinctly. Thus, the Ni-MnO_2 and Co-MnO_2 also showed the disordered network structure assembled by nanowires.

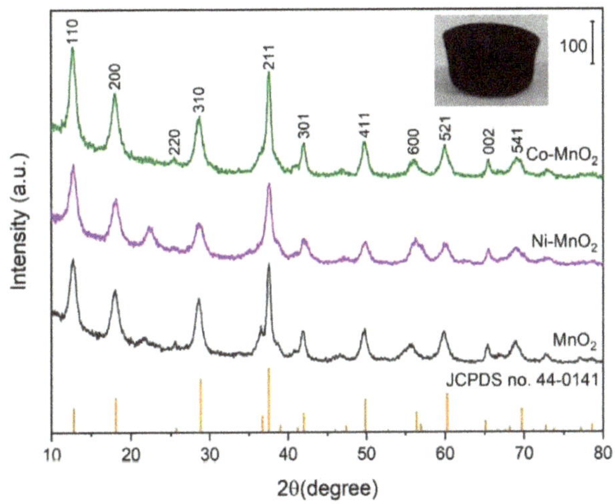

Figure 2. The X-ray diffractometer (XRD) patterns of pristine MnO_2, Ni-MnO_2, and Co-MnO_2 samples. The inset is the appearance of the macroscopic self-standing MnO_2 architecture.

On the basis of the aforesaid results, the fabrication process of macroscopic self-standing Mn-based architectures included the formation of wet gel with nanowires via hydrothermal reaction and the removal of water via ambient pressure drying maintaining the self-standing architecture simultaneously. In previous research, the MnO_2 short nanofibers were prepared and the wet gel was not formed without the addition of ammonium sulfate during the hydrothermal reaction [12,13]. The role of ammonium sulfate included: (1) modification of the nanostructure unit; the nanowires with large length-to-diameter ratio were obtained with the addition of ammonium sulfate and this morphology was favourable for the assembly, and (2) assistance to assembly; the ammonium sulfate functioned as bridging ligands and promoted the assembling of MnO_2 nanowires. It has been proposed that sulfate ions have two coordination sites which can bond water molecules via hydrogen bonds forming a three-dimensional network, and simultaneously coordinate metal nanoparticles, thus promoting nanoparticles self-assembly [14–16]. Therefore, the solvent of water also participated in the building process of the macroscopic assembly.

The mechanical properties of samples were investigated and the stress-strain curves were shown in Figure 4. The curves of all samples had a similar shape and contained elastic and plastic regions. The calculated Young's moduli in the linear region of pristine MnO_2, Ni-MnO_2, and Co-MnO_2 samples were 127, 73.8, and 172 MPa, suggesting they had good resistance to elastic deformation under load. The yield strengths of pristine MnO_2, Ni-MnO_2, and Co-MnO_2 samples reached 4.5, 3.2, and 4.2 MPa, respectively. In the plastic regions, when the samples were compressed to strain of 50%, the corresponding stresses of pristine MnO_2, Ni-MnO_2, and Co-MnO_2 samples were 6.2, 4.1, and 5.7

MPa. It suggested that all samples were robust, compared with manganese oxide sponges in previous research [10].

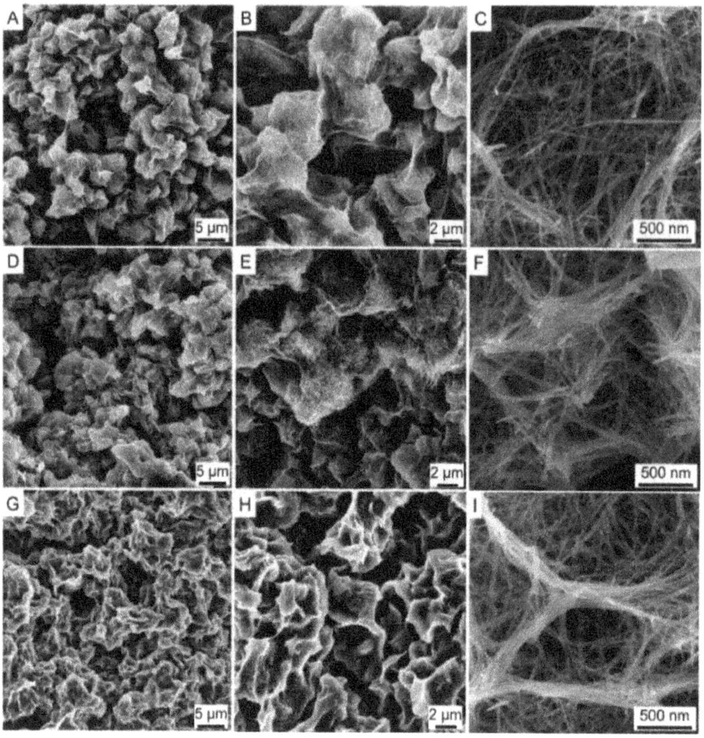

Figure 3. The Scanning Electron Microscopy (SEM) images of pristine MnO_2 (**A–C**), Ni-MnO_2 (**D–F**), and Co-MnO_2 (**G–I**) samples.

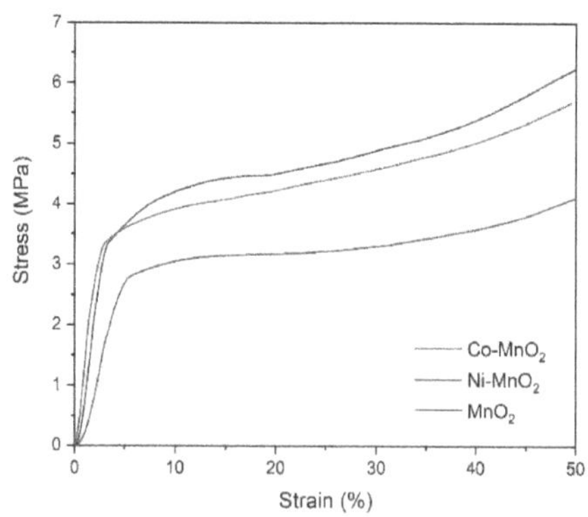

Figure 4. The mechanical properties of pristine MnO_2, Ni-MnO_2, and Co-MnO_2 samples.

The catalytic activities of propane oxidation over pristine MnO_2, Ni-MnO_2, and Co-MnO_2 samples were investigated and the results were shown in Figure 5. The T_{50} (the temperature at which 50% C_3H_8 was converted) was usually used to compare the performance of different samples. It was found that after incorporation of Ni or Co, the T_{50} shifted towards lower temperatures. Moreover, the T_{50} of Co-MnO_2 (278 °C) was lower than that of Ni-MnO_2 (289 °C), suggesting that Co-MnO_2 showed higher activity for propane oxidation than Ni-MnO_2. When the temperature reached 400 °C, the conversion of propane over Co-MnO_2, Ni-MnO_2, and MnO_2 samples was 85.9, 84.3, and 58.6%, respectively. The results verified the promotional effect of Ni or Co doping on the catalytic activity for propane oxidation over MnO_2. After the reactions, the self-standing architectures did not deform or collapse, suggesting they had good resistance to thermal shock.

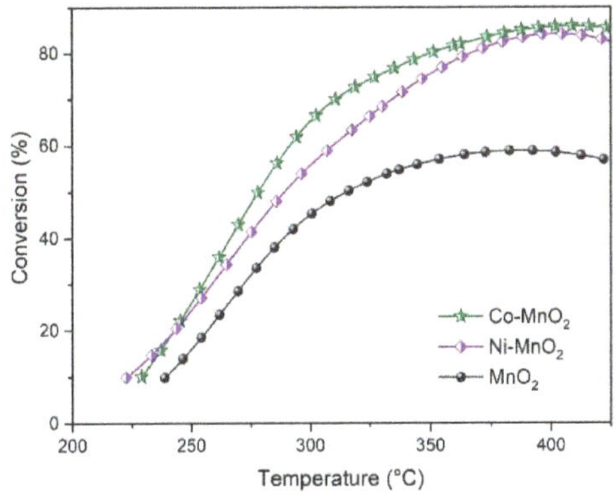

Figure 5. The catalytic activities for propane oxidation of pristine MnO_2, Ni-MnO_2 and Co-MnO_2 samples.

Furthermore, the turnover frequency (TOF) values were calculated to compare the catalytic activities of the three different catalysts. The detailed calculation method was described in previous research [17]. Based on the light-off curves and surface areas, the calculated results were shown in Table 1. It was found that the TOFs of samples increased in the order of MnO_2 (4.20×10^{-10} mol m^{-2} s^{-1}) < Ni-MnO_2 (4.53×10^{-10} mol m^{-2} s^{-1}) < Co-MnO_2 (5.10×10^{-10} mol m^{-2} s^{-1}), which suggested that the incorporation of Ni or Co remarkably enhanced the catalytic activity for propane oxidation.

Table 1. The surface areas and TOFs of the three catalysts.

	T_{10}	Surface Area (m^2 g^{-1})	TOF × 10^{10} (mol m^{-2} s^{-1})
MnO_2	239	70.8	4.20
Ni-MnO_2	223	67.8	4.53
Co-MnO_2	229	59.5	5.10

TOF = turnover frequency.

4. Conclusions

Macroscopic self-standing MnO_2 architecture was fabricated via hydrothermal reaction followed by ambient pressure drying. The addition of ammonium sulfate played an important role in the assembling of MnO_2 nanowires. After the Ni or Co doping, the macroscopic self-standing architectures could be maintained with excellent mechanical properties, which showed enhanced catalytic activities for propane oxidation. This study demonstrated a facile strategy to develop macroscopic self-standing Mn-based architectures, having great possibilities for practical applications.

Author Contributions: L.C. performed the experiments, analyzed the data and editted the paper; X.S. contributed review and supervision.

Funding: This research received no external funding.

Acknowledgments: We thank the ecomaterials laboratory of the School of Materials Science and Engineering, Tsinghua University, for performing the catalytic activity for propane oxidation tests.

Conflicts of Interest: The authors declare no conflict of interest.

References

1. Kim, S.C.; Shim, W.G. Catalytic combustion of VOCs over a series of manganese oxide catalysts. *Appl. Catal. B Environ.* **2010**, *98*, 180–185. [CrossRef]
2. Xie, Y.; Guo, Y.; Guo, Y.; Wang, L.; Zhan, W.; Wang, Y.; Gong, X.; Lu, G. A highly effective Ni-modified MnO_x catalyst for total oxidation of propane: The promotional role of nickel oxide. *Rsc. Adv.* **2016**, *6*, 50228–50237. [CrossRef]
3. Pahalagedara, L.; Kriz, D.A.; Wasalathanthri, N.; Weerakkody, C.; Meng, Y.; Dissanayake, S.; Pahalagedara, M.; Luo, Z.; Suib, S.L.; Nandi, P.; et al. Benchmarking of manganese oxide materials with CO oxidation as catalysts for low temperature selective oxidation. *Appl. Catal. B Environ.* **2017**, *204*, 411–420. [CrossRef]
4. Pahalagedara, L.R.; Dharmarathna, S.; King'ondu, C.K.; Pahalagedara, M.N.; Meng, Y.T.; Kuo, C.H.; Suib, S.L. Microwave-assisted hydrothermal synthesis of alpha-MnO_2: Lattice expansion via rapid temperature ramping and framework substitution. *J. Phy. Chem. C* **2014**, *118*, 20363–20373. [CrossRef]
5. Yang, J.; Zhou, H.; Wang, L.; Zhang, Y.; Chen, C.; Hu, H.; Li, G.; Zhang, Y.; Ma, Y.; Zhang, J. Cobalt-doped K-OMS-2 nanofibers: A novel and efficient water-tolerant catalyst for the oxidation of carbon monoxide. *ChemCatChem* **2017**, *9*, 1163–1167. [CrossRef]
6. Crespo-Biel, O.; Ravoo, B.J.; Reinhoudt, D.N.; Huskens, J. Noncovalent nanoarchitectures on surfaces: From 2D to 3D nanostructures. *J. Mater. Chem.* **2006**, *16*, 3997–4021. [CrossRef]
7. Shehzad, K.; Xu, Y.; Gao, C.; Duan, X. Three-dimensional macro-structures of two-dimensional nanomaterials. *Chem. Soc. Rev.* **2016**, *45*, 5541–5588. [CrossRef] [PubMed]
8. Long, Y.; Hui, J.F.; Wang, P.P.; Hu, S.; Xu, B.; Xiang, G.L.; Zhuang, J.; Lü, X.Q.; Wang, X. Alpha-MnO_2 nanowires as building blocks for the construction of 3D macro-assemblies. *Chem. Comm.* **2012**, *48*, 5925–5927. [CrossRef] [PubMed]
9. Jung, S.M.; Jung, H.Y.; Fang, W.; Dresselhaus, M.S.; Kong, J. A facile methodology for the production of in situ inorganic nanowire hydrogels/aerogels. *Nano Lett.* **2014**, *14*, 1810–1817. [CrossRef] [PubMed]
10. Liu, Z.; Wu, D.; Guo, X.; Fang, S.; Wang, L.; Xing, Y.; Suib, S.L. Robust macroscopic 3D sponges of manganese oxide molecular sieves. *Chem. Eur. J.* **2017**, *23*, 16213–16218. [CrossRef] [PubMed]
11. Rong, S.; Zhang, P.; Yang, Y.; Zhu, L.; Wang, J.; Liu, F. MnO_2 framework for instantaneous mineralization of carcinogenic airborne formaldehyde at room temperature. *ACS Catal.* **2017**, *7*, 1057–1067. [CrossRef]
12. Chen, L.; Ding, J.; Jia, J.; Ran, R.; Zhang, C.; Song, X. Cobalt-doped MnO_2 nanofibers for enhanced propane oxidation. *Acs Appl. Nano Mater.* **2019**, *2*, 4417–4426. [CrossRef]
13. Chen, L.; Ding, J.; Jia, J.; Ran, R.; Zhang, C.; Song, X. Nickel doping MnO_2 with abundant surface pits as highly efficient catalysts for propane deep oxidation. *Chem. Eng. J.* **2019**, *369*, 1129–1137. [CrossRef]
14. Fuchigami, T.; Kimata, R.; Haneda, M.; Kakimoto, K.I. Complex Three-dimensional Co_3O_4 nano-raspberry: Highly stable and active low-temperature CO oxidation catalyst. *Nanomaterials* **2018**, *8*, 662. [CrossRef] [PubMed]
15. Yao, F.; Chen, Y.G.; Salimi, A.R.; Mirzaei, M. Self-Assembly, Crystal structure and analysis of intermolecular interactions of the supramolecular compound based on hexamolybdochromate (III), sulfate and piperazine. *J. Clust. Sci.* **2011**, *22*, 309–318. [CrossRef]
16. Jeazet, H.B.T.; Gloe, K.; Doert, T.; Kataeva, O.N.; Jäger, A.; Geipel, G.; Bernhard, G.; Büchner, B.; Gloe, K. Self-assembly of neutral hexanuclear circular copper(II) meso-helicates: Topological control by sulfate ions. *Chem. Commun.* **2010**, *46*, 2373–2375. [CrossRef] [PubMed]
17. Hu, Z.; Qiu, S.; You, Y.; Guo, Y.; Guo, Y.; Wang, L.; Zhan, W.; Lu, G. Hydrothermal synthesis of $NiCeO_x$ nanosheets and its application to the total oxidation of propane. *Appl. Catal. B Environ.* **2018**, *225*, 110–120. [CrossRef]

© 2019 by the authors. Licensee MDPI, Basel, Switzerland. This article is an open access article distributed under the terms and conditions of the Creative Commons Attribution (CC BY) license (http://creativecommons.org/licenses/by/4.0/).

MDPI
St. Alban-Anlage 66
4052 Basel
Switzerland
Tel. +41 61 683 77 34
Fax +41 61 302 89 18
www.mdpi.com

Technologies Editorial Office
E-mail: technologies@mdpi.com
www.mdpi.com/journal/technologies

www.ingramcontent.com/pod-product-compliance
Lightning Source LLC
Chambersburg PA
CBHW040834250926
51850CB00007BA/1310
* 9 7 8 3 0 3 6 5 5 6 9 4 9 *